Springer Theses

Recognizing Outstanding Ph.D. Research

Aims and Scope

The series "Springer Theses" brings together a selection of the very best Ph.D. theses from around the world and across the physical sciences. Nominated and endorsed by two recognized specialists, each published volume has been selected for its scientific excellence and the high impact of its contents for the pertinent field of research. For greater accessibility to non-specialists, the published versions include an extended introduction, as well as a foreword by the student's supervisor explaining the special relevance of the work for the field. As a whole, the series will provide a valuable resource both for newcomers to the research fields described, and for other scientists seeking detailed background information on special questions. Finally, it provides an accredited documentation of the valuable contributions made by today's younger generation of scientists.

Theses may be nominated for publication in this series by heads of department at internationally leading universities or institutes and should fulfill all of the following criteria

- They must be written in good English.
- The topic should fall within the confines of Chemistry, Physics, Earth Sciences, Engineering and related interdisciplinary fields such as Materials, Nanoscience, Chemical Engineering, Complex Systems and Biophysics.
- The work reported in the thesis must represent a significant scientific advance.
- If the thesis includes previously published material, permission to reproduce this must be gained from the respective copyright holder (a maximum 30% of the thesis should be a verbatim reproduction from the author's previous publications).
- They must have been examined and passed during the 12 months prior to nomination.
- Each thesis should include a foreword by the supervisor outlining the significance of its content.
- The theses should have a clearly defined structure including an introduction accessible to new PhD students and scientists not expert in the relevant field.

Indexed by zbMATH.

Jonathan Yong

Optimising NMR Spectroscopy Through Method and Software Development

Doctoral Thesis accepted by
University of Oxford, Oxford, UK

Springer

Author
Dr. Jonathan Yong
The Alan Turing Institute
London, UK

Supervisors
Prof. Tim Claridge
Department of Chemistry
University of Oxford
Oxford, UK

Mohammadali Foroozandeh
Department of Chemistry
University of Oxford
Oxford, UK

ISSN 2190-5053 ISSN 2190-5061 (electronic)
Springer Theses
ISBN 978-3-031-46683-0 ISBN 978-3-031-46684-7 (eBook)
https://doi.org/10.1007/978-3-031-46684-7

This Springer imprint is published by the registered company Springer Nature Switzerland AG
The registered company address is: Gewerbestrasse 11, 6330 Cham, Switzerland

Paper in this product is recyclable.

Supervisor's Foreword

Solution-state nuclear magnetic resonance (NMR) spectroscopy is one of the most important analytical techniques in modern chemistry and finds widespread use across academia and industry. Its primary application is in the structure characterisation of novel chemical compounds, for which myriad experiments have been developed that provide increasing levels of structural detail and chemical insight. Despite its undoubted power and popularity, there remains room for further development in NMR, especially considering the increased complexity of molecules which are being synthesised and increasing demands for higher sample throughput.

This thesis describes multiple approaches towards improving the quality, speed and efficiency of NMR data acquisition. These encompass advances in pulsed NMR techniques themselves as well as new routines for experiment parameter optimisation and pulse sequence scripting. The methodology and techniques developed provide advanced tools for use across the chemical, pharmaceutical and agrochemical sciences where structure characterisation is a primary requirement. One of the notable features of the work is also the usability of the new methodology. In particular, the development of a website for the automated generation of pulse programs for immediate use on spectrometers, together with the provision of all necessary automation scripts, provides a very user-friendly interface with which to make these techniques readily available to the wider NMR community.

Some of the outstanding features of the thesis are the depth, quality and clarity of the written work. Theory is described concisely yet with great precision, and experimental results made clear with beautifully crafted figures. Every reference is provided with an active link to its associated DOI, adding to the readability of the thesis. The research carried out by Jonathan Yong has generated six journal publications, in addition to which he has contributed to a major review article on parallel NMR methodology and a book chapter describing fast NMR techniques. These endeavours and the exceptional work presented herein are certainly worthy of the Springer Thesis Prize.

Oxford, UK
August 2023

Prof. Tim Claridge

Abstract

Solution-state nuclear magnetic resonance (NMR) spectroscopy is one of the most important analytical techniques in modern organic chemistry. A series of 'core' NMR techniques are widely used for the routine characterisation of organic molecules, ranging from 1D ^{1}H and ^{13}C experiments to more complex 2D experiments such as COSY, HSQC, HMBC and more. Nonetheless, there remains much room for developments in NMR, especially considering the increased *complexity* and *quantity* of molecules which are being studied. This thesis describes multiple approaches towards improving the quality, as well as the speed, of NMR data acquisition.

In Chap. 1, I first set out the quantum mechanical formalisms required for the analysis of NMR pulse sequences. The work which follows may be divided into three sections:

- *Improved techniques for pure shift NMR* (Chap. 2). In a pure shift experiment, the effects of all homonuclear scalar couplings are suppressed, leading to highly-resolved spectra where each multiplet is collapsed into a singlet. A variety of theoretical and experimental approaches are used to search for pure shift methods which have both high sensitivity and low artefact intensity.
- *On-the-fly optimisation of NMR experiments* (Chap. 3). Here, I couple NMR acquisition with derivative-free optimisation algorithms which seek to minimise a cost function measured using spectral data. The resulting software, called POISE, allows experiment parameters to be tailored specifically for each sample and spectrometer in a fully automated fashion. The use of POISE to improve spectral sensitivity and purity is demonstrated in a wide-ranging series of NMR experiments.
- *Accelerated 2D NMR data collection* (Chap. 4). NOAH supersequences allow multiple 2D experiments ('modules') to be acquired in the time of one, through the careful manipulation of different magnetisation pools. This chapter contains work on pulse sequence development, including new NOAH-compatible modules, as well as improvements to various pre-existing modules. The GENESIS website, which creates Bruker pulse programs for arbitrary NOAH supersequences, is also described here.

Preface

> Miss Dashwood at this point, turning her eyes on him with the most angry contempt, stopped him, by saying, 'It is hardly worth while, Mr. Willoughby, for you to relate, or for me to listen any longer. Such a beginning as this cannot be followed by any thing.—Do not let me be pained by hearing any thing more on the subject'.
>
> —JANE AUSTEN, *Sense and Sensibility*

Pulse Sequence Notation

NMR pulse sequences in this thesis are depicted in a reasonably conventional manner: Fig. 1 shows an example of this. To avoid any ambiguity, I describe this notation here. Filled black rectangular bars indicate 90° pulses; empty bars indicate 180° pulses. Pulses with other flip angles are depicted using filled grey bars, typically with the Greek letter β above it representing the flip angle. Delays are variously represented by the letters Δ and τ; exact values of these are given in the respective captions. Pulses without explicit phases are assumed to be applied along the (rotating-frame) $+x$-axis; pulses with phases labelled as ϕ_i are typically phase-cycled, details of which are specified in the respective captions. Grey boxes labelled 'dec.' represent periods of heteronuclear decoupling.

In this thesis, I use the word *gradient* to refer to pulsed field gradients; this is more concise and reflects common usage. Where relevant, z-gradient amplitudes are given as percentages of the maximum gradient amplitude, which is probe-dependent (see Table 1). This maximum amplitude is unlikely to substantially affect the performance of any of the pulse sequences; consequently, in the text I quote gradient amplitudes only as percentages. Nevertheless, if necessary, absolute gradient amplitudes may still be determined for each individual dataset by referring to Table 1. Gradients which are inverted for echo–antiecho selection are depicted as a pair of positive and negative gradients with different shading (e.g. g_2 above). ε is always used to denote the time required for a gradient, plus the subsequent recovery delay (200 μs throughout this work).

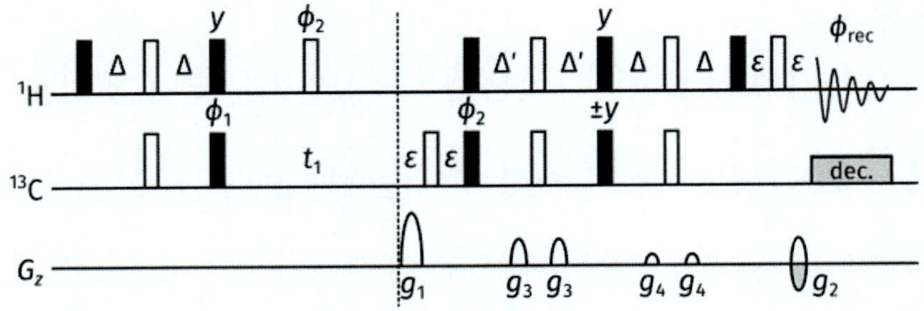

Fig. 1 An example of a pulse sequence (the sensitivity-enhanced HSQC), used here to illustrate the notation used in this thesis

Table 1 Spectrometers used in this thesis. A more complete description may be accessed via the links in the 'internal name' column

Code	Internal name	Details
7	AV700	700 MHz ^{1}H resonance frequency 5 mm TCI ^{1}H/^{13}C/^{15}N inverse cryoprobe 53 G cm^{-1} maximum z-gradient amplitude AVANCE III console, TopSpin 3.6.2
6	AV600	600 MHz ^{1}H resonance frequency 5 mm Prodigy N_2 broadcast cryoprobe (^{1}H/^{19}F outer coil) 66 G cm^{-1} maximum z-gradient amplitude AVANCE III HD console, TopSpin 3.6.2
5	AVX500	500 MHz ^{1}H resonance frequency 5 mm TBO ^{1}H/^{19}F/X broadband room-temperature probe 50 G cm^{-1} maximum z-gradient amplitude AVANCE III HD, TopSpin 3.6.2
4	AVB400	400 MHz ^{1}H resonance frequency 5 mm broadband room-temperature SmartProbe 50 G cm^{-1} maximum z-gradient amplitude AVANCE NEO console, TopSpin 4.0.8

In practice, the implementation of a pulse sequence may differ in tiny ways: for example, delays may be modified to accommodate finite pulse widths and other technicalities. Furthermore, shaped pulses may be used in place of hard pulses in order to optimise the pulse sequence, e.g. by allowing more efficient refocusing: as a rule, adiabatic pulses are used for all ^{13}C 180° rotations.

For some experiments, particularly in the NOAH chapter (Chap. 4), a product operator analysis is added above the pulse sequence. Typically, these show how a ^{1}H-X spin pair and an isolated ^{1}H spin evolve under during pulse sequence: these directly correspond to the so-called $^{1}\mathrm{H}^{\mathrm{X}}$ and $^{1}\mathrm{H}^{!\mathrm{X}}$ magnetisation pools (i.e. protons directly coupled and not directly coupled to ^{13}C, see Sect. 4.1.2 for further explanation). One-letter terms m ($m \in \{x, y, z\}$) are shorthand for single-spin terms on proton, i.e. I_m. Two-letter terms mn are shorthand for two-spin terms on both

Table 2 Samples used in this thesis. Note that concentrations are approximate and not necessarily constant, as samples were remade over time due to, e.g. decomposition. However, it is reasonable to assume that the variation in concentration is below 10%. None of the results in this thesis depend on the exact concentrations of the samples used. See Fig. 2 for chemical structures

Code	Compound	Solvent	Concentration
A	Andrographolide	DMSO-d_6	40 mM
B	(3-Fluorophenyl)boronic acid	DMSO-d_6	120 mM
C	Cyclosporin A	C_6D_6	50 mM
E	Ethanol	D_2O	1 M
F	Ferulic acid	DMSO-d_6	50 mM
G	Gramicidin S	DMSO-d_6	40 mM
P	Rodent urine	D_2O	-
S	Sucrose	90% H_2O/10% D_2O	22 mM
T	Ethyl ferulate	DMSO-d_6	200 mM
X	Brucine	$CDCl_3$	50 mM
Z	Zolmitriptan	DMSO-d_6	50 mM

the proton and heteronucleus, i.e. $2I_m S_n$. The effects of gradients are ignored in the product operator analysis: this can be justified because the coherence transfer pathways (CTPs) of interest are ultimately rephased, meaning that the gradients have no net effect (this is illustrated in greater detail in Sect. 1.4.5). For simplicity, any modulation acquired during t_1 is also omitted in these analyses; it will be discussed in the text where relevant.

Software

All NMR data were processed using TopSpin 3 or 4. Quantum mechanical NMR simulations were done in MATLAB R2021a or R2021b. This thesis was written using the LATEX typesetting system: specifically, I used the LuaLATEX engine.

Pulse sequences and other diagrams are drawn using the vector graphics program Inks-cape. Plots are generated using Python 3, using a number of packages, namely: numpy, scipy, matplotlib, seaborn, and penguins, which was written by me.

The LATEXsources for this thesis, as well as all figures, can be accessed at GitHub: https://github.com/yongrenjie/thesis.

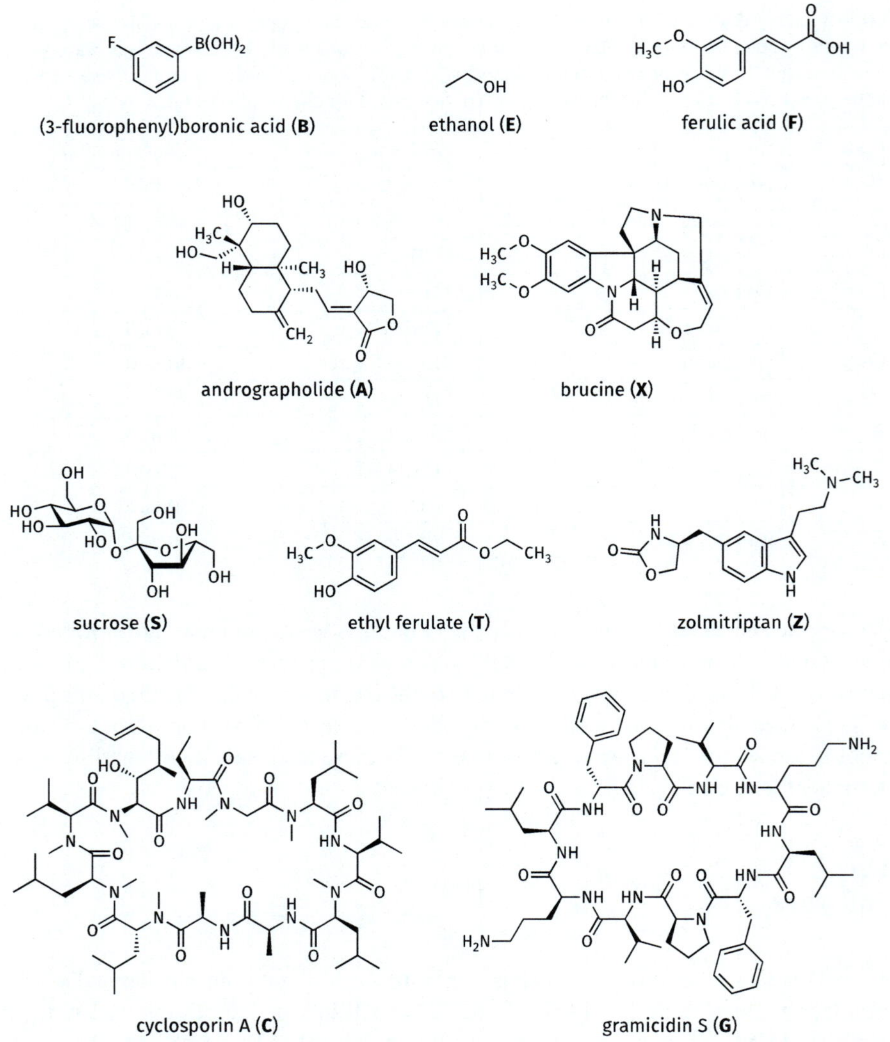

Fig. 2 Chemical structures of samples used in this thesis. See Table 2 for more information

Samples Used

The caption of every figure containing experimental data includes a 'data code' at the end, which specifies the spectrometer and sample used for the data, as well as the date it was acquired (in YYMMDD format). The spectrometers and samples used in this thesis are enumerated in Tables 1 and 2 and Fig. 2. Therefore, for example, the code 7Z-211225 would represent data acquired on the 700 MHz spectrometer, using the zolmitriptan sample, on Christmas Day 2021.

Peak assignments and other physical data for some of the samples are given in Appendix A.

London, UK
Jonathan Yong

Publications

Parts of this thesis have been published in the following journal articles:

- Yong, J. R. J.; Hansen, A. L.; Kupče, Ē.; Claridge, T. D. W. Increasing sensitivity and versatility in NMR supersequences with new HSQC-based modules. *J. Magn. Reson.* **2021**, *329*, 10.1016/j.jmr.2021.107027
- Yong, J. R. J.; Foroozandeh, M. On-the-Fly, Sample-Tailored Optimization of NMR Experiments. *Anal. Chem.* **2021**, *93*, 10735–10739, 10.1021/acs.analchem.1c01767
- Kupče, Ē.; Yong, J. R. J.; Widmalm, G.; Claridge, T. D. W. Parallel NMR Supersequences: Ten Spectra in a Single Measurement. *JACS Au* **2021**, *1*, 1892–1897, 10.1021/jacsau.1c00423
- Yong, J. R. J.; Kupče, Ē.; Claridge, T. D. W. Modular Pulse Program Generation for NMR Supersequences. *Anal. Chem.* **2022**, *94*, 2271–2278, 10.1021/acs.analchem.1c04964
- Verstraete, J.-B.; Yong, J. R. J.; Goodwin, D. L.; Myers, W. K.; Foroozandeh, M. Towards fully optimised and automated ESR spectroscopy. *Chem. Commun.* **2022**, *58*, 10715–10718, 10.1039/d2cc02742a
- Yong, J. R. J.; Kupče, Ē.; Claridge, T. D. W. A general scheme for generating NMR supersequences combining high- and low-sensitivity experiments. *Chem. Commun.* **2023**, *59*, 7827–7830, 10.1039/d3cc01472j

Acknowledgements

> 'I cannot make speeches, Emma:'—he soon resumed; and in a tone of such sincere, decided, intelligible tenderness as was tolerably convincing.—'If I loved you less, I might be able to talk about it more'.
>
> —JANE AUSTEN, *Emma*

First of all, I wish to thank my supervisors, Tim Claridge and Mohammadali Foroozandeh. Throughout my studies, their guidance has been understated yet unerring; I have become a better scientist and person for it, and hope that this thesis bears witness to the quality of their counsel. I especially appreciate being afforded the freedom to explore other things which did not directly feed back into my D.Phil. output. This includes the short period where I worked in the NMR team, as well as all the time I have spent on programming, which explains—but does not justify—many a week where no work was produced. Yet, because of this, I have gotten so much more out of these 4 years than just a D.Phil. and can look towards my future career with much optimism.

I am grateful to the Clarendon Fund, as well as the SBM CDT, for funding this work. They have made it possible for me to stay in Oxford—a place now very dear to me—for far longer than I had expected to.

I am also deeply indebted to my industrial mentors, Peter Howe, Philip Sidebottom, Iain Swan and Harry Mackenzie. They have always taken a keen interest in all aspects of my work, and I shall remember with much fondness the very kind invitation to the Jealott's Hill Syngenta site. Here, I wish to also thank my professional collaborators in the world of magnetic resonance, Ēriks Kupče and Jean-Nicolas Dumez: it has been nothing short of a privilege to work with and learn from them.

Much credit is due to the people whom I have been fortunate to work with in Oxford. In particular, I wish to thank the NMR staff: Caitlin, James, Maria, Nader, Nick and Tina (and Tim, again); and various members of the NMR group past and present, including Ali, David, Fay, JB, Jos, Noelle, Sam and Simon. By at least occasionally creating in me a desire to go into the office, you have balanced out my naturally sedentary lifestyle, which is no mean feat. It is my sincere hope that we may continue to stay in touch, now as friends rather than colleagues.

On a more personal level, learning the flute has been my greatest undertaking in Oxford outside of my work: one which has been sometimes stressful, but always joyful. For this, I must thank my teacher Jean, who has been kind to me well beyond her duty as an instructor, as well as her husband Robin.

I am also very grateful for the friends I have had during my time in Oxford, including (but not limited to) Alvin, Belinda, Jing Long, Liv, Steve, Jieyan, Sheryl and Wearn Xin; and the rest of my SBM cohort, especially Henry and Kate. Many of you will already have moved on to other adventures, but I am so glad to have crossed paths with you at some point. Distinguished most in both constancy and intimacy, however, is Marie. I have always enjoyed your company over the last few years: there are few people who listen to and understand me as you do, and on such a wide range of topics. It means a great deal to me, certainly more than I let on—thank you so much for everything.

Finally, I come to the people who have immeasurably shaped my life for the better. My parents have sacrificed much of their freedom to guarantee mine: the work I have done here is entirely due to theirs. I cannot even begin to repay their unconditional love, but I pray that through my life I may do them proud and thereby vindicate their choices. Last and dearest is my partner Bernie, who has been the source of every felicity during my time in Oxford. None of the achievements in this work can equal the profound joy which I derive from her constant presence and unfailing support.

Contents

Acronyms

2BOB	Two-bond and one-bond correlation
2DJ	(2D) J-resolved
ADEQUATE	Adequate sensitivity double-quantum spectroscopy
AP	Antiphase
API	Application programming interface
ASAP	Acceleration by sharing adjacent polarisation
BANGO	β-TANGO (i.e. small flip angle)
BEBOP	Broadband excitation by optimised pulses
BFGS	Broyden–Fletcher–Goldfarb–Shanno
BIBOP	Broadband inversion by optimised pulses
BIG-BIRD	Biselective independent gyrations BIRD
BIRD	Bilinear rotation decoupling
BOBYQA	Bound optimisation by quadratic approximation
BURBOP	Broadband universal rotation by optimised pulses
CHORUS	Chirped, ordered pulses for ultra-broadband spectroscopy
CLIP	Clean in-phase
COSY	Correlation spectroscopy
CPU	Central processing unit
CRK	Cavanagh–Rance–Kay (seHSQC)
CTP	Coherence transfer pathway
DEER	Double electron–electron resonance
DFT	Discrete Fourier transform
DIPSI	Decoupling in the presence of scalar interactions
DMSO	Dimethyl sulfoxide
DOSY	Diffusion-ordered spectroscopy
dPSYCHE	Discrete PSYCHE
DQF-COSY	Double-quantum filtered COSY
DSE	Double spin echo
DUMBO	Decoupling using mind-boggling optimisation
EA	Echo–antiecho
ECOSY	Exclusive COSY

eDUMBO	Experimental DUMBO
EPSI	Echo-planar spectroscopic imaging
ES	Excitation sculpting
ESCALADE	Efficient spin control using analytical Lie algebraic derivatives
ESR	Electron spin resonance
FE	Function evaluation
FID	Free induction decay
FT	Fourier transform
GEMSTONE	Gradient-enhanced multiplet-selective targeted-observation NMR experiment
GENESIS	Generation of supersequences in silico
GPU	Graphics processing unit
H2OBC	Heteronuclear two- and one-bond correlation
HMBC	Heteronuclear multiple-bond correlation
HMQC	Heteronuclear multiple-quantum correlation
HSQC	Heteronuclear single-quantum correlation
iDOSY	Internal DOSY
INADEQUATE	Incredible natural-abundance double-quantum transfer experiment
INEPT	Insensitive nuclei enhanced by polarisation transfer
IP	In-phase
IPAP	In-phase/antiphase
JRE	J-refocusing element
JRSE	J-refocused spin echo
JSON	JavaScript object notation
LP	Linear prediction
LPJF	Low-pass J-filter
MDS	Multidirectional search
MLEV	Malcolm Levitt
NM	Nelder–Mead
NMR	Nuclear magnetic resonance
NOAH	NMR by ordered acquisition using ^{1}H detection
NOE	Nuclear Overhauser effect
NOESY	Nuclear Overhauser effect spectroscopy
NUS	Non-uniform sampling
PEP	Preservation of equivalent pathways
POISE	Parameter optimisation by iterative spectral evaluation
PSE	Pure shift element
PSYCHE	Pure shift yielded by chirp excitation
RF	Radiofrequency
ROESY	Rotating-frame Overhauser effect spectroscopy
rSNOB	Refocusing selective excitation for biochemical applications
SAPPHIRE	Sideband averaging by periodic phase incrementation of residual J-evolution
seHSQC	Sensitivity-enhanced HSQC

SERF	Selective refocusing
SNR	Signal-to-noise ratio
SW	Spectral width/Spectral window
TANGO	Testing for Adjacent Nuclei with a Gyration Operator
TIG-BIRD	Triselective independent gyrations BIRD
TOCSY	Total correlation spectroscopy
TPPI	Time-proportional phase incrementation
TS	Time-shared
TSE	Triple spin echo
URL	Uniform resource locator
URP	Universal rotation pulse
ZF	Zero filling
ZIP	*zz* isotope-selective pulse
ZQF	Zero-quantum filter
ZS	Zangger-Sterk

List of Figures

List of Tables

List of Code Listings

Chapter 1
NMR Theory

> I have seen a great many lists of her drawing-up at various times of books that she meant to read regularly through—and very good lists they were—very well chosen, and very neatly arranged ... But I have done with expecting any course of steady reading from Emma. She will never submit to any thing requiring industry and patience.
>
> — JANE AUSTEN, *Emma*

In the opening chapter of this thesis, I provide an overview of basic NMR theory, specifically, the dynamics of quantum systems containing one or more spin-$\frac{1}{2}$ particles. Starting from the Schrödinger equation, I progressively develop the rotating frame and density operator formalisms used in the analysis and simulation of simple NMR experiments. The important product operator formalism, used throughout this thesis, is exemplified through a selection of 1D and 2D experiments. Since 2D experiments form a very large part of this thesis, I also discuss a number of general principles in 2D NMR.

Note that this chapter is not intended to be an exhaustive account of magnetic resonance theory; for more complete coverage, the reader is directed to a suitable textbook [1–5].

1.1 Quantum Mechanics

The most fundamental equation in (non-relativistic) quantum mechanics, which governs the time evolution of a quantum state $|\Psi(t)\rangle$ under a Hamiltonian H, is the time-dependent Schrödinger equation:

$$\frac{\partial|\Psi(t)\rangle}{\partial t} = -\frac{\mathrm{i}}{\hbar} H\,|\Psi(t)\rangle\,. \tag{1.1}$$

For a Hamiltonian which is constant over a period of time $t_1 \leq t \leq t_2$ (i.e. is *time-independent*), this can be integrated to yield an explicit solution:

J. Yong, *Optimising NMR Spectroscopy Through Method and Software Development*, Springer Theses, https://doi.org/10.1007/978-3-031-46684-7_1

$$|\Psi(t_2)\rangle = \exp\left[-\frac{\mathrm{i}H(t_2 - t_1)}{\hbar}\right]|\Psi(t_1)\rangle\,. \tag{1.2}$$

In NMR, it is conventional to use units of angular frequencies instead of energies, for example, by replacing $H/\hbar \to H$; this will henceforth be assumed. The term $\exp[-\mathrm{i}H(t_2 - t_1)]$ is called the *propagator* of the system and denoted $U(t_2, t_1)$; this is often further simplified to $U(\tau)$ where $\tau = t_2 - t_1$ is the duration of the evolution. For a Hamiltonian which varies with time but is piecewise constant, in that it can be broken up into several finite periods within which H is time-independent, the time evolution of the state is simply given by successive application of propagators:

$$|\Psi(t_n)\rangle = U(t_n, t_{n-1}) \cdots U(t_2, t_1)U(t_1, t_0)\,|\Psi(t_0)\rangle\,, \tag{1.3}$$

where $t_n > t_{n-1} > \cdots > t_0$. The case where H continuously varies with time is more complicated, but we will not need to consider it in this thesis.

In NMR spectroscopy, we manipulate the *spin angular momentum* of atomic nuclei in order to obtain information about chemical structure and dynamics. The present work is restricted to nuclei with spin quantum number $I = 1/2$. These are two-level systems, where the eigenstates of I_z (denoted as $|\alpha\rangle$ and $|\beta\rangle$ for $m_I = +1/2$ and $-1/2$, respectively) are used as a standard basis, called the *Zeeman basis*. Primarily for mathematical convenience, the z-axis is conventionally chosen as the quantisation axis in textbook treatments of angular momentum. However, in the context of NMR, the z-axis bears even more significance as we define it to be the axis along which the static magnetic field is aligned. Since the matrix elements of an operator O are given by $O_{mn} = \langle m|O|n\rangle$, we can work out the matrix representations of the angular momentum operators in the Zeeman basis:

$$I_x = \frac{1}{2}\begin{pmatrix}0 & 1\\ 1 & 0\end{pmatrix}; \quad I_y = \frac{1}{2}\begin{pmatrix}0 & -\mathrm{i}\\ \mathrm{i} & 0\end{pmatrix}; \quad I_z = \frac{1}{2}\begin{pmatrix}1 & 0\\ 0 & -1\end{pmatrix}. \tag{1.4}$$

Their commutators are given by

$$[I_i, I_j] = \sum_k \mathrm{i}\varepsilon_{ijk} J_k, \tag{1.5}$$

where ε_{ijk} is the Levi-Civita symbol. We also define the following linear combinations:

$$\begin{aligned} I_+ &= I_x + \mathrm{i}I_y = \begin{pmatrix}0 & 1\\ 0 & 0\end{pmatrix}; \quad I_\alpha = \frac{1}{2}E + I_z = \begin{pmatrix}1 & 0\\ 0 & 0\end{pmatrix};\\ I_- &= I_x - \mathrm{i}I_y = \begin{pmatrix}0 & 0\\ 1 & 0\end{pmatrix}; \quad I_\beta = \frac{1}{2}E - I_z = \begin{pmatrix}0 & 0\\ 0 & 1\end{pmatrix}, \end{aligned} \tag{1.6}$$

where E is the 2×2 identity matrix. The *coherence order* of an operator, denoted p, is defined by the Zeeman basis states it connects, i.e. the nonzero elements in its matrix form when expressed in this basis: an operator $O = |m_2\rangle\,\langle m_1|$ would represent

$(m_2 - m_1)$-order coherence, since $\langle m_2|O|m_1\rangle \neq 0$. Thus, in the above equations, $I_+ = |\alpha\rangle\langle\beta|$ represents a coherence order of $+1$; I_- a coherence order of -1; I_x and I_y are both a mixture of ± 1-coherence; and the remainder have coherence order 0.

States (and operators) for composite systems are formally defined as tensor products of single-spin states (and operators) [6]. Operators on the same spin commute as per Eq. (1.5), and operators on different spins fully commute. The Kronecker product allows these operators to be expressed in matrix form [5]. For example, the operator $2I_xS_z$ can be represented as follows[1]:

$$2I_xS_z = 2 \cdot \frac{1}{2} \cdot \frac{1}{2}\left[\begin{pmatrix} 0 & 1 \\ 1 & 0 \end{pmatrix} \otimes \begin{pmatrix} 1 & 0 \\ 0 & -1 \end{pmatrix}\right] = \frac{1}{2}\begin{pmatrix} 0 & 0 & 1 & 0 \\ 0 & 0 & 0 & -1 \\ 1 & 0 & 0 & 0 \\ 0 & -1 & 0 & 0 \end{pmatrix}. \tag{1.7}$$

The Hamiltonians H for nuclear spin interactions, which will be encountered frequently in this chapter, are formed from such operators [3]. In solution-state NMR, these interactions include

$$H_{\text{cs}} = \sum_i \omega_{0,i} I_{iz} \qquad \text{(chemical shift);} \tag{1.8}$$

$$H_{\text{J}} = \sum_{i>j} 2\pi J_{ij} (\mathbf{I}_i \cdot \mathbf{I}_j) \qquad \text{(scalar coupling);} \tag{1.9}$$

$$H_{\text{pulse}} = \sum_i \omega_{i,x} I_{ix} + \sum_i \omega_{i,y} I_{iy} \qquad \text{(radiofrequency pulses);} \tag{1.10}$$

$$H_{\text{grad}} = \sum_i \gamma_i G z I_{iz} \qquad \text{(pulsed field gradients on } z\text{).} \tag{1.11}$$

Pulsed-field gradients (henceforth shortened to *gradients*) can in principle be applied along any axis, not just z, but this is dependent on hardware: all the work in this thesis was done on z-gradient probes. In the above expressions:

- γ_i is the magnetogyric ratio of spin i;
- $\omega_{0,i}$ refers to the Larmor, or precession, frequency of spin i (usually on the order of MHz). The Larmor frequency is defined as

$$\omega_{0,i} = -\gamma_i B_0, \tag{1.12}$$

  where B_0 is the strength of the external (static) magnetic field;
- J_{ij} is the scalar coupling constant between spins i and j (expressed in units of Hz);

[1] This representation is not unique; it is perfectly possible to reverse the order of the Kronecker product, and as long as this is consistently done, any physically measurable quantities calculated using this alternative will be the same.

- ω_x and ω_y are amplitudes of radiofrequency (RF) pulses along the x- and y-axes, which are in general time-dependent, and are related to the so-called B_1 by a factor of γ_i.
- G is the amplitude of the gradient, typically in units of G/cm; and
- z is the position of the spin along the z-axis, typically in units of cm.

Finally, note that in the *weak coupling* regime where

$$\omega_{0,i} - \omega_{0,j} \gg J_{ij}, \tag{1.13}$$

the scalar coupling Hamiltonian may be simplified (the *secular approximation*[2]) to

$$H_{\text{J,secular}} = \sum_{i>j} 2\pi J_{ij} I_{iz} I_{jz}. \tag{1.14}$$

This condition is always satisfied whenever spins i and j are different nuclides.

Throughout the course of an NMR experiment, RF pulses and gradients are turned on and off, and thus H_{pulse} and H_{grad} are time-dependent—although they will always satisfy the 'piecewise constant' criterion which allows us to use Eq. (1.3). The 'free precession' (or simply 'free') Hamiltonian, H_{free}, refers to the Hamiltonian which is operative whenever no pulses or gradients are being applied:

$$H_{\text{free}} = H_{\text{cs}} + H_{\text{J}}. \tag{1.15}$$

1.2 The Rotating Frame

The Hamiltonians described above refer to the 'laboratory frame' or the *Schrödinger picture*, where spins precess about the z-axis at their intrinsic frequencies and obey the equation of motion (1.1). However, this proves to often be unwieldy, in particular when analysing the effects of RF pulses. It is standard procedure to transform the frame of reference to a 'rotating frame', specifically, one which rotates about the z-axis at a defined rotation frequency ω_{rot} which is close to the Larmor frequencies ω_0.

The rotating frame can be formalised using the *interaction picture* of quantum mechanics, [6] which involves the separation of the Hamiltonian into two parts, with the first typically being completely time-independent:

[2] This result comes from the use of time-independent nondegenerate perturbation theory: it is based on the assumption that the eigenstates $\{|n\rangle\}$ of the main Hamiltonian H_0 are unchanged by the perturbation V (since the first-order correction varies as $\sum_m V_{mn}/(\omega_m - \omega_n) \ll 1$), and only the first-order correction to the energies $E_n^{(1)} = \langle n|V|n\rangle$ is retained. In this context, H_0 and V are, respectively, H_{cs} and H_{J}. When the condition Eq. (1.13) does not hold, the nondegenerate treatment fails; see e.g. Sakurai [6].

$$H(t) = H_0 + H_1(t). \tag{1.16}$$

In this case, the static part H_0 simply corresponds to precession of the spins at a particular frequency:

$$H_0 = \sum_i \omega_{\text{rot},i} I_{iz}. \tag{1.17}$$

(Generally, each instance of the same nuclide (e.g. ^{1}H or ^{13}C) will share the same ω_{rot}, so the subscript i in $\omega_{\text{rot},i}$ is useful only for distinguishing different nuclear species.) This allows us to define H_1 as

$$\begin{aligned} H_1 &= H_\text{J} + H_\text{pulse} + H_\text{grad} + (H_\text{cs} - H_0) \qquad (1.18) \\ &= H_\text{J} + H_\text{pulse} + H_\text{grad} + \sum_i \Omega_i I_{iz} \\ &= H_\text{J} + H_\text{pulse} + H_\text{grad} + H_\text{offset}, \end{aligned}$$

where $\Omega_i = \omega_{0,i} - \omega_{\text{rot},i}$ is the *offset* of spin i. For reasons which will become clear later, the frequency ω_{rot} is chosen to be the centre of the spectral window for the given nuclide.

Having split up our Hamiltonian, we then define an *interaction-picture ket*:

$$|\Psi\rangle_I = \exp(\mathrm{i}H_0 t)\,|\Psi\rangle\,. \tag{1.19}$$

The time evolution of this ket is given by a transformation of the Schrödinger equation:

$$\begin{aligned} \frac{\partial\,|\Psi\rangle_I}{\partial t} &= \mathrm{i}H_0\exp(\mathrm{i}H_0 t)\,|\Psi\rangle + \exp(\mathrm{i}H_0 t)\frac{\partial|\Psi\rangle}{\partial t} \\ &= \mathrm{i}H_0|\Psi\rangle_I + \exp(\mathrm{i}H_0 t)(-\mathrm{i}H|\Psi\rangle) \\ &= \mathrm{i}H_0|\Psi\rangle_I - \mathrm{i}\exp(\mathrm{i}H_0 t)(H_0 + H_1)\exp(-\mathrm{i}H_0 t)\,|\Psi\rangle_I \\ &= \mathrm{i}H_0|\Psi\rangle_I - \mathrm{i}H_0|\Psi\rangle_I - \mathrm{i}\exp(\mathrm{i}H_0 t)H_1\exp(-\mathrm{i}H_0 t)\,|\Psi\rangle_I \\ &= -\mathrm{i}\exp(\mathrm{i}H_0 t)H_1\exp(-\mathrm{i}H_0 t)\,|\Psi\rangle_I \\ &= -\mathrm{i}H_{1,I}|\Psi\rangle_I\,, \qquad (1.20) \end{aligned}$$

where

$$H_{1,I} = \exp(\mathrm{i}H_0 t)H_1\exp(-\mathrm{i}H_0 t). \tag{1.21}$$

The underlying principle here is that the 'interesting' behaviour should be contained in H_1, and instead of explicitly considering the time evolution under the 'uninteresting' H_0, it is just used to transform H_1 into $H_{1,I}$.

We now turn our attention to how the various NMR Hamiltonians (Eqs. (1.8)–(1.11)) are transformed in the interaction picture; that is to say, what the individual terms in the right-hand side of

$$\begin{aligned} H_{1,I} &= \exp(\mathrm{i}H_0t)H_\mathrm{J}\exp(-\mathrm{i}H_0t) + \exp(\mathrm{i}H_0t)H_\mathrm{pulse}\exp(-\mathrm{i}H_0t) \\ &\quad + \exp(\mathrm{i}H_0t)H_\mathrm{grad}\exp(-\mathrm{i}H_0t) + \exp(\mathrm{i}H_0t)H_\mathrm{offset}\exp(-\mathrm{i}H_0t) \end{aligned} \tag{1.22}$$

are. We first note that H_0 (and hence $\exp(\pm\mathrm{i}H_0t)$) is a function only of the I_{iz} operators; thus, any Hamiltonian which commutes with all I_{iz}'s will be untouched by this transformation. This is trivially true of H_offset and H_grad, which are themselves both functions of the I_{iz}'s. It can also be shown that H_J (in the homonuclear case) and $H_\mathrm{J,secular}$ (heteronuclear case) fully commute with H_0. So, for three of the four terms in Eq. (1.22) we simply have the result that $\exp(\mathrm{i}H_0t)H\exp(-\mathrm{i}H_0t) = H$. This allows us to immediately write down the free precession Hamiltonian in the interaction picture:

$$H_{\mathrm{free},I} = H_\mathrm{offset} + H_\mathrm{J}. \tag{1.23}$$

The fourth term, which does not commute with H_0, is H_pulse. In the laboratory frame, *hard pulses* are applied as oscillating RF fields, which are time-dependent. For a pulse acting on a single spin, we have

$$H_\mathrm{pulse}(t) = \omega_1[\cos(\omega_\mathrm{tx}t + \phi)I_x + \sin(\omega_\mathrm{tx}t + \phi)I_y]. \tag{1.24}$$

Here, ω_1 represents the *amplitude* of the pulse, and ϕ its *phase*. This expression is similar to the expression in Eq. (1.10), but here ω_1 and ϕ are both constants, with the sinusoidal time dependence explicitly specified using the *transmitter frequency* (or just *frequency*) of the pulse, ω_tx. In the rotating frame, using that $H_0 = \omega_\mathrm{rot}I_z$, we then have the following interaction Hamiltonian:

$$\begin{aligned} H_{\mathrm{pulse},I}(t) &= \omega_1\big[\exp(\mathrm{i}\omega_\mathrm{rot}tI_z)I_x\cos(\omega_\mathrm{tx}t + \phi)\exp(-\mathrm{i}\omega_\mathrm{rot}tI_z) \\ &\quad + \exp(\mathrm{i}\omega_\mathrm{rot}tI_z)I_y\sin(\omega_\mathrm{tx}t + \phi)\exp(-\mathrm{i}\omega_\mathrm{rot}tI_z)\big], \end{aligned} \tag{1.25}$$

and using the formulae

$$\exp(\mathrm{i}\theta I_z)I_x\exp(-\mathrm{i}\theta I_z) = I_x\cos\theta - I_y\sin\theta; \tag{1.26}$$

$$\exp(\mathrm{i}\theta I_z)I_y\exp(-\mathrm{i}\theta I_z) = I_y\cos\theta + I_x\sin\theta \tag{1.27}$$

(see Appendix A.2 of Levitt [3] for a derivation), Eq. (1.25) simplifies to

$$H_{\mathrm{pulse},I}(t) = \omega_1\big\{I_x\cos[(\omega_\mathrm{tx} - \omega_\mathrm{rot})t + \phi] + I_y\sin[(\omega_\mathrm{tx} - \omega_\mathrm{rot})t + \phi]\big\}. \tag{1.28}$$

The transmitter frequency ω_tx is a parameter which can be controlled by the user and is typically placed in the centre of the spectrum of the sample under study, in order to make the most use of its *bandwidth* (the region of frequencies over which the pulse is effective). For convenience, it is typical to then choose the rotating-frame frequency to be exactly the same frequency: $\omega_{\text{rot}} = \omega_{\text{tx}}$. This allows us to simplify the rotating-frame Hamiltonian to

$$H_{\text{pulse},I} = \omega_1(I_x \cos\phi + I_y \sin\phi), \tag{1.29}$$

which is time-*independent*. Occasionally, I will also use the Cartesian components:

$$(c_x, c_y) = (\omega_1 \cos\phi, \omega_1 \sin\phi), \tag{1.30}$$

instead of the amplitude and phase, to describe the pulse.

Consider now the application of this pulse to an isolated spin for which $\omega_0 = \omega_{\text{rot}}$ and thus has an offset $\Omega = 0$. We have that $H_{\text{offset}} = H_{\text{J}} = H_{\text{grad}} = 0$, and the only active Hamiltonian is that of the pulse, which causes *nutation* of the spin magnetisation vector around the axis of the pulse; in this case, the pulse (or the spin) is said to be *on-resonance*.[3] If a duration for the pulse τ_{p} is further specified, this also allows us to define a *flip angle* $\beta = \omega_1\tau_{\text{p}}$. On the other hand, spins which are *off-resonance* ($\omega_0 - \omega_{\text{tx}} \neq 0$) evolve under not only the pulse Hamiltonian but also the offset; this leads to a different effective flip angle and axis of rotation. Off-resonance effects may be neglected when considering an idealised, infinitely hard pulse, where $\omega_1 \to \infty$ and $\tau_{\text{p}} \to 0$. However, this is of course not possible on a spectrometer, and in practice, off-resonance effects are noticeable even for hard pulses as short as several microseconds.

In general, RF pulses are more complicated than the simple case of the hard pulse shown here. For example, they may be constructed such that even in the rotating frame there is still a time dependence in the amplitudes and/or the phases; these are variously referred to as *shaped*, *amplitude-modulated* or *frequency-modulated* pulses depending on the context. In such cases, both ω_1 and ϕ can in principle be continuous functions of time; however, for ease of construction and implementation, pulses are typically generated in a *piecewise* or discrete method using n points each of time δt, within which ω_1 and ϕ are constant. The total length of the pulse is then simply $n(\delta t)$; δt is sometimes called the *timestep* of the pulse.

1.3 Density Operators

NMR experiments are not executed on one single spin at a time; instead, the samples used typically contain on the order of 10^{20} spins. Furthermore, each of these spins may have its own wavefunction: it is impossible to force every spin to possess the same

3 Strictly speaking, the rotating frame is just a mathematical formalism, so the resonance condition does not necessitate $\Omega = 0$ or $\omega_{\text{tx}} = \omega_{\text{rot}} = \omega_0$. We only need that $\omega_{\text{tx}} = \omega_0$, or in other words, that the pulse is applied at the frequency of the spin—which may or may not be the same as the rotating-frame frequency. Practically, such a situation may arise in (for example) the application of selective pulses to a specific spin which is not at the centre of the spectrum.

state. Since we are only interested in the *ensemble* behaviour such as expectation values, rather than the dynamics of each individual spin, we can use the *density operator* formalism instead of dealing with a composite wavefunction of many spins. The density operator, ρ, is defined (in the Schrödinger picture) as

$$\rho = \sum_j p_j \left|\psi_j\right\rangle\!\left\langle\psi_j\right|, \tag{1.31}$$

where p_j is the probability that a spin is in the state $\left|\psi_j\right\rangle$ (and the $\left|\psi_j\right\rangle$'s are assumed to form a complete set of states).[4] The use of ρ actually represents a loss of information, in that while Eq. (1.31) gives us a straightforward recipe for constructing ρ from a given distribution of states $\{p_j, \left|\psi_j\right\rangle\}$, the reverse is not possible: given a known ρ, it is generally not possible to determine a unique distribution of states. This is not a problem, however, because ρ contains all the information required for the calculation of expectation values, in that for any operator A,

$$\langle A\rangle = \sum_j \langle\psi_j|A\rho|\psi_j\rangle. \tag{1.32}$$

If A and ρ are expressed as matrices (through any choice of basis), then this is more easily expressed as the trace of the matrix product:

$$\langle A\rangle = \mathrm{Tr}(A\rho). \tag{1.33}$$

Other properties of the density operator are not discussed here, but can be found in virtually any textbook covering their use [6–8].

The time evolution of a Schrödinger-picture density operator is governed by the Liouville–von Neumann equation, which can be derived from Eq. (1.1):

$$\begin{aligned}
\frac{\mathrm{d}\rho}{\mathrm{d}t} &= \sum_j p_j \left(\frac{\mathrm{d}\left|\psi_j\right\rangle}{\mathrm{d}t}\left\langle\psi_j\right| + \left|\psi_j\right\rangle\frac{\mathrm{d}\left\langle\psi_j\right|}{\mathrm{d}t}\right) \\
&= \sum_j p_j \left(-\mathrm{i}H\left|\psi_j\right\rangle\!\left\langle\psi_j\right| + \left|\psi_j\right\rangle \mathrm{i}\left\langle\psi_j\right| H\right) \\
&= -\mathrm{i}H\left(\sum_j p_j \left|\psi_j\right\rangle\!\left\langle\psi_j\right|\right) + \mathrm{i}\left(\sum_j p_j \left|\psi_j\right\rangle\!\left\langle\psi_j\right|\right)H \\
&= -\mathrm{i}[H, \rho].
\end{aligned} \tag{1.34}$$

[4] This probability is a *classical* probability: that is, it is purely statistical in nature and should not be confused with the probability amplitudes associated with quantum superposition (i.e. $|c_j|^2$ in a single-spin wavefunction $\sum_j c_j \left|j\right\rangle$).

Note here that the weights p_j are time-independent, as the time evolution is contained entirely in the kets and bras.[5] For a time-independent H, this can be integrated to yield the solution:

$$\rho(t_2) = \exp(-\mathrm{i}H\tau)\rho(t_1)\exp(\mathrm{i}H\tau), \tag{1.35}$$

where $\tau = t_2 - t_1$.

In the interaction picture, the density operator is instead defined using interaction-picture states $\{|\psi_i\rangle_I\}$

$$\begin{aligned}\rho_I = \sum_j p_j\,|\psi_j\rangle_I\langle\psi_j|_I &= \sum_j p_j \exp(\mathrm{i}H_0t)\,|\psi_j\rangle\langle\psi_j|\exp(-\mathrm{i}H_0t)\\ &= \exp(\mathrm{i}H_0t)\left(\sum_j p_j\,|\psi_j\rangle\langle\psi_j|\right)\exp(-\mathrm{i}H_0t)\\ &= \exp(\mathrm{i}H_0t)\rho\exp(-\mathrm{i}H_0t)\end{aligned} \tag{1.36}$$

(note the similarity to Eq. (1.21)). Using a very similar proof as in Eq. (1.34), it can be shown that ρ_I obeys a modified Liouville–von Neumann equation:

$$\frac{\mathrm{d}\rho_I}{\mathrm{d}t} = -\mathrm{i}[H_{1,I}, \rho_I], \tag{1.37}$$

and analogously, for a time-independent $H_{1,I}$, we have that

$$\rho_I(t_2) = \exp(-\mathrm{i}H_{1,I}\tau)\rho_I(t_1)\exp(\mathrm{i}H_{1,I}\tau) = U\rho_I(t_1)U^\dagger, \tag{1.38}$$

where $U = \exp(-\mathrm{i}H_{1,I}\tau)$. Multiple propagators may be chained in a similar fashion to Eq. (1.3). This result means that from a practical point of view, the effects of H_0 can be completely ignored when analysing or simulating NMR experiments using density operators.

Finally, a mention of the *equilibrium* or *thermal* density operator is in order. For a canonical ensemble, this is given by

$$\rho_0 = \frac{\exp(-\beta\hbar H)}{\mathrm{Tr}[\exp(-\beta\hbar H)]}, \tag{1.39}$$

where $\beta = 1/(k_\mathrm{B}T)$ and the Hamiltonian H is in units of angular momentum, as has been consistently used here. At equilibrium, no pulses or gradients are being applied, so the appropriate Hamiltonian is the free (Schrödinger-picture) Hamiltonian H_free

[5] Strictly speaking, this only applies to a *closed* quantum system, which implies that effects such as relaxation are ignored (or at least, treated in only an empirical manner). The discussion of open quantum systems is beyond the scope of this work but can be found elsewhere [9, 10].

(Eq. (1.15)).[6] Consider the case of a single spin: we have that $H_{\mathrm{J}} = 0$, and hence $H_{\mathrm{free}} = H_{\mathrm{cs}} = \omega_0 I_z$. Thus,

$$\rho_0 = \frac{\exp(cI_z)}{\mathrm{Tr}[\exp(cI_z)]} \approx \frac{E + cI_z}{\mathrm{Tr}(E + cI_z)} = E + cI_z, \tag{1.40}$$

where $c = -\beta\hbar\omega_0 = -\hbar\omega_0/(k_{\mathrm{B}}T)$ and E is the identity matrix; the approximation $\exp(cI_z) \approx E + cI_z$ is justified here as c is typically very small (on the order of 10^{-5}).

Throughout this thesis, I consider only linear transformations of the form in Eq. (1.38), which use unitary propagators of the form $U = \exp(-\mathrm{i}H\tau)$:

$$U\rho_0 U^\dagger = U(E + cI_z)U^\dagger = UEU^\dagger + c(UI_zU^\dagger) = E + c(UI_zU^\dagger). \tag{1.41}$$

When describing NMR experiments, it is typical to simply ignore both the E term as well as the proportionality factor c, and focus only on the transformation of the I_z term. Thus, one may define a 'simplified' equilibrium density operator[7]:

$$\rho_0' = I_z. \tag{1.42}$$

The E term is in fact truly inconsequential, as it cannot ever be transformed into detectable magnetisation. However, the constant c is still relevant: it is manifested in the magnitude of the NMR signal which is ultimately detected. It should be mentioned that ρ_0' is not a true density operator: for example, $\mathrm{Tr}(\rho_0') = 0$ and not 1 as is required for a density operator. Nonetheless, all the physically interesting dynamics of the system such as expectation values are fully contained within ρ_0' (at least up to the proportionality constant c).

1.4 Pulse Sequences

It is impossible to provide a full overview of all, or even most, of the major NMR experiments in widespread use; the reader is directed to other books for this purpose [2, 4, 5, 13, 14]. I seek only to (somewhat briefly) explain the general structure of 1D and 2D Fourier transform (FT) experiments, and in particular, how the formalisms developed in previous sections can be used to analyse and simulate such experiments.

[6] The interaction-picture $H_{\mathrm{free},I}$ would not be appropriate here, as its entire existence is merely a mathematical formalism. If that were not the case, it would imply that we can change the equilibrium state ρ_0 by simply *choosing* a different H_0 to factor out.

[7] This is similar to the 'deviation' density operator [11, 12] which measures how far a density operator deviates from the identity; but I have gone one step further in dropping the factor of c. Note that the alternative term 'reduced density operator' has a different meaning (it refers to the density operator of a subsystem, obtained by taking a partial trace over all other degrees of freedom).

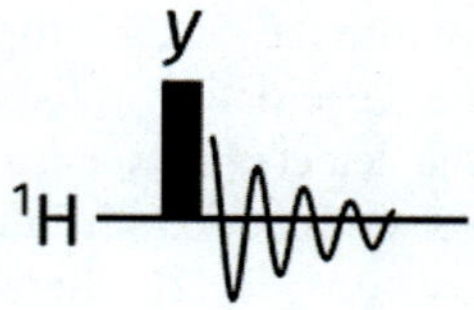

Fig. 1.1 1D ^{1}H pulse–acquire experiment

1.4.1 1D Pulse–Acquire

Consider the simplest NMR experiment, a 1D ^{1}H pulse–acquire spectrum (Fig. 1.1). This consists of a 90° pulse, immediately followed by detection; for convenience, we will first consider the pulse as being applied along the $+y$-axis, i.e. with a phase of $\phi = \pi/2$.

To understand this, we begin with the thermal density operator $\rho'_0 = I_z$ (Eq. (1.42)) and assume that there is only one spin in the sample, and that the pulse is applied on-resonance. The corresponding Hamiltonian during the pulse is simply $\omega_1 I_y$ (Eq. (1.29)). If the duration of the pulse is τ_p, then the density operator immediately following the pulse is given by

$$\rho = \exp(-\mathrm{i}\omega_1 I_y \tau_\text{p}) I_z \exp(\mathrm{i}\omega_1 I_y \tau_\text{p}) = \cos(\omega_1 \tau_\text{p}) I_z + \sin(\omega_1 \tau_\text{p}) I_x. \tag{1.43}$$

In this case, to obtain a 90° pulse, τ_p is specifically calibrated to ensure that $\omega_1 \tau_\text{p} = \pi/2$, which yields

$$\rho = I_x. \tag{1.44}$$

During the detection period, this term evolves under $H_\text{free} = H_\text{cs} = \omega_0 I_z$. (We use the Schrödinger-picture free Hamiltonian here because the measurement of the NMR signal takes place in the laboratory frame.) At a time t after detection has begun, the density operator is thus

$$\rho(t) = \exp(-\mathrm{i}\omega_0 I_z t) I_x \exp(\mathrm{i}\omega_0 I_z t) = \cos(\omega_0 t) I_x + \sin(\omega_0 t) I_y. \tag{1.45}$$

The NMR signal derives from both x- and y-magnetisation (M_x and M_y), which are in turn proportional to I_x and I_y by a factor of γ. (If multiple spins are present, then each spin induces its own magnetisation: we would have that $M_x = \sum_i \gamma_i I_{ix}$, and likewise for M_y.) These are then combined to form a complex signal (this process is known as *quadrature detection*):

$$\begin{aligned} s(t) &= M_x(t) + \mathrm{i}M_y(t) && (1.46) \\ &\propto \langle I_x(t)\rangle + \mathrm{i}\langle I_y(t)\rangle \\ &= \mathrm{Tr}[I_x \rho(t)] + \mathrm{i}\,\mathrm{Tr}[I_y \rho(t)] \\ &\propto \cos(\omega_0 t) + \mathrm{i}\sin(\omega_0 t) \\ &= \exp(\mathrm{i}\omega_0 t). && (1.47) \end{aligned}$$

Before the signal is digitised, the NMR spectrometer mixes this with a *reference* RF field oscillating at the transmitter frequency ω_{tx}. This results in downconversion of the detected frequencies by ω_{tx}, such that the actual digitised signal oscillates at the offset frequency Ω rather than ω_0 (recall we have chosen $\omega_{\text{rot}} = \omega_{\text{tx}}$, so $\omega_0 - \omega_{\text{tx}} = \omega_0 - \omega_{\text{rot}} = \Omega$). Therefore, instead of Eq. (1.47), the signal we really see is

$$s(t) \propto \exp(\mathrm{i}\Omega t). \tag{1.48}$$

This result is the same as if we had pretended that during the detection period, ρ evolved under the *interaction-picture* free Hamiltonian $H_{\text{free},I} = H_{\text{offset}}$; we will henceforth adopt this simplification, even though it is not physically accurate.

In practice, relaxation causes this signal to decay with time; this is frequently modelled as an exponential, in accordance with the Bloch equations [15]:

$$s(t) = \exp(\mathrm{i}\Omega t)\exp(-t/T_2), \tag{1.49}$$

where T_2 is the transverse relaxation time constant.[8] The NMR signal is thus often called a *free induction decay* (FID). Fourier transformation of the FID then yields a spectrum with absorption- and dispersion-mode lineshapes in the real and imaginary parts, respectively (Fig. 1.2):

$$\begin{aligned} S(\omega) = \mathcal{F}[s(t)] &= \frac{1}{\sqrt{2\pi}}\int_0^\infty s(t)\exp(-\mathrm{i}\omega t)\,\mathrm{d}t \\ &= \underbrace{\frac{k}{k^2 + (\omega - \Omega)^2}}_{\text{absorption: } A(\omega;\Omega)} + \mathrm{i}\underbrace{\frac{\Omega - \omega}{k^2 + (\omega - \Omega)^2}}_{\text{dispersion: } D(\omega;\Omega)}, \end{aligned} \tag{1.50}$$

where $k = 1/T_2$. The notation $A(\omega;\,\Omega)$ here means that the spectrum is a function of the frequency ω, but is parametrised by the peak offset Ω. Conventionally, only the real part of the spectrum is displayed, so it is desirable for the real part to contain the absorption-mode lineshape. This provides better resolution due to the narrower lineshape and is also less affected by cancellation when multiple peaks overlap.

Strictly speaking, the Lorentzian lineshapes above are only obtained when there is nonzero relaxation during the FID. For example, in the limit where $k \to 0$ (i.e. no relaxation at all), $A(\omega;\,\Omega)$ tends to a delta function $\delta(\omega = \Omega)$. However, for simplicity, in this thesis, I will drop the relaxation term $\exp(-kt)$ unless absolutely necessary; I will simply pretend that a signal of the form $s(t) = \exp(\mathrm{i}\Omega t)$ is directly Fourier transformed to give $A(\omega;\,\Omega) + \mathrm{i}D(\omega;\,\Omega)$.

Consider now changing the initial pulse such that it is applied along the $+x$-axis instead ($\phi = 0$). Repeating the above analysis, we find that the resulting signal will have a phase shift:

---

[8] Transverse (and longitudinal) relaxation is sometimes called spin–spin (and spin–lattice) relaxation, although the continued usage of these terms has been criticised [3, 4, 16].

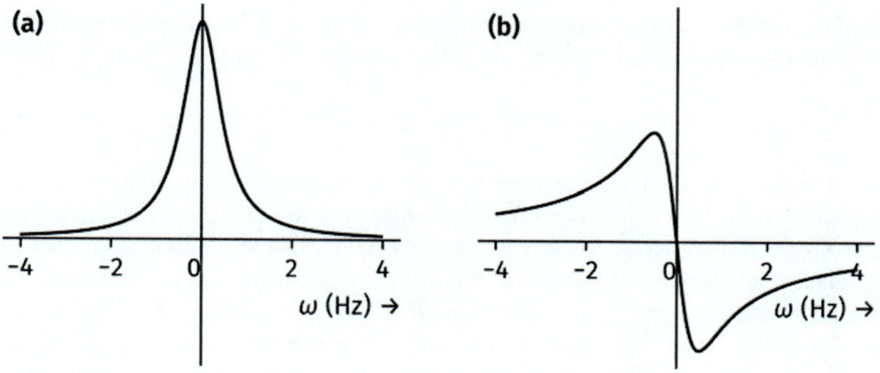

Fig. 1.2 **a** Absorption-mode lineshape $A(\omega;\, \Omega = 0)$. **b** Dispersion-mode lineshape $D(\omega;\, \Omega = 0)$. Both lines have been plotted using $k = \pi\,\mathrm{rad\,s^{-1}}$

$$s'(t) = -\mathrm{i}\exp(\mathrm{i}\Omega t) \tag{1.51}$$

$$\Rightarrow \quad S'(\omega) = \mathcal{F}[s'(t)] = D(\omega;\, \Omega) - \mathrm{i}A(\omega;\, \Omega). \tag{1.52}$$

If we were to plot the real part of the spectrum here, we would obtain the undesired dispersion-mode lineshape $D(\omega;\, \Omega)$. There are two ways of removing this phase shift. The first is to shift the *receiver phase* by ϕ_{rec}, which introduces an extra factor of $\exp(-\mathrm{i}\phi_{\mathrm{rec}})$ to the detected signal: we can thus choose $\phi_{\mathrm{rec}} = 3\pi/2$ in order to cancel out the $-\mathrm{i}$ term in $s'(t)$. Alternatively, the spectrum can be processed through *phase correction*, in which $S(\omega)$ is directly multiplied by a term $\exp(\mathrm{i}\phi_{\mathrm{corr}})$, where ϕ_{corr} is a linear function of the frequency ω:

$$\phi_{\mathrm{corr}} = \phi_{\mathrm{corr}}^{(0)} + \omega\phi_{\mathrm{corr}}^{(1)}. \tag{1.53}$$

$\phi_{\mathrm{corr}}^{(0)}$ and $\phi_{\mathrm{corr}}^{(1)}$ are, respectively, termed the *zeroth-* and *first-order phase corrections*: in this idealised case, we can simply choose $(\phi_{\mathrm{corr}}^{(0)}, \phi_{\mathrm{corr}}^{(1)}) = (\pi/2, 0)$ to again remove the unwanted phase shift. More realistically, due to instrumental imperfections, both of these values will have to be nonzero in order to ensure that every peak in the spectrum has the correct phase, i.e. is displayed in absorption mode.

An alternative framework for analysing pulse sequences is to use the ladder operators I_+ and I_- (Eq. (1.6)). Using the original example with our initial pulse on $+y$, the density operator immediately after the pulse is

$$\rho = I_x = \frac{1}{2}(I_+ + I_-), \tag{1.54}$$

and during detection, this evolves as

$$\rho(t) = \cos(\Omega t)I_x + \sin(\Omega t)I_y = \frac{1}{2}\left[\exp(-\mathrm{i}\Omega t)I_+ + \exp(\mathrm{i}\Omega t)I_-\right]. \tag{1.55}$$

(Notice that the $+1$-coherence I_+ actually evolves at the negative frequency $-\Omega$.) To obtain the same signal as previously done in Eq. (1.47), we 'detect' the I_- term:

$$s(t) \propto \mathrm{Tr}[I_- \rho(t)] \propto \exp(\mathrm{i}\Omega t), \tag{1.56}$$

which leads to the common assertion that *only quantum coherences of order –1 are detectable*. It is true that coherences with orders $p = 0, \pm 2, \pm 3, \ldots$ can never be detected in an FID. However, it is worth pointing out that the 'uniqueness' of -1-coherence is merely a result of how the x- and y-magnetisation are combined to form the complex signal (Eq. (1.46)). We do not *physically* detect I_-: we detect I_x and I_y, and combine them to form a complex signal which is mathematically equal to detecting I_-. If we had instead chosen to combine them in a different way, such as $s(t) = M_x(t) - \mathrm{i}M_y(t)$, this would give us $s(t) \propto \exp(-\mathrm{i}\Omega t)$—corresponding to 'detection' of $+1$-coherence—although this alternative does come with the drawback that frequencies must be reversed after Fourier transformation. In any case, we will stick to the established convention of detecting -1-coherence here.

To end this section, it should be pointed out that the complex signal is not obtained as an infinitely long, continuous function of time, as the treatment above implies. The complex-valued signal is digitised at an interval called the *dwell time*, τ_{dw}, and detection must be stopped after a finite period called the *acquisition time*, τ_{aq}. The transform being used is actually a discrete Fourier transform (DFT), which yields a periodic function $S(\omega)$; its period (in Hz) is given by $1/\tau_{\mathrm{dw}}$.[9] The NMR spectrum displayed to the user corresponds to one single period of $S(\omega)$, and thus the *spectral width* is also equal to $1/\tau_{\mathrm{dw}}$.[10] In principle, the periodicity of the DFT means that signals which would ordinarily fall outside of the spectral width would appear at incorrect frequencies in the spectrum [17]. On modern instrumentation, this is no longer the case for direct detection; peaks outside of the spectral width are removed using digital filters. However, *folding* or *aliasing* of peaks in the indirect dimension(s) of multidimensional NMR spectra still occur.

The DFT $S(\omega)$ is also a discrete function itself, and its resolution is given by $1/\tau_{\mathrm{aq}}$. It is possible to extend the effective acquisition time (and thus improve spectral resolution) without actually acquiring more data: this can be done either by *forward linear prediction* of the signal, or by simply adding zeros onto the end of the signal (*zero-filling*).

---

[9] The periodicity property of the DFT is equivalent to the Nyquist theorem, which is usually formulated as follows: the sampling rate required to correctly digitise a signal containing frequencies in the range $[0, f_{\max}]$ is $1/(2f_{\max})$. In the main text, it appears as if we have dropped the factor of 2 in the denominator; but in truth, this statement of the Nyquist theorem is applicable to *real-valued* signals, and here we have a *complex-valued* signal $s(t)$, which effectively doubles the range of correctly sampled frequencies.

[10] Frustratingly, the `DW` parameter in Bruker's TopSpin software is actually equal to $\tau_{\mathrm{dw}}/2$. The reason is because this parameter corresponds to the interval between which *real* data points are sampled, which is effectively twice as fast as complex-valued sampling.

Fig. 1.3 INEPT pulse sequence. The delay Δ is set to $1/(4 \cdot {}^1J_{\text{CH}})$

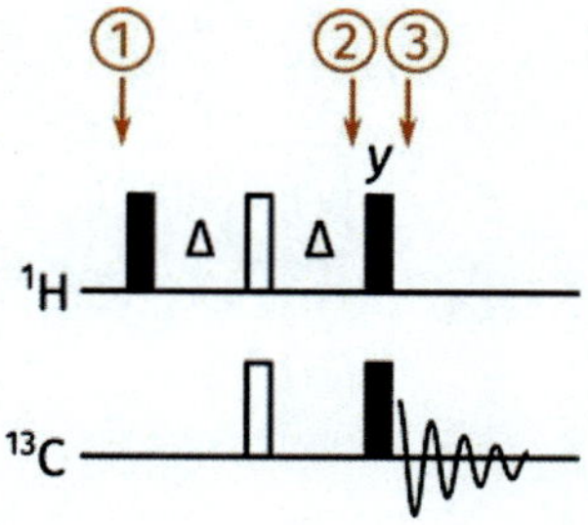

1.4.2 INEPT and Product Operators

Having tackled a simple single-spin case, we now move to the analysis of coupled spin systems and the development of the so-called 'product operator formalism' [18]. In particular, we look at the INEPT experiment [19, 20], in which magnetisation is transferred from a nuclide with a high magnetogyric ratio to one with a low magnetogyric ratio through a scalar coupling: for example, from ^{1}H to ^{13}C using the one-bond coupling constant, $^1J_{\text{CH}}$ (Fig. 1.3). Following tradition, the two nuclei are, respectively, labelled I and S.[11] The Schrödinger-picture free Hamiltonian for a weakly coupled system (cf. Eqs. (1.13) and (1.14)) is $H_{\text{free}} = \omega_{0,I} I_z + \omega_{0,S} S_z + 2\pi J_{IS} I_z S_z$. At the very beginning of the sequence (point ①), we formally have the equilibrium density operator

$$\rho_0 = \frac{\exp(-\beta\hbar H_{\text{free}})}{\text{Tr}[\exp(-\beta\hbar H_{\text{free}})]} \approx E - \beta\hbar(\omega_{0,I} I_z + \omega_{0,S} S_z + 2\pi J_{IS} I_z S_z), \qquad (1.57)$$

using the same approximations as in Eq. (1.40). The scalar coupling term can be safely neglected as $2\pi J_{IS}$ is several orders of magnitude smaller than the Larmor frequencies ω_0. After removing the physically irrelevant E term and factoring out a constant of $\beta\hbar B_0$, we end up with

$$\rho_0' = \gamma_I I_z + \gamma_S S_z. \qquad (1.58)$$

This represents equilibrium magnetisation (or *polarisation*) on both spins I and S, in proportion to their magnetogyric ratios. In general, an NMR experiment may manipulate—and ultimately detect—both of these terms. Since unitary evolution according to the Liouville–von Neumann equation is *linear*, in that $U(\rho_1 + \rho_2)U^\dagger = U\rho_1 U^\dagger + U\rho_2 U^\dagger$, we can treat these two terms separately: we focus first on the spin-I polarisation, $\rho_I = \gamma_I I_z$. The first 90°_x ^{1}H pulse tips this magnetisation into the transverse plane (ignoring off-resonance effects):

---

[11] This may seem insensible since I is the *sensitive* and S the *insensitive* nucleus, and indeed, in the original INEPT literature [19], the meanings of I and S were swapped. However, this usage has not been universal [21], and in modern usage, the identification of I as the sensitive nucleus seems to have prevailed.

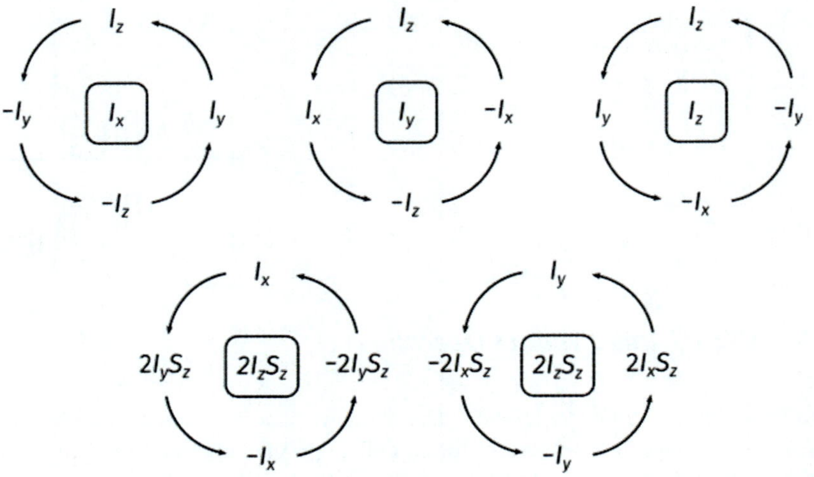

Fig. 1.4 Simplified rules for the evolution of product operators under different common Hamiltonians (offset, weak/secular J-coupling and pulses). These Hamiltonians often have the form ωM where M is some 'base operator', and are applied for a time τ. The boxed operators in the centre of each group refer to M; the initial state is then 'rotated' about this by an angle of $\omega\tau$ to obtain the final state, or more formally, it is transformed into itself times $\cos(\omega\tau)$, plus the next term in the cycle times $\sin(\omega\tau)$. For example, a 90°_x pulse has the 'base' operator I_x and the angle $\omega\tau = \pi/2$; thus, the initial state I_z would be rotated to $I_z \cos(\pi/2) - I_y \sin(\pi/2) = -I_y$

$$\rho_I \to \exp[-\mathrm{i}(\pi/2)I_x]\gamma_I I_z \exp[\mathrm{i}(\pi/2)I_x] = -\gamma_I I_y. \tag{1.59}$$

In principle, we could continue in this manner through repeated application of the 'sandwich' formulae (Eqs. (1.26) and (1.27)), as well as an analogous version for the I_zS_z term). For example, in the Δ delay which follows, we have that

$$\begin{aligned}\rho_I &\to -\gamma_I \exp(-\mathrm{i}H_{\text{free},I}\Delta)I_y \exp(\mathrm{i}H_{\text{free},I}\Delta)\\ &= -\gamma_I \exp(-\mathrm{i}H_\mathrm{J}\Delta)\exp(-\mathrm{i}H_\text{offset}\Delta)I_y \exp(\mathrm{i}H_\text{offset}\Delta)\exp(\mathrm{i}H_\mathrm{J}\Delta)\\ &= \ldots\end{aligned} \tag{1.60}$$

When performing simulations of NMR experiments, such as those in later chapters, this is precisely what happens, with the slight difference that the Liouville–von Neumann equation (Eq. (1.38)) is evaluated numerically rather than symbolically. Note that in going from the first to the second line, we can only 'split up' $H_{\text{free},I}$ into its constituent components H_offset and H_J because they commute.

When analysing pulse sequences by hand, however, it is far more convenient to use a set of heuristics which summarise the effects of various pulse sequence elements. For example, Fig. 1.4 summarises the evolution of a density operator under a single term of the Hamiltonian: as above, since $[H_\text{offset}, H_\mathrm{J}] = 0$, we only need to consider one term at a time. More high-level rules may be devised as well: for example, during the Δ–$180^\circ_x(I, S)$–Δ spin echo which comes next, the J_{IS} interaction in $H_{\text{free},I}$ is allowed to evolve for a period of 2Δ, but the offset term is *refocused* and can be ignored. (The sign inversion caused by the 180° pulses must also be included.) As per Fig. 1.4, this transforms the $-I_y$ term to $-2I_xS_z$ at point ②: the Hamiltonian is $\pi J_{IS} 2I_zS_z$ for a total time of $2\Delta = 1/(2J_{IS})$, so the 'angle' rotated through is $\pi J_{IS}/(2J_{IS}) = \pi/2$. Immediately after this, the $90^\circ_x(I, S)$ pair of pulses rotates this magnetisation to $-2I_zS_y$ (point ③). These transformations are often denoted with simpler notation:

$$\gamma_I I_z \xrightarrow{90^\circ_x(I)} -\gamma_I I_y \xrightarrow{\Delta-180^\circ_x(I,S)-\Delta} -2\gamma_I I_x S_z \xrightarrow{90^\circ_y(I),90^\circ_x(S)} -2\gamma_I I_z S_y. \tag{1.61}$$

During the detection period, the term $-2\gamma_I I_z S_y$ evolves as

$$\begin{aligned} -2\gamma_I I_z S_y \xrightarrow{H_{\text{free},I}} & -2\gamma_I I_z S_y \cos(\Omega_S t)\cos(\pi J t) + \gamma_I S_x \cos(\Omega_S t)\sin(\pi J t) \\ & -2\gamma_I I_z S_x \sin(\Omega_S t)\cos(\pi J t) + \gamma_I S_y \sin(\Omega_S t)\sin(\pi J t), \end{aligned} \tag{1.62}$$

from which we extract the complex signal

$$s_I(t) = \langle S_x(t)\rangle + \mathrm{i}\langle S_y(t)\rangle = \frac{\gamma_I}{2\mathrm{i}}\big\{\exp[\mathrm{i}(\Omega_S + \pi J_{IS})t] - \exp[\mathrm{i}(\Omega_S - \pi J_{IS})t]\big\}. \tag{1.63}$$

After Fourier transformation, the resulting spectrum has two peaks with frequencies $\Omega_S \pm \pi J_{IS}$ and opposite phases; because of the factor of $1/(2\mathrm{i})$, the real part of the spectrum will contain dispersion-mode signals (Fig. 1.5a). If desired, zeroth-order phase correction can be performed here, yielding instead a pair of absorption-mode signals still with opposite phases (Fig. 1.5b). In either case, this is termed an *antiphase* doublet; the product operators which give rise to it ($2I_zS_x$ and $2I_zS_y$) are said to be antiphase with respect to spin I. Importantly, the amplitude of the signal scales as γ_I rather than γ_S; since $\gamma_I > \gamma_S$, this represents a sensitivity enhancement compared to the direct excitation of S-magnetisation.

Of course, this is only half of the picture; we have not considered what happens to the other part of the magnetisation, namely $\rho_S = \gamma_S S_z$. Clearly, this is unaffected by the initial $90^\circ_x(I)$ pulse and the first Δ delay. The $180^\circ_x(S)$ pulse inverts it, and the final $90^\circ_x(S)$ pulse in fact transforms it into observable S-magnetisation:

$$\gamma_S S_z \xrightarrow{90^\circ_x(I)-\Delta} \gamma_S S_z \xrightarrow{180^\circ_x(I),180^\circ_x(S)-\Delta} -\gamma_S S_z \xrightarrow{90^\circ_y(I),90^\circ_x(S)} \gamma_S S_y. \tag{1.64}$$

This term produces *in-phase* spin-S magnetisation during the detection period (where the two components of the doublet have the same phase):

$$s_S(t) = \frac{\mathrm{i}\gamma_S}{2}\big\{\exp[\mathrm{i}(\Omega_S + \pi J_{IS})t] + \exp[\mathrm{i}(\Omega_S - \pi J_{IS})t]\big\}, \tag{1.65}$$

Because of the factor of i, the real part of the spectrum will contain a dispersion-mode doublet (Fig. 1.5c). The signal actually measured by the spectrometer is

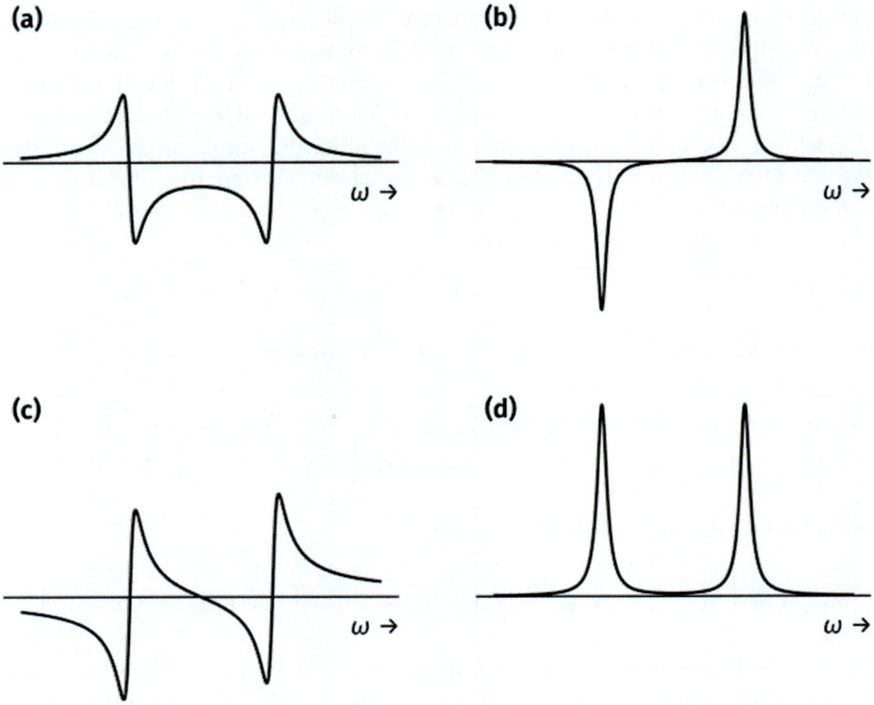

Fig. 1.5 Peak shapes of a doublet. In all cases, the separation between the two peaks is $2\pi J_{IS}$. **a** Antiphase, dispersion-mode. **b** Antiphase, absorption-mode. **c** In-phase, dispersion-mode. **d** In-phase, absorption-mode

$s(t) = s_I(t) + s_S(t)$; and the spectrum is a weighted sum of in-phase and antiphase magnetisation. This leads to potentially unwanted phase distortions in the spectrum, which one would prefer to suppress.

This can be accomplished through the technique of *phase cycling*, where pulse and receiver phases are changed in concert and the resulting FIDs summed in order to select for a particular signal. In this case, the INEPT experiment is performed twice, once with the phases as given in Fig. 1.3, and once where the initial $90^\circ_x(I)$ pulse is replaced with a $90^\circ_{-x}(I)$ pulse. The first of these gives us the same signals as above. However, in the second, inverting the initial I pulse leads to s_I acquiring a minus sign, because the initial I_z term is rotated to I_y instead of $-I_y$. On the other hand, the signal component s_S is unaffected by this pulse and thus does not experience a change of sign. The two FIDs we record are thus as follows:

$$s_1(t) = s_I(t) + s_S(t); \tag{1.66}$$

$$s_2(t) = -s_I(t) + s_S(t). \tag{1.67}$$

Simply taking the difference of these two FIDs yields a signal where the desired s_I has been accumulated and s_S has been cancelled out. In practice, instead of subtracting the two signals, it is typical to shift the receiver phase ϕ_{rec} by π (i.e. 180°) in the second experiment: this introduces a phase shift of $\exp(-\mathrm{i}\pi) = -1$ to the signal, and the two signals can now be *added* together instead of subtracted to cancel out s_S. Since both ϕ_1 and ϕ_{rec} are 0 on the first experiment and π on the second experiment, we can express this as $\phi_1 = \phi_{\text{rec}} = (0, \pi)$. This is more commonly denoted as $\phi_1 = \phi_{\text{rec}} = (x, -x)$, because the phases $(0, \pi)$ correspond to the $+x$- and $-x$-axes, respectively.

The 'simplified' analysis of pulse sequences shown in Eqs. (1.61) and (1.64) is often called *'product operator'* analysis [18], because the underlying two-spin operators are products of single-spin Cartesian operators. Although this is often touted as being 'simpler' than full-density operator calculations, it is really just a shorthand which masks the quantum mechanical theory developed in this chapter:

$$\underbrace{I_z \xrightarrow{90^\circ_x(I)} -I_y}_{\textit{product operator}} \iff \underbrace{\exp(-\mathrm{i}I_x\pi/2)I_z\exp(\mathrm{i}I_x\pi/2) = -I_y}_{\textit{density operator}}. \tag{1.68}$$

Since the operators $\{E, I_x, I_y, I_z\}$ form a complete basis for a single-spin system, their products (i.e. product operators) likewise form a complete basis for multiple-spin systems, and so *any* density matrix for a multiple-spin system may be expressed as a linear combination of product operators. Thus, strictly speaking, the use of product operators therefore does not actually sacrifice any power in and of itself. However, the heuristics such as those in Fig. 1.4 *are* limiting, in that the evolution under some Hamiltonians—for example, strong coupling $\mathbf{I} \cdot \mathbf{S}$, or pulses for off-resonance spins where H is a sum of I_x and I_z—cannot be neatly captured in such a pictorial form.

1.4.3 2D NMR: General Principles

Much of this thesis is concerned with two-dimensional (2D) NMR experiments. Before considering an explicit example of a 2D experiment, we will first describe some general principles [22, 23]. 2D experiments contain one *indirect* and one *direct* time dimension, traditionally labelled t_1 and t_2. The t_1 period is a variable period in the pulse sequence which starts at 0[12] and is incremented by a constant amount $\delta(t_1)$ on every iteration of the sequence. For each value of t_1 (or each t_1 *increment*), one complex FID $s'(t_2)$ is obtained. Putting this all together, the raw data thus take the form $s(t_1, t_2)$, which can be viewed as a 2D data matrix. Fourier transformation in both dimensions leads to a spectrum $S(\omega_1, \omega_2)$. In experimental contexts, the

[12] Or *as close to 0 as possible*, considering that pulse elements during t_1—such as the $180^\circ(I)$ pulse in Fig. 1.6—require finite amounts of time. In some cases, it is possible to arrange spin echoes around t_1 such that the t_1 evolution on the first increment is refocused.

frequency dimensions are often referred to as F_1 and F_2, but in this chapter, I stick to the more mathematically consistent ω.

2D experiments generally comprise four components: *preparation*, *evolution*, *mixing* and *detection*. The t_1 and t_2 periods correspond to evolution and detection, respectively. The relevant product operators broadly follow this pattern:

$$\rho_0' \xrightarrow{\text{preparation}} P \xrightarrow{\text{evolution}} P\cos(\Omega_P t_1) + P'\sin(\Omega_P t_1) \xrightarrow{\text{mixing}} Q\cos(\Omega_P t_1) + \cdots \tag{1.69}$$

The role of the preparation period is to generate some operator P from the equilibrium magnetisation ρ_0'. During the t_1 period, P evolves into P' at a frequency of Ω_P (as per Fig. 1.4): this leads to two terms which are *amplitude-modulated* in t_1. The mixing period simply transforms P into some other operator Q while preserving its t_1 modulation; for now, we will assume that the sine-modulated term P' is turned into unobservable magnetisation.

During the detection period, the x- and y-magnetisation generated by the term Q is recorded. As discussed in Sect. 1.4.1, Q must therefore be (or at least contain) -1-quantum coherence on some spin, and thus evolves at the offset of that spin, Ω_Q. The complex signal $s(t_1, t_2)$ is therefore of the form $\cos(\Omega_P t_1)\exp(\mathrm{i}\Omega_Q t_2)$. Note that, since the term P is not directly detected, it can have *any* coherence order; 2D NMR thus allows us to 'detect' frequencies of otherwise unobservable coherences.

We can equivalently express this signal as a sum of complex exponentials, which is easier to Fourier transform:

$$s(t_1, t_2) = \cos(\Omega_P t_1)\exp(\mathrm{i}\Omega_Q t_2) = \frac{1}{2}\big[\exp(\mathrm{i}\Omega_P t_1)\exp(\mathrm{i}\Omega_Q t_2) + \exp(-\mathrm{i}\Omega_P t_1)\exp(\mathrm{i}\Omega_Q t_2)\big]. \tag{1.70}$$

Fourier transformation of the first term reveals a peak centred at $(\omega_1, \omega_2) = (\Omega_P, \Omega_Q)$:

$$\begin{aligned}\mathcal{F}[\exp(\mathrm{i}\Omega_P t_1)\exp(\mathrm{i}\Omega_Q t_2)] &= [A_1(\Omega_P) + \mathrm{i}D_1(\Omega_P)][A_2(\Omega_Q) + \mathrm{i}D_2(\Omega_Q)] \\ &= A_1(\Omega_P)A_2(\Omega_Q) - D_1(\Omega_P)D_2(\Omega_Q) \\ &\quad + \mathrm{i}[A_1(\Omega_P)D_2(\Omega_Q) + D_1(\Omega_P)A_2(\Omega_Q)],\end{aligned} \tag{1.71}$$

where we have used the shorthand $A_i(\Omega)$ to denote what should properly be written as $A(\omega_i; \Omega)$. We should ideally like our 2D peaks to be absorption-mode in both dimensions, i.e. $S(\omega_1, \omega_2) = A_1(\Omega_P)A_2(\Omega_Q)$. Unfortunately, if we take the real part of this spectrum, we have an undesirable mixture of double absorption and double dispersion: this peak shape is called a *phase twist*. It is not possible to extract only the double-absorption part through phase correction.

This is not the only problem, however: if we perform the same Fourier transformation on the second term in Eq. (1.70), we get *another* phase twist peak but this time centred at $(\omega_1, \omega_2) = (-\Omega_P, \Omega_Q)$. So, merely recording the complex signal $s(t) = \cos(\Omega_P t_1)\exp(\mathrm{i}\Omega_Q t_1)$ is clearly not good enough:

- firstly, our spectra are not *pure phase* or *phase-sensitive*, in that the peaks are an inseparable mixture of absorptive and dispersive lineshapes;
- secondly, we have lost *quadrature detection* in the indirect dimension, in that we cannot distinguish positive and negative offsets.

There are multiple different ways of solving these dual issues, which are extensively covered in NMR textbooks [1–4, 14]. Among these are the States method [24], the time-proportional phase incrementation(TPPI) method [25], and the echo–antiecho (EA) method. I will briefly cover the States and EA methods; the TPPI method can be shown to be essentially mathematically equivalent to the States method [26].[13]

In the States method, the cosine-modulated signal described above (and denoted $s_{\cos}(t_1, t_2)$ here) forms only one part of the signal. It is also necessary to acquire a *sine-modulated signal* of the form $s_{\sin}(t_1, t_2) = \sin(\Omega_P t_1) \exp(\mathrm{i}\Omega_Q t_2)$. Once we have these two datasets, we can perform a Fourier transform along ω_2 first to get two intermediate signals:

$$s'_{\cos}(t_1, \omega_2) = \cos(\Omega_P t_1)[A_2(\Omega_Q) + \mathrm{i}D_2(\Omega_Q)]; \tag{1.72}$$

$$s'_{\sin}(t_1, \omega_2) = \sin(\Omega_P t_1)[A_2(\Omega_Q) + \mathrm{i}D_2(\Omega_Q)]. \tag{1.73}$$

These idealised signals have the absorption-mode component A_2 as the real part in the ω_2 dimension. However, as before, an experimentally measured signal will typically require phase correction, which can be performed at this point by multiplication with a phase factor $\exp(\mathrm{i}\phi_{\mathrm{corr},2})$. After any phase correction, we discard the imaginary parts of these signals and use their real parts to construct another complex signal:

$$s''(t_1, \omega_2) = \Re\{s'_{\cos}(t_1, \omega_2)\} + \mathrm{i}\Re\{s'_{\sin}(t_1, \omega_2)\} = \exp(\mathrm{i}\Omega_P t_1)A_2(\Omega_Q). \tag{1.74}$$

Fourier transformation along ω_1 yields

$$S(\omega_1, \omega_2) = [A_1(\Omega_P) + \mathrm{i}D_1(\Omega_P)]A_2(\Omega_Q), \tag{1.75}$$

which represents a peak only at the correct frequency (Ω_P, Ω_Q), and the real part of which is the desired double-absorption lineshape. Phase correction along ω_1 can now be carried out using a different phase factor $\exp(\mathrm{i}\phi_{\mathrm{corr},1})$. (In practice, the imaginary

[13] There is a slight difference in that the TPPI method pushes *axial peaks*—artefacts arising at $\omega_1 = 0$—to the edge of the spectrum, whereas with the States method, these artefacts appear in the middle of the spectrum, potentially obscuring useful peaks. The reader is referred to the references cited in the main text for a discussion of this. In practice, both the States and EA methods can be easily adapted to incorporate this shifting of axial peaks by phase shifting the t_1 modulation as well as the receiver by 180° every time t_1 is incremented (in the former case it is creatively known as the *States–TPPI method*). Thus, I do not consider the TPPI method any further here.

parts of $s'_{\cos}$ and $s'_{\sin}$ are not actually discarded: they are retained to allow phase correction in ω_2 to be performed at a later stage if necessary.)[14]

Naturally, the obvious question is how the sine-modulated signal $s_{\sin}(t_1, t_2)$ can be obtained. Returning to the product operators in Eq. (1.69), we see that we can obtain $s_{\sin}$ if we change the mixing period to transform $P' \rightarrow Q$ instead of $P \rightarrow Q$. This can usually be done by phase shifting one or more pulses after the t_1 period by 90°.[15] Alternatively (in fact, more commonly), we can modify the preparation period such that it produces an operator $-P'$ which rotates into P during t_1. That way, after the t_1 evolution period we have a density operator of the form $-P' \cos(\omega_P t_1) + P \sin(\omega_P t_1)$; if we keep the mixing period the same then we will obtain the desired sine-modulated data. This can be accomplished by phase shifting one or more pulses *before* the t_1 period by –90°.

On the other hand, the EA method seeks instead to measure two signals which are *phase-modulated* in t_1 (instead of *amplitude-modulated* as in the States method):

$$s_{\text{echo}}(t_1, t_2) = \frac{1}{2} \exp(-\mathrm{i}\Omega_P t_1) \exp(\mathrm{i}\Omega_Q t_2); \tag{1.76}$$

$$s_{\text{antiecho}}(t_1, t_2) = \frac{1}{2} \exp(\mathrm{i}\Omega_P t_1) \exp(\mathrm{i}\Omega_Q t_2). \tag{1.77}$$

Once recorded, these signals can be added and subtracted to obtain the cosine- and sine-modulated signals of the States method; the standard States processing can then follow. The echo and antiecho signals can be obtained through the use of pulsed field gradients during the t_1 period, though a concrete example is deferred until the next section. The factor of 1/2 in Eqs. (1.76) and (1.77) requires an explanation: this arises because the 'original' signal is cosine- or sine-modulated, which

[14] This treatment of the States method stems from the original paper [24] and is consistently reproduced in textbooks, although I have personally always found the steps rather contrived. I argue instead that a more coherent formulation can be given in terms of *quaternions*, a type of *hypercomplex number*, expressed as $a + \mathrm{i}b + \mathrm{j}c + \mathrm{k}d$ where $a, b, c, d \in \mathbb{R}$, $\mathrm{i}^2 = \mathrm{j}^2 = \mathrm{k}^2 = -1$, and $\mathrm{ij} = -\mathrm{ji} = \mathrm{k}$. We can write that $s_{\cos} = \cos(\Omega_P t_1) \exp(\mathrm{j}\Omega_Q t_2)$, and that $s_{\sin} = \sin(\Omega_P t_1) \exp(\mathrm{j}\Omega_Q t_2)$; then, we can directly form a quaternionic 2D matrix $s(t_1, t_2) = s_{\cos} + \mathrm{i}s_{\sin} = \exp(\mathrm{i}\Omega_P t_1) \exp(\mathrm{j}\Omega_Q t_2)$. At this point, we can use a quaternion Fourier transform to *directly* obtain the data: $S(\omega_1, \omega_2) = (2\pi)^{-1} \iint \exp(-\mathrm{i}\omega_1 t_1) s(t_1, t_2) \exp(-\mathrm{j}\omega_2 t_2)\, \mathrm{d}t_1\, \mathrm{d}t_2$, and phase correction in *both* dimensions can be carried out at will and simultaneously using $\exp(\mathrm{i}\phi_{\text{corr},1}) S(\omega_1, \omega_2) \exp(\mathrm{j}\phi_{\text{corr},2})$, where the ϕ_{corr}'s represent the phase corrections in the two respective frequency dimensions. (Note that quaternion multiplication is non-commutative, so the ordering of terms in these expressions matters.) Indeed, in the representation of 2D data as used by Bruker instruments, the four files `2rr`, `2ri`, `2ir`, and `2ii` essentially correspond to the four components (or *quadrants*) of a quaternionic $S(\omega_1, \omega_2)$. Of course, this is merely nice notation: none of the underlying science is changed. But the beauty of this is that we can consider the steps given in the main text to be an *implementation* of something more fundamental (the quaternion Fourier transform), a generalisation which naturally follows from the 1D case and immediately suggests its extension to the 3D case, rather than a *prescription* which—at least to a novice—seems to work only as if by magic. This treatment has been proposed before by Delsuc [27], although it does not appear to have caught on much.

[15] Actually, if double-quantum coherence is sampled in t_1, then the phase shift used must be halved, i.e. 45°; and likewise for higher coherence orders. But we will not encounter such cases in this thesis.

can be thought of as a sum of two opposite phase modulations, i.e. $\cos(\Omega_P t_1) = [\exp(-\mathrm{i}\Omega_P t_1) + \exp(\mathrm{i}\Omega_P t_1)]/2$. The gradients effectively select for only one sense of the phase modulation and reject the other. Although this factor of 1/2 can be cancelled out when combining the echo and antiecho datasets, in that:

$$s_{\text{cos,EA}} = s_{\text{echo}} + s_{\text{antiecho}} = \cos(\Omega_P t_1)\exp(\mathrm{i}\Omega_Q t_2) = s_{\cos} \tag{1.78}$$

(and likewise for the sine component), the process of adding up two separate datasets leads to a $\sqrt{2}$ increase in the noise level when compared to measuring $s_{\cos}$ directly. Therefore, EA spectra have a $\sqrt{2}$ decrease in signal-to-noise ratio (SNR) compared to their States counterparts. This point is explored more fully in an article by Keeler and coworkers [28]. Despite this loss in SNR, the use of gradients typically leads to spectra with far better artefact suppression, completely obviating the need for long phase cycles. Consequently, a significant proportion of modern 2D experiments—especially heteronuclear experiments—use EA selection.

1.4.4 The States HSQC Experiment

To illustrate the ideas developed in the previous section, we now turn our attention to two typical implementations of the 2D HSQC experiment, where both phase cycling as well as gradients are used to select for particular product operators. Figure 1.6 shows an HSQC experiment where quadrature detection in the indirect dimension is performed using the States method.

The HSQC experiment seeks to only detect protons directly bonded to ^{13}C; the signals from all other protons must be suppressed. We first consider the simplest possible implementation of the HSQC, shown in Fig. 1.6; as before, we will illustrate this with an IS spin pair. It can be shown that the equilibrium S-magnetisation cannot be transformed into observable I-magnetisation at the end of the sequence, so we will simply take $\rho_0' = I_z$. Consider first the 'basic' case where $\phi_1 = x$. The *preparation*

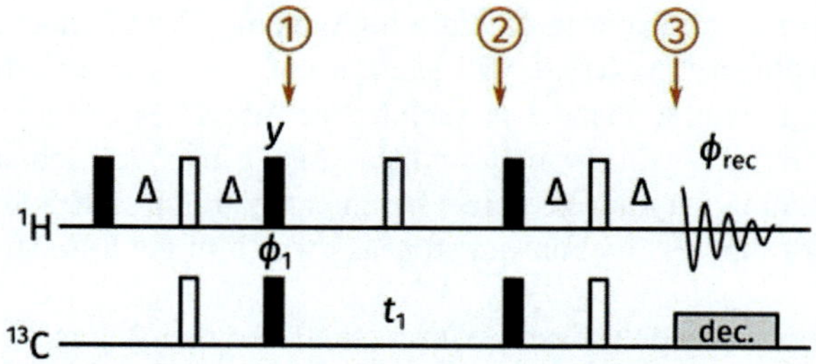

Fig. 1.6 A typical HSQC pulse sequence utilising the States method for quadrature detection in ω_1. The delay Δ is set to $1/(4 \cdot {}^1J_{\text{CH}})$. To record the cosine-modulated dataset $s_{\cos}(t_1, t_2)$, we set $\phi_1 = \phi_{\text{rec}} = (x, -x)$. The sine-modulated dataset $s_{\sin}(t_1, t_2)$, on the other hand, is obtained using $\phi_1 = (-y, y)$ and $\phi_{\text{rec}} = (x, -x)$

period is just the same INEPT element as analysed in Sect. 1.4.2, so at point ① we already know that we have

$$\rho = -2I_zS_y. \tag{1.79}$$

During t_1, the J-coupling is refocused, but the offset of spin S evolves to give:

$$\rho = 2I_zS_y\cos(\Omega_S t_1) - 2I_zS_x\sin(\Omega_S t_1), \tag{1.80}$$

at point ②, bearing in mind that the $180^\circ_x(I)$ pulse inverts the I_z component of both terms. Immediately after this, the reverse INEPT element (the 'mixing' segment) transfers the amplitude-modulated $2I_zS_y$ coherence I_x for detection. However, the $2I_zS_x$ term is turned into a mixture of unobservable zero- and double-quantum coherence, such that immediately before detection (point ③) we have:

$$\rho = I_x\cos(\Omega_S t_1) = \frac{1}{2}(I_+ + I_-)\cos(\Omega_S t_1). \tag{1.81}$$

Performing the usual quadrature detection in the direct t_2 dimension gives us the desired cosine-modulated signal of

$$s_{\cos}(t_1, t_2) = \frac{1}{2}\cos(\Omega_S t_1)\exp(\mathrm{i}\Omega_I t_2), \tag{1.82}$$

In a similar way, it can be shown that setting $\phi_1 = -y$ yields the following after t_1 (point ②)

$$\rho = 2I_zS_x\cos(\Omega_S t_1) + 2I_zS_y\sin(\Omega_S t_1), \tag{1.83}$$

which after the reverse INEPT gives us the desired sine-modulated signal

$$s_{\sin}(t_1, t_2) = \frac{1}{2}\sin(\Omega_S t_1)\exp(\mathrm{i}\Omega_I t_2). \tag{1.84}$$

These can be combined in the way described in Sect. 1.4.3 to yield an HSQC spectrum with double absorption-mode lineshapes $A_1(\Omega_S)A_2(\Omega_I)$.

This alone is not sufficient to obtain a high-quality HSQC spectrum, however. Not every ^{1}H spin in a molecule will share a one-bond coupling to a ^{13}C spin: for example, in a natural-abundance sample, the HSQC experiment only detects around 1% of ^{1}H spins. Even small artefacts arising from the remaining 99% of proton magnetisation may have comparable intensity to the signals of interest. We therefore need to suppress any unwanted peaks which might arise from this *'bulk'* magnetisation.

The simplest way to do this is to perform (at least) a two-step phase cycle, where the pulse phase ϕ_1 and receiver phase ϕ_{rec} are simultaneously inverted. This has no effect on the desired signal (it picks up two minus signs which cancel out), but will lead to cancellation of any signal arising from the bulk magnetisation, which does not evolve under H_J and therefore cannot be affected by the pulse on spin S.

Note the similarity to the INEPT phase cycling in Sect. 1.4.2, where we chose to cycle a pulse which the undesired pathway did not 'experience': this is not the only possible choice, but is one that is conceptually simple to understand. Naturally, this phase cycling must be carried out for both the cosine- and sine-modulated datasets. In practice, even longer phase cycles are typically required to deal with experimental imperfections.

1.4.5 The Echo–Antiecho HSQC: Gradients and Coherence Selection

The EA version of the HSQC experiment (Fig. 1.7) is very similar to the States version discussed above. The only real difference is that two gradients are added: one immediately after t_1, and one directly before detection. As we will see, the

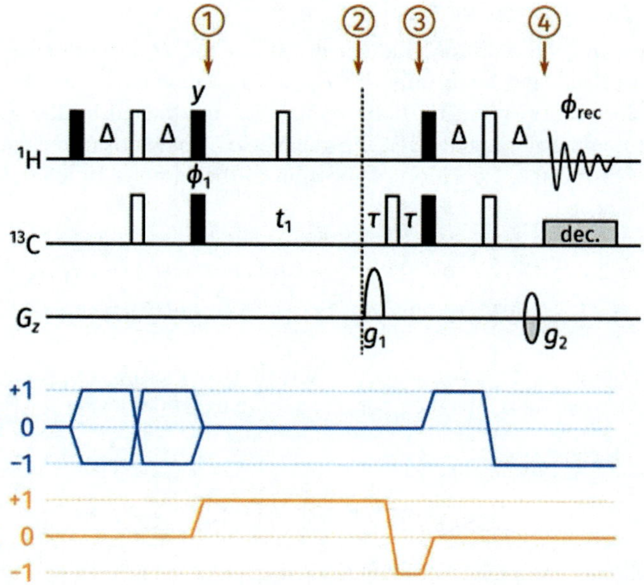

Fig. 1.7 A typical EA HSQC pulse sequence. The delay Δ is set to $1/(4 \cdot {}^1J_{CH})$. The gradient amplitudes are chosen such that $|g_1/g_2| = \gamma_H/\gamma_C \approx 4$. Specifically, the echo dataset is obtained by setting $(g_1, g_2) = (80\%, 20\%)$, and the antiecho dataset by setting $(g_1, g_2) = (80\%, -20\%)$, where gradient amplitudes are quoted as a percentage of the maximum gradient amplitude. Unlike in the States HSQC (Fig. 1.6), phase cycling of ϕ_1 and ϕ_{rec} is no longer mandatory as the gradients g_1 and g_2 dephase unwanted magnetisation (although it can still be performed to attain *even* better artefact suppression). τ represents the duration of both gradients, and is usually on the order of 1 ms. The lines below the pulse sequence illustrate the coherence orders selected for during the echo experiment, which are collectively referred to as a coherence transfer pathway (CTP): for the antiecho experiment the ^{13}C CTP must be inverted

effect of this is to enforce a relationship between the coherence orders during the two gradients, or in other words, to select for a specific *coherence transfer pathway* (CTP), as illustrated by the lines beneath the pulse sequence in Fig. 1.7.

Brute-Force Analysis

Recall that the Hamiltonian caused by a gradient of strength G is given by $H_{\text{grad}} = \sum_i \gamma_i G z I_{iz}$ (Eq. (1.11)), where z is the z-position of the spin. In the case of our two-spin system, we can more explicitly write this as:

$$H_{\text{grad}} = \gamma_I G z I_z + \gamma_S G z S_z. \tag{1.85}$$

Points ① and ② are the same as in the States HSQC (Fig. 1.6), so from Eq. (1.80) we know that the density operator at point ② is:

$$\rho = 2I_z S_y \cos(\Omega_S t_1) - 2I_z S_x \sin(\Omega_S t_1). \tag{1.86}$$

In the next spin echo with the gradient g_1, H_{offset} and H_{J} are refocused, so we can ignore their effects and focus only on H_{grad}. We assume here that the gradients are applied with duration τ.[16] The I_z terms in ρ are unaffected by the gradient since they commute with H_{grad}; however, the transverse S-magnetisation acquires a phase which depends on the z-position of the spin in the sample. Immediately after the gradient, we have:

$$\begin{aligned}\rho(z) = {} & 2I_z S_y \cos(\Omega_S t_1)\cos(\gamma_S g_1 z\tau) - 2I_z S_x \cos(\Omega_S t_1)\sin(\gamma_S g_1 z\tau) \\ & - 2I_z S_x \sin(\Omega_S t_1)\cos(\gamma_S g_1 z\tau) - 2I_z S_y \sin(\Omega_S t_1)\sin(\gamma_S g_1 z\tau),\end{aligned} \tag{1.87}$$

where the notation $\rho(z)$ reminds us that this density matrix is spatially dependent. The $180^\circ_x(S)$ pulse then flips the S_y terms to give us, at point ③,

$$\begin{aligned}\rho(z) = {} & -2I_z S_y \cos(\Omega_S t_1)\cos(\gamma_S g_1 z\tau) - 2I_z S_x \cos(\Omega_S t_1)\sin(\gamma_S g_1 z\tau) \\ & - 2I_z S_x \sin(\Omega_S t_1)\cos(\gamma_S g_1 z\tau) + 2I_z S_y \sin(\Omega_S t_1)\sin(\gamma_S g_1 z\tau).\end{aligned} \tag{1.88}$$

We know already from the States HSQC analysis that the mixing period (i.e. reverse INEPT) causes the transformation $2I_z S_y \to I_x$, and that the $2I_z S_x$ term is lost. So, *if* the gradient g_2 were absent, we would have the following terms at point ④:

$$\begin{aligned}\rho(z) &= -I_x \cos(\Omega_S t_1)\cos(\gamma_S g_1 z\tau) + I_x \sin(\Omega_S t_1)\sin(\gamma_S g_1 z\tau) \\ &= -I_x \cos(\Omega_S t_1 + \gamma_S g_1 z\tau).\end{aligned} \tag{1.89}$$

[16] In practice, we also need to include a gradient recovery delay immediately after the gradient to allow for the dissipation of eddy currents, which causes the spin echo to be lengthened slightly. This is inconsequential to the analysis and will be ignored here.

This is of course not the case. In principle, the last Δ delay should be split up into two parts: one of duration $(\Delta - \tau)$ where only $H_{\mathrm{free},I}$ is active, and one of duration τ where the Hamiltonian $H_{\mathrm{free},I} + H_{\mathrm{grad}}$ operates. Thankfully, we know that H_{grad} commutes with $H_{\mathrm{free},I}$, so we can in fact separate the relevant propagators:

$$\begin{aligned}
&\exp[-\mathrm{i}(H_{\mathrm{free},I} + H_{\mathrm{grad}})\tau]\exp[-\mathrm{i}H_{\mathrm{free},I}(\Delta - \tau)] \\
&\quad = \exp(-\mathrm{i}H_{\mathrm{grad}}\tau)\exp(-\mathrm{i}H_{\mathrm{free},I}\tau)\exp[-\mathrm{i}H_{\mathrm{free},I}(\Delta - \tau)] \\
&\quad = \exp(-\mathrm{i}H_{\mathrm{grad}}\tau)\exp(-\mathrm{i}H_{\mathrm{free},I}\Delta),
\end{aligned} \tag{1.90}$$

where the first term is just the gradient on its own, and the second is just the delay without a gradient. So, we can directly apply the propagator $\exp(-\mathrm{i}H_{\mathrm{grad}}\tau)$ to the density operator in Eq. (1.89) to get[17]:

$$\rho(z) = -I_x \cos(\Omega_S t_1 + \gamma_S g_1 z\tau)\cos(\gamma_I g_2 z\tau) - I_y \cos(\Omega_S t_1 + \gamma_S g_1 z\tau)\sin(\gamma_I g_2 z\tau) \tag{1.91}$$

$$\begin{aligned}
&= -\frac{1}{2}I_x \cos[\Omega_S t_1 + (\gamma_S g_1 + \gamma_I g_2)z\tau] - \frac{1}{2}I_x \cos[\Omega_S t_1 + (\gamma_S g_1 - \gamma_I g_2)z\tau] \\
&\quad - \frac{1}{2}I_y \sin[\Omega_S t_1 + (\gamma_S g_1 + \gamma_I g_2)z\tau] + \frac{1}{2}I_y \sin[\Omega_S t_1 + (\gamma_S g_1 - \gamma_I g_2)z\tau].
\end{aligned} \tag{1.92}$$

(As a sanity check, it can be verified that this reduces to Eq. (1.89) if $g_2 = 0$.) The signal which we detect stems from the entire sample length, so we in fact need to perform an integration over z:

$$\rho = \frac{1}{L} \cdot \int_{-L/2}^{L/2} \rho(z)\,\mathrm{d}z, \tag{1.93}$$

where the sample length is L and $z = 0$ is assumed to be the middle of the sample.

For the *echo* experiment, we choose the ratio $g_1/g_2 = \gamma_I/\gamma_S$; this means that $\gamma_S g_1 - \gamma_I g_2 = 0$. The second and fourth terms in Eq. (1.92) thus *lose* their z-dependence; when integrated over z these terms are unchanged. On the other hand, the first and third terms are attenuated by a factor proportional to $\int_{-L/2}^{L/2} \cos[(\gamma_S g_1 + \gamma_I g_2)z\tau]\,\mathrm{d}z$ (or one with the equivalent sine-modulated integrand). The term in square brackets here can be identified as the spatially dependent phase caused by the evolution under the gradient pulse; we say that the gradients *dephase* coherences. (In the case of the second and fourth terms, they also *rephase* coherences.) If the gradient strengths (g_1, g_2) and/or their durations τ are sufficiently large, the integral evaluates to a small number which is effectively zero, corresponding to complete dephasing. The result is that, just before detection, we have:

[17] We have implicitly assumed here that the z-coordinate of the IS spin pair during g_1 is the same as its z-coordinate during g_2, or in other words, that there is no *diffusion* or *convection* between the two gradients. In general, this is not true, and these effects will lead to a loss of signal as the rephasing by the second gradient is not perfect.

$$\rho_{\text{echo}} = -\frac{1}{2}I_x \cos(\Omega_S t_1) + \frac{1}{2}I_y \sin(\Omega_S t_1), \tag{1.94}$$

or, extracting the detectable I_- components,

$$\rho_{\text{echo}} = -\frac{1}{4}I_- \cos(\Omega_S t_1) + \frac{\mathrm{i}}{4}I_- \sin(\Omega_S t_1) = -\frac{1}{4}I_- \exp(-\mathrm{i}\Omega_S t_1). \tag{1.95}$$

Detection of this in t_2 gives us the echo signal

$$s_{\text{echo}}(t_1, t_2) = -\frac{1}{4}\exp(-\mathrm{i}\Omega_S t_1)\exp(\mathrm{i}\Omega_I t_2), \tag{1.96}$$

in accordance with Eq. (1.76). Note that the prefactor here is $-1/4$, instead of $1/2$ in the States method (Eqs. (1.82) and (1.84)): this accounts for the decreased SNR in the EA method as previously described. (The minus sign comes from the extra $180°(S)$ pulse after t_1, but is inconsequential as it can be removed by phase correction.)

In a similar way, it can be shown that if we invert the sign of g_2, we have that $g_1/g_2 = -\gamma_I/\gamma_S$. Now, the second and fourth terms in Eq. (1.92) are dephased, and we get the antiecho spectrum from the first and third terms:

$$s_{\text{antiecho}}(t_1, t_2) = -\frac{1}{4}\exp(\mathrm{i}\Omega_S t_1)\exp(\mathrm{i}\Omega_I t_2). \tag{1.97}$$

The Shift Basis

In the above treatment, I have only used the 'rules' developed so far for Cartesian product operators to explain the effects of gradients. *This is clearly very tedious!* When gradients are involved, it proves simpler to use a different basis, specifically $\{E, I_z, I_+, I_-\}$: this is called the *shift basis*.[18] The coherence orders of these, and their products, are easily read off from the terms involved: for example, I_-S_- is double-quantum coherence with $p = -2$. The evolution of these terms under various Hamiltonians is summarised by Keeler [4], but two cases are particularly simple and important here:

1. 180°_x pulses invert I_z and interchange $I_+ \leftrightarrow I_-$;
2. An operator I_{ip}, on a spin i and with coherence order p, evolves under the Hamiltonian ωI_z for a time t to give:

$$I_{ip} \longrightarrow I_{ip}\exp(-\mathrm{i}p\omega t). \tag{1.98}$$

[18] The normalisation factors of each of these operators may be modified slightly for consistency, as in Cavanagh et al. [2], but this is largely inconsequential for the purposes of this section.

We have seen examples of the latter before: compare, for example, Eqs. (1.54) and (1.55). Being flexible in switching between the two bases can greatly simplify the mathematics involved, as terms which do not survive can be immediately dropped, and the simpler phase modulation $\exp(\mathrm{i}\omega t)$ can be used instead of unwieldy $\cos(\omega t)$ and $\sin(\omega t)$ terms.

Consider now how a spatially dependent phase is imparted to a coherence as it proceeds through the HSQC sequence. We assume that between the two gradients the coherence I_{ip} is transferred to I_{jq}, i.e. q-order coherence on some other spin j:

$$I_{ip} \xrightarrow{g_1} I_{ip}\exp(-\mathrm{i}p\gamma_i g_1 z\tau) \xrightarrow{\text{mixing}} I_{jq} \xrightarrow{g_2} I_{jq}\exp(-\mathrm{i}q\gamma_j g_2 z\tau)\exp(-\mathrm{i}p\gamma_i g_1 zt). \tag{1.99}$$

For the coherence to be rephased, we require that

$$p\gamma_i g_1\tau + q\gamma_j g_2\tau = 0, \tag{1.100}$$

and in the specific case of the HSQC, we know that spins i and j are, respectively, S and I, so

$$p\gamma_S g_1 + q\gamma_I g_2 = 0. \tag{1.101}$$

For the echo experiment, we choose $g_1/g_2 = \gamma_I/\gamma_S$, which means that Eq. (1.101) is satisfied if and only if $p + q = 0$. Since I_q is only detectable if $q = -1$, this means that $p = +1$.

With this knowledge, we can work through the pulse sequence analysis with much greater facility. At the start of t_1, we have the operator $-2I_zS_y = \mathrm{i}(I_zS_+ - I_zS_-)$: of these two terms, the gradient combination selects for the I_zS_+ term during t_1 (for which $p = +1$), and rejects the I_zS_- term ($p = -1$). This evolves during t_1 (and the $180°(I)$ pulse) to give

$$-\mathrm{i}I_zS_+\exp(-\mathrm{i}\Omega_S t_1). \tag{1.102}$$

The gradient echo after t_1 transforms this to

$$-\mathrm{i}I_zS_-\exp(-\mathrm{i}\Omega_S t_1)\exp(-\mathrm{i}\gamma_S g_1 z\tau) = -\mathrm{i}I_z(S_x - \mathrm{i}S_y)\exp(-\mathrm{i}\Omega_S t_1)\exp(-\mathrm{i}\gamma_S g_1 z\tau). \tag{1.103}$$

Again, the mixing period only transforms $2I_zS_y \to I_x$, so we can discard the I_zS_x term and get

$$-\frac{1}{2}I_x\exp(-\mathrm{i}\Omega_S t_1)\exp(-\mathrm{i}\gamma_S g_1 z\tau) = -\frac{1}{4}(I_+ + I_-)\exp(-\mathrm{i}\Omega_S t_1)\exp(-\mathrm{i}\gamma_S g_1 z\tau) \tag{1.104}$$

just before applying the g_2 gradient. The rest of the story is the same as before: the I_- component is rephased by g_2 (it picks up another $\exp(\mathrm{i}\gamma_I g_2 z\tau)$ term, which cancels out with the phase from the first gradient), and is then detected to yield $s_{\text{echo}}(t_1, t_2)$ (Eq. (1.96)). This *CTP selection* process is frequently depicted in the diagrammatic form of Fig. 1.7, where solid lines indicate the coherence orders which are selected for at each stage of the pulse sequence. In this case, during the gradient g_1, we have $+1$-coherence on ^{13}C; and during the gradient g_2, we have -1-coherence on ^{1}H.

Conversely, for the antiecho experiment, the gradient ratio $g_1/g_2 = -\gamma_I/\gamma_S$ necessitates instead that $p - q = 0$. This therefore selects for the $I_z S_-$ term during t_1, and the remaining analysis can be carried out along the same lines. The relevant CTP diagram is similar to that in Fig. 1.7, except that the ^{13}C component (orange line) is inverted.

Finally, we note here that any signal generated from the bulk magnetisation (protons not directly coupled to ^{13}C) cannot be rephased by these gradients. Since such magnetisation is always on spin I, this would require that

$$p\gamma_I g_1 + q\gamma_I g_2 = 0 \Leftrightarrow p\gamma_I + q\gamma_S = 0, \tag{1.105}$$

which cannot be satisfied for any sensible integer values of p and q except for $p = q = 0$, which is not observable during t_2 anyway. So, the CTP gradients effectively remove all unwanted signals arising from the bulk magnetisation: the cycling of ϕ_1 and ϕ_{rec} done in the States experiment is rendered unnecessary.

The points developed in this chapter are neatly demonstrated in Fig. 1.8. The States–TPPI experiment used here is the same as the States experiment in Fig. 1.6, except that on every t_1 increment ϕ_1 and ϕ_{rec} are inverted, causing the artefacts to be shifted to the edge of the spectrum. Figure 1.8a and b show the States–TPPI and EA HSQC spectra obtained with one scan per increment. Comparing the projections of these two spectra (Fig. 1.8d), it can be seen that the SNR of the States–TPPI version is greater by a factor of $\sqrt{2}$. However, the spectral quality of the States–TPPI spectrum is far poorer: presentable data can only be obtained when using a two-step phase cycle, as was done in Fig. 1.8c. In this case, although the EA spectrum sacrifices some SNR, its sensitivity is still perfectly usable: the improved spectral quality thus makes EA the preferred implementation for many heteronuclear experiments. For homonuclear experiments which do not have such stringent artefact suppression requirements, States–TPPI quadrature detection is still commonly used.

Coherence Transfer Pathways

To end this chapter, we generalise the CTP refocusing requirement introduced in Eq. (1.100). The treatment here is similar to that of Mitschang et al. [29]. In general, during a pulse sequence we may have n gradients in total, with amplitudes $g^{(1)}, g^{(2)}, \ldots, g^{(n)}$. We assume that their durations $\tau^{(i)}$ are all the same, such that they can be factored out of the equation. We express the coherence during the i-th gradient as a product of single-spin coherences

$$M^{(i)} = \prod_{j}^{\text{spins}} I_j(p_j^{(i)}), \tag{1.106}$$

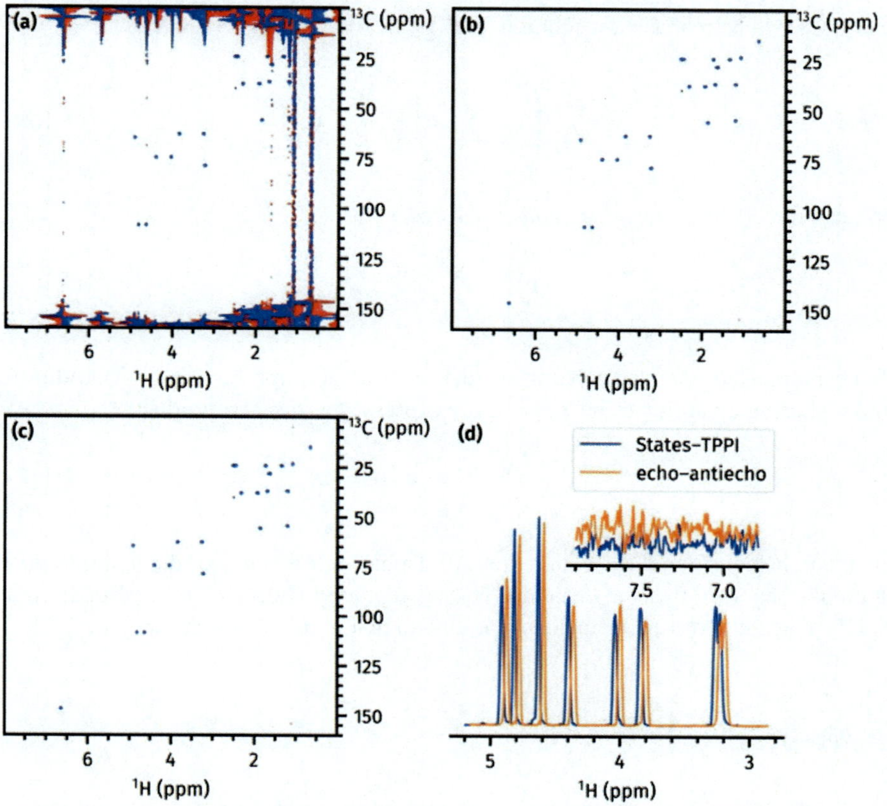

Fig. 1.8 A comparison of HSQC data acquired using various quadrature detection schemes. **a** States–TPPI HSQC with one scan (i.e. no phase cycling) applied. **b** EA HSQC with one scan. **c** States–TPPI HSQC with two scans, using the phase cycling shown in Fig. 1.6. The artefacts are essentially completely removed. **d** Projections of the States–TPPI and EA spectra in **a** and **b** onto the F_2 axis (to avoid picking up the artefacts, only the region between 60 and 110 ppm in F_1 was projected). The two projections are slightly horizontally offset for visual clarity. The signal intensity is the same, but the noise level in the EA spectrum is larger (by a factor of $\sqrt{2}$, as described in the text) *Data code:* `6A-220809`.

where $I_j(p_j^{(i)})$ represents $p_j^{(i)}$-order coherence on spin j during gradient i. The spatially dependent phase imparted by the gradient $g^{(i)}$ is the sum of the phases acquired by each individual coherence:

$$\phi^{(i)} = -zg^{(i)} \sum_j p_j^{(i)} \gamma_j. \tag{1.107}$$

For a CTP to be rephased, we require that $\sum_i \phi^{(i)} = 0$, or equivalently

$$\sum_i \left(g^{(i)} \sum_j p_j^{(i)} \gamma_j \right) = 0. \tag{1.108}$$

We can now define a *weighted coherence order* [30] as

$$p^{(i)} = \sum_j p_j^{(i)} \gamma_j. \tag{1.109}$$

For example, the weighted coherence order for I_+S_- is simply $\gamma_I - \gamma_S$.[19] This allows the rephasing condition to be very simply expressed as a scalar product:

$$\sum_i g^{(i)} p^{(i)} = \mathbf{g} \cdot \mathbf{p} = 0. \tag{1.110}$$

$\mathbf{g} \cdot \mathbf{p}$ can be considered to be an 'extent of dephasing': ideally, gradient amplitudes $\mathbf{g}$ are chosen such that the desired CTP $\mathbf{p}$ is rephased (Eq. (1.110)), and undesired CTPs $\mathbf{p}'$ are suppressed as much as possible in that $\mathbf{g} \cdot \mathbf{p}'$ is maximised.

References

1. Ernst RR, Bodenhausen G, Wokaun A (1987) Principles of nuclear magnetic resonance in one and two dimensions. Clarendon Press, Oxford, U.K
2. Cavanagh J, Fairbrother WJ, Palmer AG III, Rance M, Skelton NJ (2007) Protein NMR spectroscopy: principles and practice, 2nd edn. Academic Press, Burlington, Mass
3. Levitt MH (2008) Spin dynamics: basics of nuclear magnetic resonance, 2nd edn. Wiley, Chichester, U.K
4. Keeler J (2010) Understanding NMR spectroscopy, 2nd edn. Wiley, Chichester, U.K
5. Hore PJ, Jones JA, Wimperis S (2015) NMR: The toolkit, How pulse sequences work, 2nd edn. Oxford University Press, Oxford, U.K
6. Sakurai JJ, Napolitano J (2021) Modern quantum mechanics. Cambridge University Press, Cambridge, U.K
7. Blum K (2012) Density matrix theory and applications. Springer, Heidelberg
8. Cohen-Tannoudji C, Diu B, Laloë F (2020) Quantum mechanics, 2nd edn. Wiley, Weinheim, Germany
9. Breuer H-P, Petruccione F (2002) The theory of open quantum systems. Oxford University Press, Oxford, U.K

[19] Mitschang et al. define a *composite coherence order* as $\sum_j p_j^{(i)} \gamma_j / \gamma_I$, using some nuclide I as a reference. This follows an earlier paper [30] where I is explicitly chosen to be ^{1}H, and has the advantage that if the experiment is homonuclear (i.e. all the spins j are simply I), the composite coherence order is the same as the coherence order. Here, I prefer not to tie the analysis to a particular nuclide as the choice of I may, in principle, depend on the experiment under consideration. This different definition naturally induces a different choice of terminology.

10. Lidar DA (2019) Lecture notes on the theory of open quantum systems. https://doi.org/10.48550/arxiv.1902.00967
11. Chuang IL, Gershenfeld N, Kubinec MG, Leung DW (1998) Bulk quantum computation with nuclear magnetic resonance: theory and experiment. Proc R Soc Lond A 454:447–467. https://doi.org/10.1098/rspa.1998.0170
12. Jones JA (2011) Quantum computing with NMR. Prog Nucl Magn Reson Spectrosc 59:91–120. https://doi.org/10.1016/j.pnmrs.2010.11.001
13. Findeisen M, Berger S (2013) 50 and more essential NMR experiments: a detailed guide. Wiley, Weinheim
14. Claridge TDW (2016) High-resolution NMR techniques in organic chemistry, 3rd edn. Elsevier, Amsterdam
15. Bloch F (1946) Nuclear induction. Phys Rev 70:460–474. https://doi.org/10.1103/physrev.70.460
16. Gupta A, Stait-Gardner T, Price WS (2021) Is it time to forgo the use of the terms "spin- lattice" and "spin-spin" relaxation in NMR and MRI?. J Phys Chem Lett 12:6305–6312. https://doi.org/10.1021/acs.jpclett.1c00945
17. Turner CJ, Hill HDW (1986) Artifacts in quadrature detection. J Magn Reson 66:410–421. https://doi.org/10.1016/0022-2364(86)90185-x
18. Sørensen OW, Eich GW, Levitt MH, Bodenhausen G, Ernst RR (1984) Product operator formalism for the description of NMR pulse experiments. Prog Nucl Magn Reson Spectrosc 16:163–192. https://doi.org/10.1016/0079-6565(84)80005-9
19. Morris GA, Freeman R (1979) Enhancement of nuclear magnetic resonance signals by polarization transfer. J Am Chem Soc 101:760–762. https://doi.org/10.1021/ja00497a058
20. Morris GA (1980) Sensitivity enhancement in nitrogen-15 NMR: polarization transfer using the INEPT pulse sequence. J Am Chem Soc 102:428–429. https://doi.org/10.1021/ja00521a097
21. Pines A, Gibby MG, Waugh JS (1972) Proton-enhanced nuclear induction spectroscopy. A method for high resolution NMR of dilute spins in solids. J Chem Phys 56:1776–1777. https://doi.org/10.1063/1.1677439
22. Aue WP, Bartholdi E, Ernst RR (1976) Two-dimensional spectroscopy. Application to nuclear magnetic resonance. J Chem Phys 64:2229–2246. https://doi.org/10.1063/1.432450
23. Jeener J, Alewaeters G (2016) "Pulse pair technique in high resolution NMR" a reprint of the historical 1971 lecture notes on two-dimensional spectroscopy. Prog Nucl Magn Reson Spectrosc 94–95:75–80. https://doi.org/10.1016/j.pnmrs.2016.03.002
24. States DJ, Haberkorn RA, Ruben DJ (1982) A two-dimensional nuclear Overhauser experiment with pure absorption phase in four quadrants. J Magn Reson 48:286–292. https://doi.org/10.1016/0022-2364(82)90279-7
25. Marion D, Wuthrich K (1983) Application of phase sensitive two-dimensional correlated spectroscopy (COSY) for measurements of $^1H^1H$ spin-spin coupling constants in proteins. Biochem Biophys Res Commun 113:967–974. https://doi.org/10.1016/0006-291x(83)91093-8
26. Keeler J, Neuhaus D (1985) Comparison and evaluation of methods for two-dimensional NMR spectra with absorption-mode lineshapes. J Magn Reson 63:454–472. https://doi.org/10.1016/0022-2364(85)90236-7
27. Delsuc MA (1988) Spectral representation of 2D NMR spectra by hypercomplex numbers. J Magn Reson 77:119–124. https://doi.org/10.1016/0022-2364(88)90036-4
28. Kontaxis G, Stonehouse J, Laue ED, Keeler J (1994) The sensitivity of experiments which use gradient pulses for coherence-pathway selection. J Magn Reson Ser A 111:70–76. https://doi.org/10.1006/jmra.1994.1227
29. Mitschang L, Ponstingl H, Grindrod D, Oschkinat H (1995) Geometrical representation of coherence transfer selection by pulsed field gradients in high-resolution nuclear magnetic resonance. J Chem Phys 102:3089–3098. https://doi.org/10.1063/1.468618
30. John BK, Plant D, Heald SL, Hurd RE (1991) Efficient detection of CαH-HN correlations in proteins using gradient-enhanced ^{15}N HMQC-TOCSY. J Magn Reson 94:664–669. https://doi.org/10.1016/0022-2364(91)90158-p

Chapter 2
Pure Shift NMR

> Then seating herself with a gentleman on each side, she ... discussed the possibility of improvements with much animation. Nothing was fixed on—but Henry Crawford was full of ideas and projects, and, generally speaking, whatever he proposed was immediately approved.
>
> —JANE AUSTEN, *Mansfield Park*

Pure shift NMR refers to the technique of acquiring *broadband homodecoupled* NMR spectra, where all multiplet structure is suppressed and each chemical environment gives rise to a singlet [1, 2]. Here, 'broadband' means that the couplings are removed from the entire spectrum, and 'homodecoupled' refers to the removal of homonuclear couplings: unlike heteronuclear couplings, these cannot be removed by applying RF pulses during the acquisition period, as that will destroy the desired signal itself. Although pure shift techniques can be applied to any nuclide, ^{1}H spectra are of greatest interest because of the narrow chemical shift range of ^{1}H which often leads to peak overlap, as well as the abundance of ^{1}H–^{1}H couplings in typical organic molecules. An example of a pure shift spectrum is provided in Fig. 2.1.

In the first two sections, I first cover the theory underpinning, and a brief history of, pure shift experiments. I then describe my work towards increasing the quality of pure shift experiments: this is measured both in terms of *sensitivity* as well as *purity*, i.e. the lack of spectral artefacts arising from imperfect decoupling. Each of the approaches described here is compared against the PSYCHE pure shift method, which is the current state of the art. I end with a section discussing the combination of pure shift diffusion spectroscopy—formally a pseudo-3D experiment—with the use of ultrafast NMR techniques to collapse the diffusion dimension. This last project was carried out in collaboration with Jean-Nicolas Dumez (University of Nantes).

The work in this chapter has not been submitted for publication.

J. Yong, *Optimising NMR Spectroscopy Through Method and Software Development*, Springer Theses, https://doi.org/10.1007/978-3-031-46684-7_2

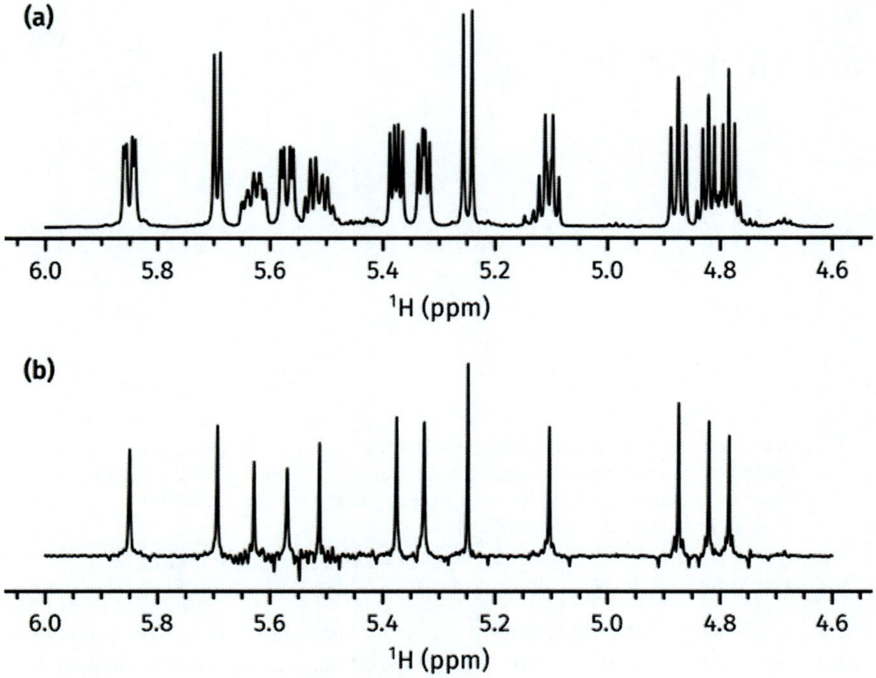

Fig. 2.1 **a** A pulse–acquire ^{1}H spectrum. **b** The corresponding PSYCHE pure shift spectrum. The PSYCHE technique is discussed in detail in Sect. 2.2.3. *Data code:* `7C-211123`

2.1 Theoretical Background

In Chap. 1, I showed how density operators could be expressed in either of the two bases $\{E, I_x, I_y, I_z\}$ or $\{E, I_z, I_+, I_-\}$, depending on which was most mathematically expedient. To analyse pure shift NMR, it turns out to be most convenient to introduce a third basis, namely $\{I_\alpha, I_\beta, I_+, I_-\}$ [3–5]. The definitions of these terms are given in Eq. (1.6): it is clear from there that, when expressed in the Zeeman basis, each matrix element of the density operator corresponds to one of these terms:

$$\rho = \begin{pmatrix} \rho_\alpha & \rho_+ \\ \rho_- & \rho_\beta \end{pmatrix} = \rho_\alpha I_\alpha + \rho_+ I_+ + \rho_- I_- + \rho_\beta I_\beta. \tag{2.1}$$

In systems containing multiple spins, the corresponding *single-element operators* are just the products of these operators. In this section, I use a system with two weakly coupled spins I_1 and I_2 to illustrate the ideas behind pure shift NMR. We have that $H_{\text{free},I} = \Omega_1 I_{1z} + \Omega_2 I_{2z} + 2\pi J I_{1z} I_{2z}$, which is diagonal in the Zeeman basis:

$$H_{\text{free},I} = \begin{pmatrix} \omega_{\alpha\alpha} & 0 & 0 & 0 \\ 0 & \omega_{\alpha\beta} & 0 & 0 \\ 0 & 0 & \omega_{\beta\alpha} & 0 \\ 0 & 0 & 0 & \omega_{\beta\beta} \end{pmatrix}. \tag{2.2}$$

Here, $\omega_{\lambda\mu} = \langle\lambda\mu|H_{\text{free},I}|\lambda\mu\rangle$ ($\lambda, \mu \in \{\alpha, \beta\}$) represents the precession frequency of the state $|\lambda\mu\rangle$. Given that $I_z|\alpha\rangle = (1/2)|\alpha\rangle$ and $I_z|\beta\rangle = -(1/2)|\beta\rangle$, these frequencies are relatively easy to work out:

$$\begin{aligned} \omega_{\alpha\alpha} &= \frac{1}{2}(\Omega_1 + \Omega_2 + \pi J); & \omega_{\alpha\beta} &= \frac{1}{2}(\Omega_1 - \Omega_2 - \pi J); \\ \omega_{\beta\alpha} &= \frac{1}{2}(-\Omega_1 + \Omega_2 - \pi J); & \omega_{\beta\beta} &= \frac{1}{2}(-\Omega_1 - \Omega_2 + \pi J). \end{aligned} \tag{2.3}$$

The corresponding propagator $U(\tau) = \exp(-\mathrm{i}H_{\text{free},I}\tau)$ is then just:

$$U(\tau) = \begin{pmatrix} \exp(-\mathrm{i}\omega_{\alpha\alpha}\tau) & 0 & 0 & 0 \\ 0 & \exp(-\mathrm{i}\omega_{\alpha\beta}\tau) & 0 & 0 \\ 0 & 0 & \exp(-\mathrm{i}\omega_{\beta\alpha}\tau) & 0 \\ 0 & 0 & 0 & \exp(-\mathrm{i}\omega_{\beta\beta}\tau) \end{pmatrix}. \tag{2.4}$$

Consequently, the evolution of the single-element operators under $H_{\text{free},I}$ is extraordinarily simple to calculate in matrix form: for example, we have that

$$I_{1+}I_{2\alpha} = \begin{pmatrix} 0 & 0 & 1 & 0 \\ 0 & 0 & 0 & 0 \\ 0 & 0 & 0 & 0 \\ 0 & 0 & 0 & 0 \end{pmatrix}, \tag{2.5}$$

so $U(\tau)I_{1+}I_{2\alpha}U^{\dagger}(\tau)$ is

$$\begin{pmatrix} 0 & 0 & \exp(-\mathrm{i}\omega_{\alpha\alpha}\tau)\exp(\mathrm{i}\omega_{\beta\alpha}\tau) & 0 \\ 0 & 0 & 0 & 0 \\ 0 & 0 & 0 & 0 \\ 0 & 0 & 0 & 0 \end{pmatrix} = \exp[-\mathrm{i}(\Omega_1 + \pi J)\tau]I_{1+}I_{2\alpha}. \tag{2.6}$$

Essentially, under $H_{\text{free},I}$, all of these operators acquire phase factors which depend on the difference between two of the frequencies $\omega_{\lambda\mu}$. I explicitly state the rules for the single-quantum operators on spin 1 here[1]:

[1] Note that Sect. 10.4.2 of Keeler's text [3] has a sign error in these equations—the sign of the πJ evolution is flipped—though it proves to be inconsequential as the sign of J cannot be observed.

$$
\begin{align}
I_{1+}I_{2\alpha} &\longrightarrow \exp[-\mathrm{i}(\Omega_1 + \pi J)\tau]I_{1+}I_{2\alpha}; \tag{2.7}\\
I_{1+}I_{2\beta} &\longrightarrow \exp[-\mathrm{i}(\Omega_1 - \pi J)\tau]I_{1+}I_{2\beta}; \tag{2.8}\\
I_{1-}I_{2\alpha} &\longrightarrow \exp[\mathrm{i}(\Omega_1 + \pi J)\tau]I_{1-}I_{2\alpha}; \tag{2.9}\\
I_{1-}I_{2\beta} &\longrightarrow \exp[\mathrm{i}(\Omega_1 - \pi J)\tau]I_{1-}I_{2\beta}. \tag{2.10}
\end{align}
$$

The rules for the corresponding operators on spin 2 can be easily obtained by permutation of labels. Notice that the evolution frequencies of the -1-quantum operators (Eqs. (2.9) and (2.10)) each correspond to one peak of the corresponding multiplet in an NMR spectrum: for example, $\Omega_1 + \pi J$ and $\Omega_1 - \pi J$ correspond to the two peaks of the spin-1 doublet.

Consider now a simple spin echo sequence: 90°x–τ–180°x–τ–detection. The initial excitation pulse acts on both spins 1 and 2 and thus generates a mixture of all eight possible single-quantum operators (the four above plus four more on spin 2). For simplicity, we consider only the $I_{1+}I_{2\alpha}$ term. This evolves in the first τ delay to give $\exp[-\mathrm{i}(\Omega_1 + \pi J)\tau]I_{1+}I_{2\alpha}$. The 180° pulse *flips* both spins 1 and 2, in that it causes the transitions $I_+ \leftrightarrow I_-$ and $I_\alpha \leftrightarrow I_\beta$; consequently, we have that

$$
\exp[-\mathrm{i}(\Omega_1 + \pi J)\tau]I_{1+}I_{2\alpha} \longrightarrow \exp[-\mathrm{i}(\Omega_1 + \pi J)\tau]I_{1-}I_{2\beta}. \tag{2.11}
$$

In the second delay, we get a second phase factor from the evolution of the $I_{1-}I_{2\beta}$ operator:

$$
\begin{aligned}
\exp[-\mathrm{i}(\Omega_1 + \pi J)\tau]I_{1-}I_{2\beta} &\longrightarrow \exp[-\mathrm{i}(\Omega_1 + \pi J)\tau]\exp[\mathrm{i}(\Omega_1 - \pi J)\tau]I_{1-}I_{2\beta}\\
&= \exp(-2\mathrm{i}\pi J\tau)I_{1-}I_{2\beta}.
\end{aligned} \tag{2.12}
$$

Detection of this gives us one of the two peaks of the spin-1 doublet, as described previously, but with a phase factor tacked on. The Ω_1 terms in the phase factor are cancelled out, which reflects the fact that the offset (or chemical shift) is refocused by the 180° pulse. However, the J-evolution is not refocused, which leads to characteristic phase distortions in the detected multiplets. The same is true of the seven other single-quantum operators.

In order to refocus the J-evolution as well as the chemical shift, we would need—instead of a 180° pulse—a pulse sequence element which simultaneously effects *all* of the following transitions on spin 1:

$$
I_{1+}I_{2\alpha} \longrightarrow I_{1-}I_{2\alpha};\quad I_{1+}I_{2\beta} \longrightarrow I_{1-}I_{2\beta};\quad I_{1-}I_{2\alpha} \longrightarrow I_{1+}I_{2\alpha};\quad I_{1-}I_{2\beta} \longrightarrow I_{1+}I_{2\beta}; \tag{2.13}
$$

and on spin 2:

$$
I_{1\alpha}I_{2+} \longrightarrow I_{1\alpha}I_{2-};\quad I_{1\beta}I_{2+} \longrightarrow I_{1\beta}I_{2+};\quad I_{1\alpha}I_{2-} \longrightarrow I_{1\alpha}I_{2+};\quad I_{1\beta}I_{2-} \longrightarrow I_{1\beta}I_{2-}. \tag{2.14}
$$

Such an element forms the basis of a pure shift technique, and I refer to it here as a *pure shift element* (PSE). The difficulty in designing a PSE is that *all* spins must be simultaneously decoupled from each other (and not just one spin). For example, if we only had to invert spin 1 and not spin 2 (i.e. only Eq. (2.13) and not Eq. (2.14)), this could be trivially accomplished by a selective 180° pulse on spin 1. However, this would not bring about the correct transitions for spin 2. Yet another complicating factor is that the spins will have offsets and couplings which are *a priori* not known; the design of the PSE therefore cannot use these parameters as inputs. These limitations mean that it is impossible to accomplish the above transitions *in full*; rather, a more realistic scenario involves

$$I_{1+}I_{2\alpha} \longrightarrow cI_{1-}I_{2\alpha} + \sum_i c_i' M_i, \tag{2.15}$$

and likewise for the other operators. Here, the desired transition probability $c < 1$ directly correlates with the sensitivity of the PSE, and the M_i's are some other undesired operators which (if detectable) lead to artefacts if not suppressed.

In the above discussion, note that the role of the PSE is to invert the I_+ and I_- terms, and to leave the I_α and I_β terms untouched. The spins with I_+ and I_- terms are referred to as *active spins*, and the I_α and I_β spins as *passive spins*. Thus, for example, in the context of Eq. (2.13), spin 1 is active and spin 2 is passive. The detected signal always arises from the active spins (unless coherence transfer between spins occurs, for example in the mixing period of a 2D pure shift experiment).

Before moving on to the discussion of how PSEs are implemented in practice, I insert a slight digression about *J-resolved* (or *2DJ*) *spectroscopy*, which is very closely related to pure shift NMR. The basic 2DJ sequence involves a spin echo of duration t_1, immediately followed by detection (Fig. 2.2). If we only consider a single operator and reuse the analysis above, setting the spin echo delay τ to be $t_1/2$ instead, then we have that

$$I_{1+}I_{2\alpha} \xrightarrow{t_1/2} \xrightarrow{180^\circ} \xrightarrow{t_1/2} \exp(-\mathrm{i}\pi J t_1) I_{1-}I_{2\beta}. \tag{2.16}$$

This yields a complex signal of the form

$$s(t_1, t_2) = \exp(-\mathrm{i}\pi J t_1)\exp[\mathrm{i}(\Omega_1 - \pi J)t_2], \tag{2.17}$$

which when Fourier transformed yields a phase twist lineshape at $(-\pi J, \Omega_1 - \pi J)$. The other component on spin 1 (starting from $I_{1+}I_{2\beta}$) likewise yields a phase twist at $(\pi J, \Omega_1 + \pi J)$. It has long been known that *shearing* this 2DJ spectrum by 45° (i.e. moving each data point (Ω_1, Ω_2) to $(\Omega_1, \Omega_2 - \Omega_1)$) generates a spectrum which only has chemical shift information in the ω_2 dimension. After magnitude-mode processing, projection of this spectrum onto the ω_2 axis, for example, would in

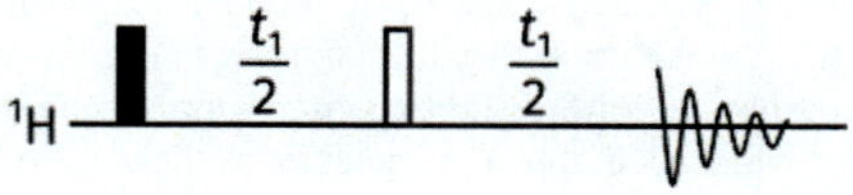

Fig. 2.2 Pulse sequence for the J-resolved (2DJ) experiment

principle yield a pure shift spectrum.[2] This is true, but in practice the phase twist lineshapes cause the resulting resolution to be very poor, which defeats the purpose of using a pure shift spectrum. To circumvent this issue, a number of special processing techniques have been proposed [7–9] (see also references therein); but more ideally, we want a phase-sensitive 2DJ spectrum, which can be obtained by recording a pair of 'echo' and 'antiecho' signals[3]:

$$s_{\text{echo}}(t_1, t_2) = \exp(\mathrm{i}\pi J t_1)\exp[\mathrm{i}(\Omega_1 - \pi J)t_2]; \tag{2.18}$$

$$s_{\text{antiecho}}(t_1, t_2) = \exp(-\mathrm{i}\pi J t_1)\exp[\mathrm{i}(\Omega_1 - \pi J)t_2]. \tag{2.19}$$

These can be processed in the same way as described in Sect. 1.4.3 to yield double absorption-mode lineshapes. The antiecho signal (Eq. (2.19)) is of course the same as in Eq. (2.17), but to obtain the echo signal in Eq. (2.18), we require a different pulse sequence with a PSE inserted just prior to detection:

$$I_{1-}I_{2\alpha} \xrightarrow{t_1/2} \exp[\mathrm{i}(\Omega_1 + \pi J)t_1/2]I_{1-}I_{2\alpha} \tag{2.20}$$

$$\xrightarrow{180^\circ} \exp[\mathrm{i}(\Omega_1 + \pi J)t_1/2]I_{1+}I_{2\beta} \tag{2.21}$$

$$\xrightarrow{t_1/2} \exp(\mathrm{i}\pi J t_1)I_{1+}I_{2\beta} \tag{2.22}$$

$$\xrightarrow{\text{PSE}} c\exp(\mathrm{i}\pi J t_1)I_{1-}I_{2\beta} \tag{2.23}$$

$$\xrightarrow{t_2} c\exp(\mathrm{i}\pi J t_1)\exp[\mathrm{i}(\Omega_1 - \pi J)t_2]. \tag{2.24}$$

(Note that we began the analysis with a different operator here, $I_{1-}I_{2\alpha}$, in order to end up with the same $I_{1-}I_{2\beta}$ operator just before detection.) In order to apply echo–antiecho processing, the decrease in sensitivity by a factor of c must also be applied to the antiecho spectrum: this can be done by simply inserting the PSE immediately after the initial excitation pulse, which simply scales all operators down by c. Thus, we see that *exactly the same PSE* allows us to generate pure shift spectra as well as absorption-mode 2DJ spectra: this has been previously demonstrated using various PSEs [10, 11].

In fact, the same formalism can be used to describe a family of small flip angle COSY experiments, including ECOSY [5, 12, 13] and z-COSY [14–16]; these are also closely related to pure shift NMR. In particular, the anti z-COSY experiment is a precursor to PSYCHE and is analysed in Sect. 2.2.3. However, a full discussion of these is beyond the scope of this thesis.

---

[2] Or equivalently, projection of the unsheared spectrum along a 45° axis [6].

[3] The terms 'echo' and 'antiecho' refer to the relative senses of the coherences evolving during t_1 and t_2: in the echo spectrum these have opposite signs, e.g. I_zS_+ and I_- in the HSQC, and in the antiecho spectrum they have the same sign. As pointed out by Pell and Keeler [10], this is not really appropriate for the 2DJ experiment since each half of t_1 has a coherence with a different sense, but we will stick to this nomenclature as the underlying concept is very similar to that of echo–antiecho processing.

2.2 Pure Shift in Practice

In the previous section, I described the underlying theory used for analysing PSEs and showed how such an element could be used to record absorption-mode 2DJ spectra. From this, one can obtain a pure shift spectrum through shearing and projection. However, this is only an *indirect* route to a pure shift spectrum. In this section, we will tackle the main question of how pure shift experiments may be *directly* acquired using a PSE. Following this, I cover several examples of PSEs reported in the literature. This is not an exhaustive survey of pure shift methods: I only choose to cover a handful of PSEs which specifically accomplish the transformations listed in Eqs. (2.13) and (2.14). Thus, for example, constant-time techniques (which are widely used to suppress ^{13}C–^{13}C couplings in labelled biomolecules) are not mentioned.

2.2.1 *Acquisition Modes*

Restating Eq. (2.15), suppose we have a PSE which accomplishes the transformation

$$I_{1+}I_{2\alpha} \longrightarrow cI_{1-}I_{2\alpha} + \sum_i c'_i M_i \tag{2.25}$$

(and likewise for the other single-quantum operators, which are not shown here). The simplest method of using this is to insert it in the middle of a t_1 period of a 2D experiment. This is actually not entirely desirable, because the PSE causes *both* chemical shifts and J-couplings to be refocused; consequently, there will be *no* frequency modulation during t_1 at all! It is more sensible to combine the PSE with a hard 180° pulse, which refocuses only chemical shifts. Together, the effect is to refocus J-couplings and allow only chemical shifts to evolve; this combination is thus called a *J-refocusing element*, or JRE (Fig. 2.3a). We can equivalently say that the JRE flips all passive spins and leaves active spins untouched. This distinction between a JRE and a PSE is important and will be referred to several times in this chapter.

Implementing a JRE in the middle of a t_1 period is simple, requiring minimal modification of existing 2D experiments. This results in a 2D spectrum which is pure shift in the ω_1, or F_1, dimension. In homonuclear 2D spectra, this pure shift 'character' may further be mapped to the F_2 dimension through indirect covariance processing [17–22]. However, the increased resolution in the F_1 dimension provided by homodecoupling cannot really be realised unless many t_1 increments are acquired. Furthermore, this does not help with acquiring a 1D pure shift spectrum, where there is no indirect dimension.

These considerations lead us to the second acquisition mode for pure shift data, called *real-time acquisition* [23–26]. Here, JREs are inserted at regular intervals throughout an acquisition period, causing the chemical shift evolution to be effec-

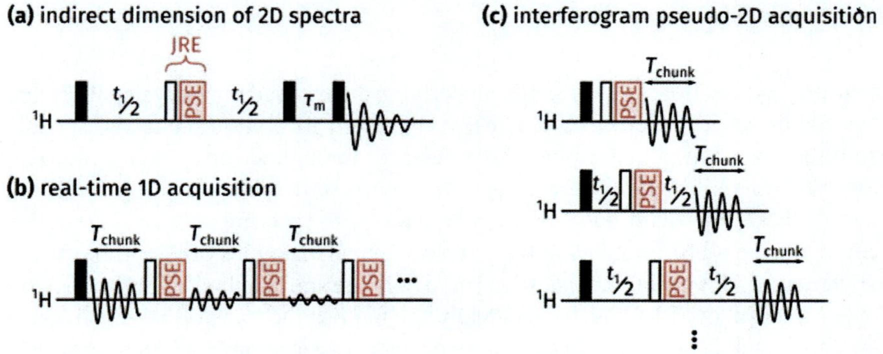

Fig. 2.3 Possible acquisition modes for pure shift spectroscopy. The red box labelled 'PSE' indicates a generic pure shift element, which can be any of those described in the main text. In practice, gradients are also used to suppress unwanted coherence transfers; these are not shown here for simplicity. **a** Insertion of a J-refocusing element (JRE) in the centre of an indirect-dimension evolution period, which leads to a spectrum which is pure shift in F_1. The $90°–\tau_m–90°$ block shown here corresponds to a NOESY experiment, but in principle any 2D experiment can be adapted in this fashion. **b** Real-time acquisition of a 1D pure shift spectrum in chunks of duration T_{chunk}. **c** Interferogram acquisition of a 1D pure shift spectrum, where t_1 is lengthened by T_{chunk} every increment

tively 'suspended' for the duration of the JRE, and the sense of J-evolution to be reversed (Fig. 2.3b).[4] This leads to a series of FID 'chunks' which must then be concatenated to form the desired FID. To prevent the J-coupling from evolving too much during a single chunk, the required spacing of the JREs, or equivalently the duration of each chunk T_{chunk}, must satisfy $T_{\text{chunk}} \ll 1/J$ (in practice, it is on the order of $1/(2J)$). Naturally, real-time acquisition still comes with the sensitivity penalty of c. However, it allows a pure shift spectrum to be acquired in effectively the same time as the original coupled spectrum; its 'single-scan' nature also allows, for example, the application of hyperpolarisation techniques which cannot be reproducibly repeated over multiple increments [27, 28].

Unfortunately, it is not always possible to perform real-time acquisition. The reason is because the JRE is applied multiple times, and each time it is, it must select for the same active and passive spins in the same molecule as it did the last time. In other words, *for any given molecule in the sample*, it must enforce this CTP:

$$I_{1-}I_{2\alpha} \xrightarrow{\text{JRE}} I_{1-}I_{2\beta} \xrightarrow{\text{JRE}} I_{1-}I_{2\alpha} \xrightarrow{\text{JRE}} I_{1-}I_{2\beta} \xrightarrow{\text{JRE}} \cdots. \tag{2.26}$$

As will be described later, the BIRD and Zangger–Sterk methods always select the same active spins in the same molecules, but the PSYCHE method does not.

[4] Relaxation during the JRE must also be taken into account for the real-time method: this causes each successive chunk to decay in intensity faster than usual, thereby leading to peak broadening, which can be an issue for very long JREs. In contrast, the interferogram method only applies one JRE per increment, so the relaxation losses in each chunk are just a constant.

Therefore, in order to acquire pure shift PSYCHE spectra, we have to resort to the *interferogram method*, where each chunk is obtained as a separate increment of a 2D experiment (Fig. 2.3c). The insertion of the JRE in the middle of the t_1 period means that when detection is started, it is 'as if' only the chemical shift has evolved for a period t_1. On each increment, one chunk—again of duration $T_{\text{chunk}} \ll 1/J$—is detected, and then t_1 is incremented by T_{chunk} so that the next chunk can be recorded. Finally, the chunks are stitched together to form the requisite FID.[5] Since the indirect dimension is not processed by a Fourier transform, this is sometimes called a *pseudo-2D* experiment (or *pseudo-3D* if this is applied to the direct dimension of a 2D experiment, and so on). In this case, the sensitivity drop c is incurred, but there is on top of that also a time penalty in that the experiment duration must be lengthened by n times in order to collect n chunks. n is typically on the order of 16–32.

2.2.2 *Pure Shift Elements*

2.2.2.1 Zangger–Sterk

We are finally now in a position to study individual PSEs and their mechanisms of action. We begin with the Zangger–Sterk (ZS, or 'slice-selective') PSE [31], in which a selective refocusing pulse and a weak gradient are simultaneously applied (Fig. 2.4a). In practice, an rSNOB pulse [37] is often used as the refocusing pulse. The effect of the gradient is to make each spin in the sample have a spatially dependent offset; therefore, in each *slice* (or cross section) of the sample, a different spin will fall within the specific bandwidth of the refocusing pulse. This spin is refocused by the PSE and therefore becomes the active spin *within that specific slice*; the bracketing pair of CTP gradients serve to destroy coherences on all the other spins which are not inverted. Each signal of the pure shift spectrum therefore derives from a specific slice of the sample; during direct detection, all slices simultaneously contribute to the signal, thus yielding a broadband pure shift spectrum.

The sensitivity of the ZS method tends to be low (the factor c tends to be on the order of 0.01–0.05), as each signal only comes from a narrow section of the sample. Nevertheless, it still finds wide usage in pure shift applications nowadays, especially because it is compatible with the real-time acquisition mode [24]: as long as the pulse and the weak gradient are the same each time, then the same active spins will always be chosen in the same slice (as long as diffusion effects are ignored). The PSE can also be customised through the bandwidth of the refocusing pulse: decreasing this improves the spatial differentiation between spins which have similar intrinsic offsets,

[5] Not included in Fig. 2.3b and c is the extra detail that scalar couplings are usually allowed to evolve for a period of $T_{\text{chunk}}/2$ at the start of the sequence by the addition of a spin echo: this amounts to a *prefocusing* of J-evolution, such that J-coupling is refocused in the middle of the chunk rather than the beginning [29]. This allows chunk sizes twice as large to be used, reducing the total duration of the experiment. This J-prefocusing can also be done in a more intelligent manner via the SAPPHIRE method [30], which is discussed in more detail in Sect. 4.3.6.

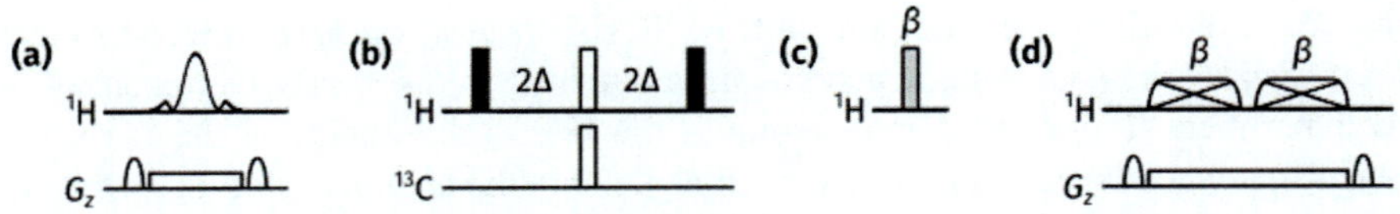

Fig. 2.4 A selection of pure shift elements. **a** Zangger–Sterk PSE [31], involving the combination of a selective 180° refocusing pulse and a weak gradient. **b** BIRD PSE [32, 33]; the delay Δ is set to $1/(4 \cdot {}^1J_{CH})$. **c** Time-reversal PSE [13], simply consisting of a hard pulse with variable flip angle β. Multiple spectra with different values of β must be co-added to suppress artefacts (though this suppression is not perfect, as discussed in the text). **d** PSYCHE PSE [34], consisting of two saltire pulses [35, 36] with flip angle β, and a weak gradient

yielding better decoupling quality, albeit at the cost of sensitivity. The ZS element can be easily—and has been—adapted for use in many experiments, including (but not limited to) absorption-mode 2DJ spectroscopy [10] and selective refocusing (SERF) experiments for the measurement of ${}^nJ_{HH}$ [38–41].

2.2.2.2 BIRD

Next up is the *bilinear rotation decoupling* (BIRD) pulse element (Fig. 2.4b). BIRD is not spatially selective like the ZS method; instead, it is *isotope-selective* in that it acts as a 180°_y pulse on ^{13}C–bound protons and does not affect ^{12}C–bound protons. Consequently, all ^{13}C–bound protons become the active spins in the context of pure shift NMR. The first report of the BIRD element [32], in 1982, was clearly ahead of its time: it reported the use of an interferogram-type approach to obtain 1D pure shift spectra. However, in the subsequent decades, this seemed to have been forgotten: BIRD found much more use as an isotope-selective rotation element in heteronuclear NMR [42], until its use as a pure shift element was 'rediscovered' [33, 43].

An immediate drawback of BIRD is that it does not decouple geminal (diastereotopic) CH2 groups, as both protons would be either both active or passive. The sensitivity penalty of BIRD is also relatively severe: the factor c derives from the natural abundance of ^{13}C, which is approximately 0.011. However, it is also compatible with real-time acquisition [23] and has found particular success as a pure shift element in the F_2 dimension of HMQC and HSQC experiments [26, 43–47]: in this case, the use of BIRD leads to no loss of sensitivity as only ^{13}C–bound protons are detected in HSQC experiments anyway.[6] It should be noted that the BIRD element does not need to be combined with a hard 180° pulse to form a JRE: the inverse effect of only flipping the passive spins can be accomplished by simply changing the phase of both internal 180° pulses to $+y$. Using the nomenclature of Uhrín et al. [42] the PSE and JRE forms of the BIRD pulse can be labelled as $\text{BIRD}^{\text{d,X}}$ and $\text{BIRD}^{\text{r,X}}$ respectively.

[6] In fact, the sensitivity is increased by the collapse of multiplet structure.

2.2.2.3 Time-Reversal

The *time-reversal* PSE (Fig. 2.4c) is even simpler: it consists only of one hard pulse with flip angle β [13]. The catch is that the experiment must be repeated with different values of β, and the results added up with specific weightings [5]. This leads to cancellation of *some*, but not all, of the unwanted coherences: for example, on its own, it does not suppress COSY-type coherence transfer such as $I_{1-}I_{2\alpha} \rightarrow I_{1\alpha}I_{2+}$. This proved to be inconsequential in the original application of F_1 decoupling in a 2D NOESY [13]; however, it is not acceptable in a 1D context as it leads to substantial artefacts. This will be discussed in more detail in Sect. 2.5, so a fuller analysis of the time-reversal PSE is deferred until then.

2.2.2.4 PSYCHE

Finally, we come to the PSYCHE method, which is generally considered the current state of the art for pure shift NMR [34]. The corresponding PSE (Fig. 2.4d) consists of two swept-frequency pulses applied during a weak gradient: for example, a pair of chirps with opposite sweep directions can be used. The signal-to-artefact ratio of PSYCHE can be further improved by using two *saltire pulses*: these are superpositions of chirps which simultaneously sweep in both frequency directions [35]. The operation of this PSE is not easy to fully explain [36]. However, we can adopt the (not fully accurate, but still useful) instant-flip approximation—that the swept-frequency pulse acts as an instantaneous 180° rotation on each frequency it passes through [48, 49]. Using this, the PSYCHE element (or strictly, the PSYCHE JRE, including a hard 180° pulse) may be viewed as a spatially parallelised version of the anti z-COSY experiment [14, 15], which we now describe.

2.2.3 *PSYCHE in Detail*

The anti z-COSY experiment utilises a β–zero-quantum filter (ZQF)–β mixing period (Fig. 2.5a), where β is a small angle, typically 10°–20°. The role of the ZQF [50] is to remove terms such as $I_{1-}I_{2+}$ between the two β pulses, retaining only population terms such as $I_{1\alpha}I_{2\alpha}$. The CTPs selected for by this mixing period ultimately give rise to peaks which lie perpendicular to the main diagonal of the spectrum. In an elegant paper, Pell et al. [15] showed that by isolating the diagonal peaks of a 2D anti z-COSY experiment and taking the 45° projection of these, a pure shift spectrum could be obtained. Here, we will go one step further and consider the *direct* use of this element as a JRE: this will reveal some problems which are neatly taken care of by the PSYCHE experiment.

We first need to introduce how the basis operators $\{I_+, I_-, I_\alpha, I_\beta\}$ evolve under a hard pulse (applied along the x-axis) with flip angle β. This is given by

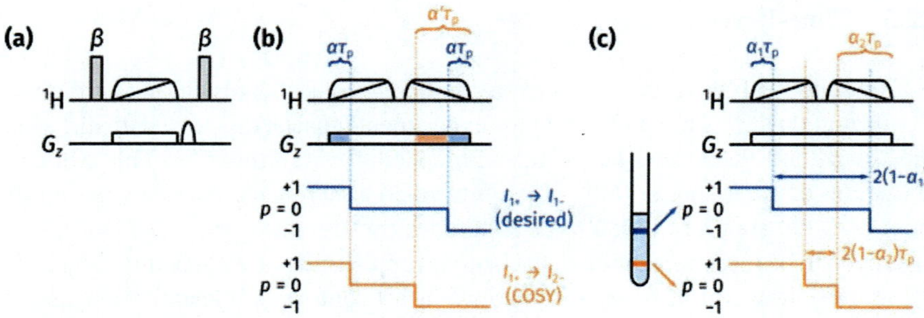

Fig. 2.5 A closer look at the mechanism of the PSYCHE PSE. **a** The β–ZQF–β mixing period used in the original anti z-COSY experiment. This has a similar action to a JRE, but does not suppress COSY-type coherence transfers from spin i to spin j. **b** An illustration of how COSY-type artefacts are suppressed by the PSYCHE pulse element. The desired CTP which remains on spin 1 is rephased by the gradients, but the COSY CTP is dephased. **c** An illustration of how zero-quantum terms are suppressed by the PSYCHE element through spatial averaging: in each slice of the sample (highlighted in blue and orange), the zero-quantum terms are allowed to evolve for a different duration

$$I_\pm \xrightarrow{\beta I_x} c^2 I_\pm + s^2 I_\mp \pm \frac{iS}{2}(I_\alpha - I_\beta); \tag{2.27}$$

$$I_\alpha \xrightarrow{\beta I_x} c^2 I_\alpha + s^2 I_\beta + \frac{iS}{2}(I_+ - I_-); \tag{2.28}$$

$$I_\beta \xrightarrow{\beta I_x} c^2 I_\beta + s^2 I_\alpha - \frac{iS}{2}(I_+ - I_-), \tag{2.29}$$

where $S = \sin\beta$, $s = \sin(\beta/2)$, and $c = \cos(\beta/2)$. Using these formulae, we can show that for a two-spin system (see Pell et al. [15] for the equivalent analysis on a three-spin system), the β–ZQF–β element converts the term $I_{1+}I_{2\alpha}$ to

$$I_{1+}I_{2\alpha} \longrightarrow \underbrace{\frac{1}{2}S^2c^4 I_{1+}I_{2\beta}}_{\text{term 1}} + \underbrace{S^2c^2s^2 I_{1+}I_{2\alpha}}_{\text{term 2}} \underbrace{- \frac{1}{4}S^2c^4 I_{1\alpha}I_{2+} + \frac{1}{4}S^2c^4 I_{1\beta}I_{2+}}_{\text{terms 3 and 4}} + \cdots, \tag{2.30}$$

where other terms with different coherence orders have been neglected (on the basis that they can be easily suppressed with bracketing CTP gradients), and terms with higher orders in s have been discarded since $s = \sin(\beta/2) \ll 1$ for small β.

The first term $I_{1+}I_{2\beta}$, corresponding to the flipping of passive spins only, is the only term we want to see from a JRE. The second term, $I_{1+}I_{2\alpha}$, corresponds to the case where neither active nor passive spins have been flipped. In the original anti z-COSY work, these give rise to 'off-diagonal' peaks which are part of a multiplet on the diagonal, but when projected at 45° generate artefacts around the pure shift peak. In the context of pure shift NMR, these are called 'recoupling artefacts', as they arise from imperfect J-refocusing. Note that the ratio of recoupling artefacts to desired signal is proportional to $S^2c^2s^2/S^2c^4 = \tan^2(\beta/2)$: using a smaller value for β therefore leads to better signal-to-artefact ratios, but also lower overall sensitivity.

The PSYCHE element is similar to the β–ZQF–β element in this regard: it does not suppress the recoupling artefacts, but instead relies on the user choosing a suitable value for β such that the artefact-to-signal ratio is tolerably small. If the sensitivity proves to be insufficient, the flip angle β may be increased instead: this leads to a larger artefact-to-signal ratio, but if the sample is not concentrated anyway, it may well be that the artefacts do not rise above the noise level.

The third and fourth terms, $I_{1\lambda}I_{2+}$ ($\lambda \in \{\alpha, \beta\}$), represent 'COSY-type' coherence transfer to a coupled spin. In the original anti z-COSY, these led to crosspeak multiplets at (Ω_1, Ω_2), which could be removed by hand before taking the projection. However, in a pure shift sequence, the peaks arising from these terms cannot simply be removed in the same way. It is precisely this issue which precludes the β–ZQF–β element from being directly used as a JRE and motivates the development of the PSYCHE PSE, which *does* suppress these coherence transfers.

To understand how this occurs, we invoke the instantaneous spin-flip assumption. Each coherence I_{i+} is converted (or 'flipped') to a population term $I_{1\lambda}$ at a specific point $\alpha\tau_\mathrm{p}$ after the start of the first chirp and can be reconverted to a coherence on the same spin I_{i-} at a time $\alpha\tau_\mathrm{p}$ before the end of the second chirp (the blue CTP in Fig. 2.5b).[7] Here, τ_p is the duration of the chirp, and $0 < \alpha < 1$. In this case, the coherence is perfectly rephased by the weak gradient applied during the chirp pulses, since the *time* it experiences these gradients for is the same on both sides of the chirps. Now, if the I_{i+} term is instead converted to a coherence on a different spin I_{j-}, it experiences the gradient for a total duration of $\alpha\tau_\mathrm{p}$ after the start of the first chirp, and also $\alpha'\tau_\mathrm{p}$ before the end of the second chirp (the orange CTP in Fig. 2.5b). In general, since spins i and j have different offsets, we have that $\alpha \neq \alpha'$. Therefore, this CTP is dephased by the gradients, resulting in suppression of the COSY-type artefacts in the spectrum.

It remains to also consider how the PSYCHE element selects for the population terms between the two-spin flips. Any terms with nonzero coherence order are of course simply dephased by the weak gradient. However, zero-quantum terms (in homonuclear systems) are not dephased by gradients, and to eliminate them in a single-scan manner, they must be spatially averaged, for example by a ZQF. It turns out that the PSYCHE element also results in a similar spatial averaging. Following on from the previous paragraph, the time *between* the spin flips (for the desired pathways, i.e. not COSY-type coherence transfer) is given by $2(1 - \alpha)\tau_\mathrm{p}$. At the same time, the weak gradient induces a range of offsets across the sample, much like in the Zangger–Sterk experiment. Thus, the offset, and thus the value of α, for a given spin depends on which slice it is in; for example, Fig. 2.5c uses values of α_1 and α_2 for two different slices (blue and orange). If zero-quantum terms are present between the spin flips, they evolve during this time and accrue a spatially dependent phase: summation of these during FID acquisition leads to a cancellation of these terms. Only population terms such as $I_{1\alpha}I_{2\alpha}$ survive during this, as they do not precess during this time.

[7] Note the change in the sign of the coherence, which differs from the analysis of the anti z-COSY experiment. This arises because we are only considering the PSYCHE PSE on its own, *not* the JRE.

The sensitivity of PSYCHE is significantly better than for other methods: depending on the flip angle β chosen, c is typically on the order of 0.05–0.15 (see also Fig. 2.6 for explicit simulations). Furthermore, it is generally more robust with respect to strong coupling compared to other pure shift methods (artefacts from strong coupling often arise due to unexpected coherence transfer [4], which is suppressed in a similar way to the COSY-type artefacts). These two factors alone have meant that PSYCHE has enjoyed substantial adoption since its introduction: a large number of 2D experiments utilising PSYCHE decoupling in either F_1 or F_2 have been developed, [22, 51–56] notably the PSYCHE-iDOSY diffusion experiment [57], where the increased resolution provided by pure shift spectroscopy translates into increased resolution in the *diffusion* dimension as well. Like the ZS element before it, the PSYCHE element has also been used for the acquisition of absorption-mode 2DJ spectra [11].

Despite this success, PSYCHE suffers from one significant drawback: it cannot be used in a real-time fashion. The PSYCHE PSE is often said to select active and passive spins in a 'statistical' manner: this is because of the c^2 and s^2 terms arising from the low-flip angle pulses. What this really means is that we do not care *exactly* which spins are active and which are passive, but that a certain proportion of the spins are active and passive. Repeated application of the PSE therefore does not select for the same active spins each time, which precludes its application to real-time acquisition.

Although the PSYCHE PSE may appear deceptively simple at first glance, the closer analysis given here (and elsewhere [35]) clearly shows that its inner workings are anything but. Along with other ingenious experiments such as the ZQF [4], ultrafast NMR [58–60], and more recently GEMSTONE [61], PSYCHE is a prime example of how *spatiotemporal averaging* and pulsed field gradients can be used to great effect in modern NMR spectroscopy [62].

At the same time, PSYCHE itself is not *perfect*: it does not fully suppress recoupling artefacts and can only be used in the interferogram mode. To improve on PSYCHE would therefore entail one of the following:

1. increasing the sensitivity (while maintaining purity);
2. increasing the purity (while maintaining sensitivity); or
3. developing a pure shift element which is compatible with real-time acquisition while giving comparable sensitivity and purity to PSYCHE.

The sections which follow describe my efforts towards objectives (1) and (2).

2.3 PSYCHE with a Variable Number of Saltires

The first attempted method was to change the number of saltire pulses used in the PSYCHE PSE. As described in Sect. 2.2.3, PSYCHE relies on spatiotemporal averaging to suppress unwanted artefacts: this crucially relies on the fact that the pulse(s) used in the PSE are symmetric. This can be accomplished with two opposing chirps,

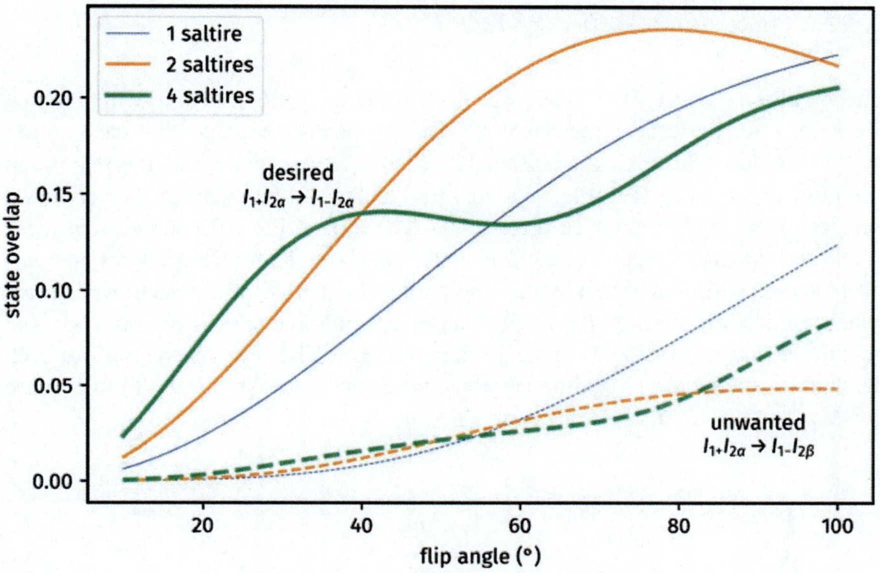

Fig. 2.6 Simulated signal and artefact intensity for various PSYCHE-style PSEs as a function of the flip angle used. Calculations were performed on a two-spin system with a coupling of 7 Hz, and an offset difference of 1.5 kHz. The total PSE duration was 30 ms (so, for example, in the four-saltire PSYCHE, each saltire was 7.5 ms). Solid lines indicate the coefficients for the desired $I_{1+}I_{2\alpha} \rightarrow I_{1-}I_{2\alpha}$ pathway which contributes to the pure shift signal; dashed lines the coefficients for the undesired $I_{1+}I_{2\alpha} \rightarrow I_{1-}I_{2\beta}$ pathway, which gives rise to recoupling artefacts

or two saltires, both of which are symmetric. However, a *single* saltire is also symmetric in itself: it is not hard to show that a single saltire can provide the requisite averaging. Likewise, the use of four saltires would also be valid.

Theoretical simulations show that the overall profile of signal and artefact versus flip angle varies with the number of saltires (Fig. 2.6). Generally, using a larger number of saltires but with a smaller flip angle accomplishes a similar sensitivity level. This may be qualitatively rationalised as more saltires providing more possible CTPs which generate the desired signal (the same idea is generally invoked when discussing the difference between unidirectional chirps and saltires [35]). However, regardless of the number of saltires, the fundamental strategy of adjusting the flip angle to control the signal-to-artefact ratio remains valid, which naturally raises the question of whether specific waveforms and flip angles can be chosen in order to obtain the best decoupling quality.

2.3.1 Quadruple-Saltire PSYCHE

The quadruple-saltire PSYCHE was first evaluated experimentally. In the first instance, I set the total duration of the PSE to 30 ms, meaning that each saltire was 7.5 ms long. In this experiment, the sensitivity was defined as the maximum height of the main peak in Fig. 2.7a, and the artefact as the mean of the maximum heights of the two artefacts surrounding it. The plot in Fig. 2.7b shows how these quantities vary as a function of the flip angle. Interestingly, for the artefacts, the profile observed is similar to that in the simulations: the double-saltire version performs better at low and very high flip angles, but in the middle there is a region where the quadruple-saltire version has lower artefact intensity. The signal intensities for both the double- and quadruple-saltire versions, however, seem to plateau off rather more quickly than the simulations suggest.

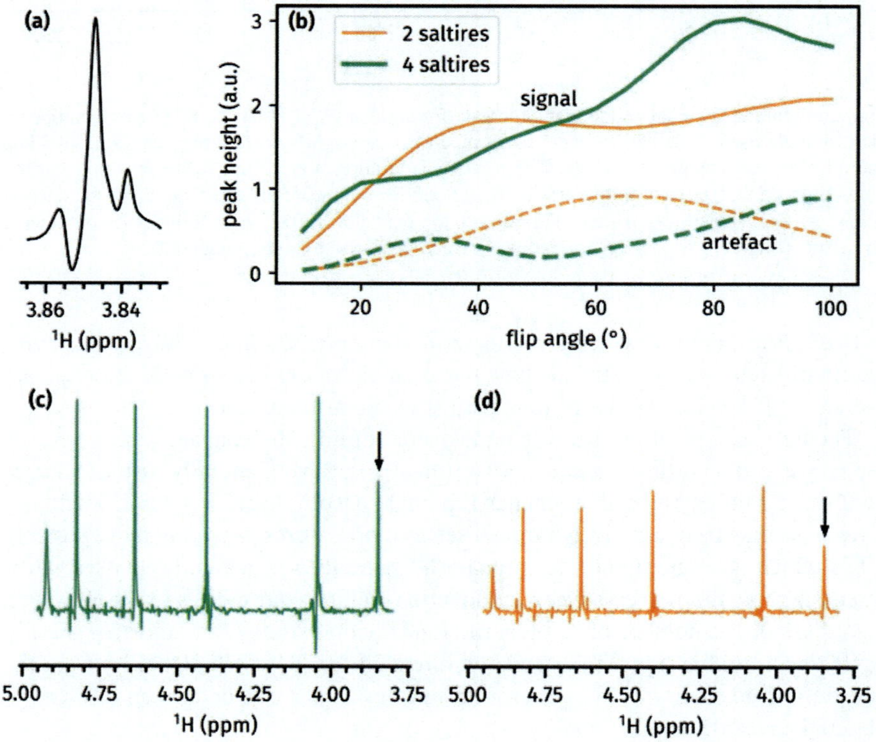

Fig. 2.7 **a** The peak in the pure shift spectrum of andrographolide is used for the calculation of signal and artefact intensity. The recoupling artefacts flanking the main peak are clearly visible. **b** Peak heights of the desired signal (the central peak, solid lines) and artefacts (the mean of the two flanking peaks, dashed lines), as a function of flip angle. **c** Quadruple-saltire PSYCHE with $\beta = 55°$. **d** The reference spectrum, a double-saltire PSYCHE with $\beta = 20°$. The peak at 3.85 ppm used for the sensitivity and purity analysis is labelled with an arrow. *Data code:* `7A-201016`

To highlight one particular data point, Fig. 2.7b suggests that the quadruple-saltire experiment with $\beta \approx 55°$ has a similar artefact level to the double-saltire experiment with $\beta \approx 20°$, but with substantially greater signal intensity. Insets from these two spectra are respectively shown in Fig. 2.7c and d. This conclusion does indeed seem to be true for the specific peak used for this analysis, which is at the right edge of the spectral insets shown here. Overall, the quadruple-saltire 55° experiment does have better sensitivity than the double-saltire 20° experiment. However, the artefacts are not always suppressed as nicely as in the chosen reference peak: for example, the peak at 4.05 ppm is significantly less clean in the quadruple-saltire experiment. Any apparent advantage over the double-saltire experiment is therefore not very clear—at least from this data alone.[8]

2.3.2 *Single-Saltire PSYCHE*

Moving on to the single-saltire case, in addition to searching for a better signal and artefact profile as before, another possible motivation would be that the duration of the PSE can be decreased. This would minimise losses due to relaxation and diffusion during the PSE, which were not taken into account in Fig. 2.6 (and the simulations there did not vary τ_p anyway). A single-saltire PSYCHE experiment was thus recorded with various combinations of flip angle ($\beta/° \in \{15, 20, 25, 28, 30\}$) and pulse duration ($\tau_p/\mathrm{ms} \in \{15, 25, 30, 35, 45, 55\}$). In a similar way to the quadruple-saltire study, the sensitivity was defined as the maximum height of the main peak in Fig. 2.8a,[9] and the artefact as the mean of the heights of the two artefacts surrounding it. (However, note that the sample used was different.) The purity, or signal-to-artefact ratio, is simply the former divided by the latter.

The results thus obtained are shown in Fig. 2.8b. In this plot, both the sensitivities as well as the signal-to-artefact ratios are normalised with respect to the reference double-saltire experiment (black dot at (1, 1), acquired using $\beta = 20°$ and a total $\tau_p = 30$ ms, i.e. 15 ms per saltire). An ideal PSE would fall in the top-right corner of this plot. Clearly, as the flip angle increases, the sensitivity increases but at the cost of the purity: this is hardly unsurprising given that the double-saltire experiment has the same property. The effect of the pulse duration is less clear-cut: in general, a shorter PSE yields to increased signal, and very usable results were obtained even with a simple 15 ms single-saltire PSE. However, a shorter PSE also leads to poorer spatiotemporal suppression of artefacts and thus lower purity.

---

[8] In fact, I also performed some preliminary experiments where the four-saltire PSE was lengthened to 60 ms. The artefact behaviour in these spectra was better than in their 30 ms counterparts, which is to be expected since a longer PSE leads to better spatiotemporal averaging. However, I unfortunately did not compare these against a 60 ms double-saltire experiment, so these results are not included in this thesis. (It would be rather unfair to compare them against the 30 ms double saltire.)

[9] As an alternative, the Bruker `sinocal` AU program was also used to measure the sensitivity of the spectrum; it yielded extremely similar results, so is not shown here. I found the `sinocal` routine to be rather unreproducible as the exact value calculated depends on random noise.

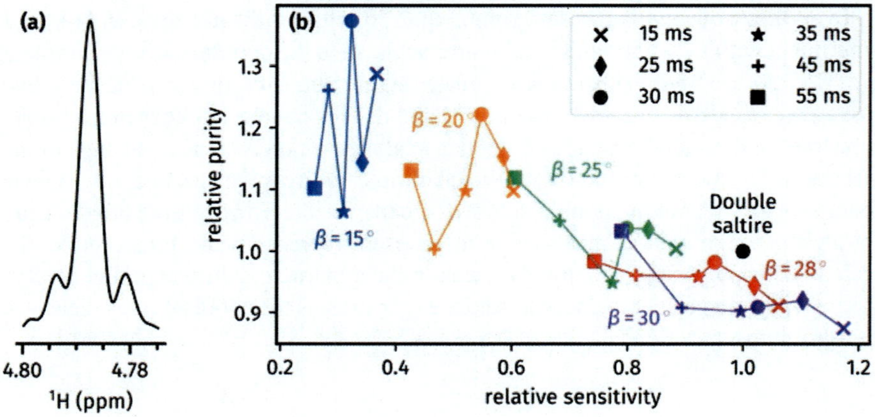

Fig. 2.8 **a** The peak in the (reference, double saltire) pure shift spectrum of cyclosporin is used for the calculation of signal intensity and signal-to-artefact ratio. **b** Plot showing the signal intensity and signal-to-artefact ratio obtained in various single-saltire PSYCHE experiments, normalised against the double-saltire experiment with $\beta = 20°$ and $\tau_p = 30$ ms. Each line represents a series of single-saltire experiments acquired with the same value of β; τ_p generally decreases going left to right (i.e. a longer pulse means less signal). *Data code:* `5C-190617`

While there is no standout candidate which is *clearly* better than the double saltire (i.e. greater sensitivity as well as purity), the single-saltire pulse with $\beta = 28°$ and $\tau_p = 30$ ms comes close in performance to the double-saltire experiment. (The use of $\beta = 28°$ here is not coincidental: this flip angle for a single saltire was chosen to (approximately) match the sensitivity of the $\beta = 20°$ double-saltire experiment.) Therefore, this does at least prove that a single saltire can be used as a PSE. However, both the single- and quadruple-saltire cases suffer from the classic dilemma of pure shift NMR: just as in the original PSYCHE, and in the Zangger–Sterk method before it, there is a compromise between sensitivity and purity, and one can only be increased at the cost of the other. Arguably, then, there is not much value in changing the number of saltire pulses as simply varying the *flip angle* of the basic double-saltire already gives the experimentalist a way to balance these competing objectives.

2.4 Direct Optimisation of PSYCHE Waveform

The changes to the PSYCHE waveform discussed in the previous section—changing the number of saltire pulses to either 1 or 4—are fairly minor, in that they do not fundamentally alter the form of the PSE. Furthermore, there is an obvious issue in that the quality of the spectrum is not measured in a particularly rigorous manner. Ideally, we would like to have a mathematical way of measuring how well the pure shift sequence has worked. Such a metric would also enable a more automated optimisation process, where a program is allowed to find the best PSE without any interference from human subjectivity.

In this section, I discuss more radical changes which depart from the tried-and-tested saltire pulse. I also show how two different *cost functions*—functions which determine how 'bad' a spectrum is—can be used to objectively evaluate PSEs. Although the work in this section did not quite yield any groundbreaking results, it provided substantial insights into the nature of pure shift optimisations, which were later used in the context of POISE (Chap. 3).

2.4.1 Techniques for Pure Shift Optimisations

Throughout this chapter (and more generally this thesis), various algorithms are used for *optimisation*: that is, to find the parameters $\mathbf{x} \in \mathbb{R}^n$ which minimise a *cost function* $f : \mathbb{R}^n \to \mathbb{R}$. These algorithms can loosely be categorised as either *derivative-based* or *derivative-free*: the former use extra information in the form of ∇f to help locate the optimum, whereas the latter do not, using only the value of $f(\mathbf{x})$. While derivative-based algorithms typically converge to an optimum more quickly, they are unsuitable for problems where the cost function f is noisy. In this section, the Nelder–Mead (NM) simplex algorithm [63] was used: it is a very popular derivative-free method, and although mathematical convergence is not guaranteed [64], in practice such cases are rather unlikely to arise. In this section, the implementation of the NM algorithm in the Python `scipy` package was simply used as-is.

The cost function measuring the performance of a pure shift experiment can be measured in one of two ways: either *theoretically*, in that the pure shift experiment is simulated using the density operator formalism, or *experimentally*, in that the experiment is run on a spectrometer. Unfortunately, the simulation of PSYCHE-type PSEs, where a shaped pulse is applied together with a gradient, requires a large amount of time. The pulse itself already has $m \sim 10000$ points, but on top of that, the application of gradients also requires splitting up the sample into multiple slices ($n \sim 100$ to 1000) such that the evolution of ρ can be simulated in each slice and the results summed up. To make matters worse, $H_{\mathrm{pulse},I}$ does not commute with H_{grad}, so a propagator

$$U(i, z) = \exp\left[-\mathrm{i}(H_{\mathrm{grad}}(z) + H_{\mathrm{free},I} + c_x^{(i)} I_x + c_x^{(i)} I_y)(\delta t)\right] \tag{2.31}$$

must be calculated for each pulse point, in each slice of the sample, for a total of mn matrix exponentials. As a result, the simulation of PSYCHE spectra in the Spinach package [65] typically requires minutes to hours.[10]

[10] With highly optimised handwritten code, exploiting the symmetry of the PSYCHE element, the fastest simulation I could do took 16 s on a 20-core computer, for a simple 2-spin system. This number increases exponentially for larger spin systems. It is possible that GPU acceleration could

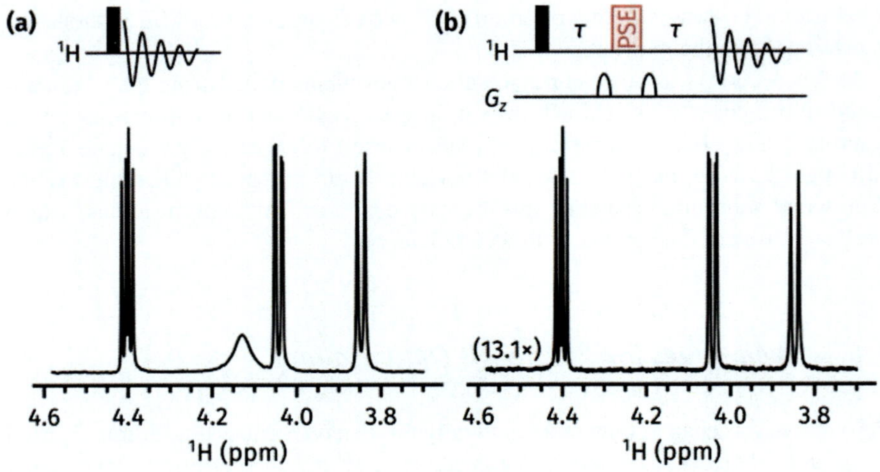

Fig. 2.9 **a** Pulse–acquire experiment and the resulting spectrum. **b** J-refocused spin echo experiment and the resulting spectrum. The PSE used was the PSYCHE double saltire, with a flip angle of 25°; the delay τ was 11 ms. The OH peak at 4.1 ppm is lost, most likely due to chemical exchange. *Data code:* `6A-200816`

This in fact makes it faster to experimentally acquire a pure shift spectrum and calculate a cost function based on that. Running an actual pure shift experiment, though, is suboptimal: firstly, the pseudo-2D interferogram method is slow, and secondly, there is no easy way to devise a cost function for the resulting spectrum without prior knowledge of where peaks and artefacts lie. Instead, we can use a simple 1D 'J-refocused spin echo' (JRSE) sequence, which has the form 90°–τ–PSE–τ–detect (Fig. 2.9b). An ideal PSE would lead to complete refocusing of both chemical shifts and J-couplings, and the signal detected after this would simply be the same as in a pulse–acquire experiment (Fig. 2.9a).

Of course, there is a sensitivity penalty which reflects that of the PSE (there are also relaxation losses during the τ delays, but these are a constant, unaffected by the form of the PSE). On top of that, if the J-refocusing is not perfect, then the multiplets in the spin echo sequence acquire a degree of phase distortion: the delay τ has to be long enough to allow for this to evolve, but its exact value is otherwise largely insignificant. These distortions are just about visible in Fig. 2.9b.

Two cost functions were designed and used in this section:

$$f_{\text{phase}} = \operatorname{Var}_i \left[\arctan\left(\frac{S_{\text{re},i}}{|S_{\text{im},i}|} \right) \right]; \tag{2.32}$$

$$f_{\text{diff}} = \left\| \frac{\mathbf{S}_{\text{re}}}{\|\mathbf{S}_{\text{re}}\|} - \frac{\mathbf{T}_{\text{re}}}{\|\mathbf{T}_{\text{re}}\|} \right\|, \tag{2.33}$$

result in substantial speedups, but I have not looked into this sufficiently. Anyway, experimental acquisition of the JRSE spectrum (to be described) takes only around 5 s.

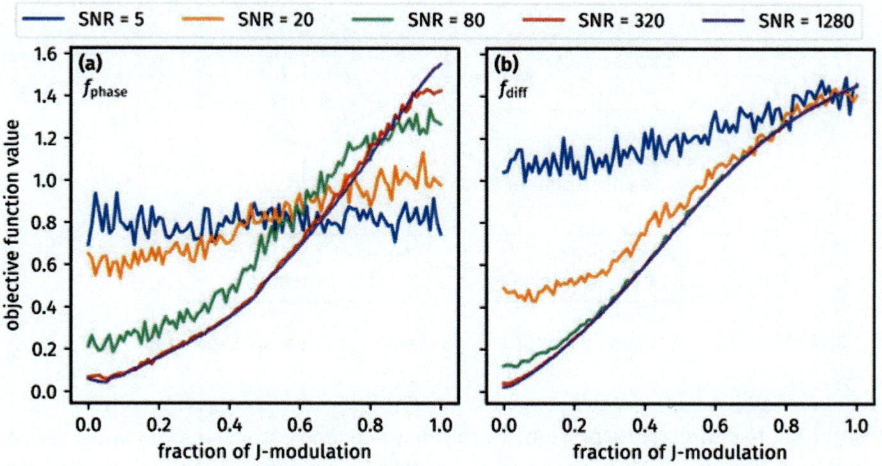

Fig. 2.10 Behaviour of the two cost functions, f_{phase} and f_{diff}, on synthetic spectra with various SNRs. Zero phase distortion refers to an in-phase absorption-mode doublet, whereas complete phase distortion refers to an antiphase dispersion-mode doublet. **a** The f_{phase} cost function. **b** The f_{diff} cost function, measured against a spectrum with no phase distortion and an SNR of 500

where the JRSE and pulse–acquire 1D spectra are treated as complex-valued vectors $\mathbf{S}$ and $\mathbf{T}$ respectively (for *'spectrum'* and *'target'*). $\mathbf{S}_{\text{re}}$ and $\mathbf{S}_{\text{im}}$ are the real and imaginary parts of the spectrum $\mathbf{S}$, and $S_{\text{re},i}$ is the i-th point of the real part of the spectrum. The operator Var_i represents the variance over all points in the spectrum (indexed by i), and $\|\mathbf{x}\|$ denotes the 2-norm of the vector $\mathbf{x}$, i.e. $\sqrt{\sum_i x_i^2}$. Python implementations of these are given in Listing 2.1.[11]

Listing 2.1 Pure shift cost functions.

```
import numpy as np
# assume S and T are complex numpy arrays which have been read in
Sr = np.real(S); Si = np.imag(S); Tr = np.real(T)
f_phase = np.var(np.arctan(Sr / np.abs(Si)))
f_diff = np.linalg.norm((Sr / np.linalg.norm(Sr))
                        - (Tr / np.linalg.norm(Tr)))
```

These two cost functions were chosen as they exhibited desirable characteristics on synthetic data (Fig. 2.10). In these simulations, the 'target' spectrum was chosen to simply be an in-phase absorption-mode doublet with an SNR of 500. Synthetic data with increasing amounts of J-modulation (i.e. spectra ranging from in-phase

[11] The use of `np.arctan` (the arctangent), and *not* `np.arctan2` (the argument of a complex number), is intentional. The behaviour shown in Fig. 2.10 isn't reproduced with `arctan2`. Of course, this means the name 'phase' is a misnomer; it's not really the phase of anything meaningful.

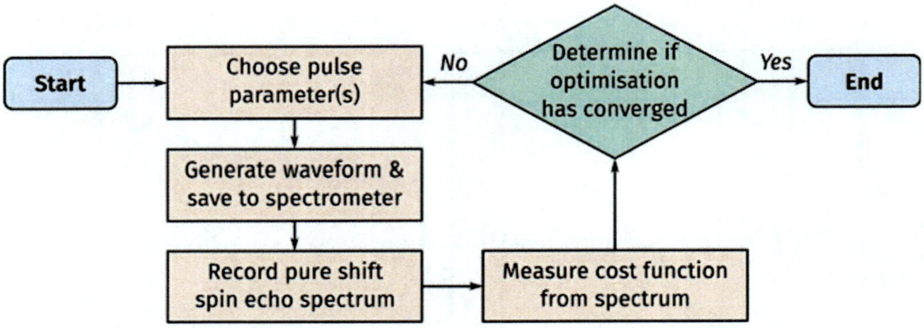

Fig. 2.11 Flowchart illustrating the steps for optimisation of a pure shift spectrum

absorption, to antiphase dispersion) were generated, and extra Gaussian noise added to mimic different SNRs. It can be seen that, for data which have little J-modulation (left edges of the plots), both f_{phase} (Fig. 2.10a) and f_{diff} (Fig. 2.10b) penalise lower SNRs. Furthermore, both of the cost functions penalise J-modulation, since they increase going from left to right. This penalty is stricter for high-SNR spectra, which is also desirable, since it is only in high-SNR spectra that the J-modulation becomes noticeable.

The cost function f_{diff} is easier to comprehend: it simply scales both the target and JRSE spectra down by their respective intensities and compares each point to determine whether the peak shapes obtained are similar. Although this seems like it should be agnostic towards signal intensity, this is only true for noiseless spectra. If a (genuine) JRSE spectrum has low SNR, $\|\mathbf{S}_{\text{re}}\|$ will be small, and the noise will be scaled down less than for the target spectrum; this difference in the *noise* (rather than the signal) then contributes towards the cost function. On the other hand, a proper rationalisation of why the cost function f_{phase} works is unfortunately not within my capabilities! It was mostly developed by trial-and-error (based on the notion that phase distortions would have something to do with $\mathbf{S}_{\text{re}}$ and $\mathbf{S}_{\text{im}}$), and I do not have a good explanation of why it works.

The general optimisation procedure is conceptually simple and largely consists of the loop shown in Fig. 2.11: this is essentially a specialised version of the POISE flowchart (Fig. 3.1). The optimisation algorithm is responsible for determining convergence, as well as choosing the new parameters based on previously obtained information; the initial parameters must be supplied by the user.

In practice, it is a technical challenge to implement this loop on a spectrometer as the cost function calculation is performed either in MATLAB or Python 3, both of which are not compatible with Bruker's TopSpin software. TopSpin instead provides Jython (Python 2.7) and C programming interfaces;[12] the former is not compatible with Python 3 packages like `numpy` or `scipy`, and the latter is too low-level to

[12] TopSpin 4.1.4 introduced a Python 3 interface which would have made much of this work simpler. Unfortunately, this was not available at the time of this work.

be worth implementing numerical algorithms in.[13] Thus, we require a means of *communication* between the spectrometer and the optimisation control program: this includes a signal from the controlling program to trigger acquisition on the spectrometer, as well as a signal from the spectrometer that acquisition is done so that the cost function can be calculated. As it turns out, the code used for the optimisations in this section was a very rudimentary and fragile form of that eventually used in POISE (for example, the aforementioned signals were transmitted via the creation and deletion of files). I therefore defer the discussion of this issue to Sect. 3.2.5, where the more robust POISE interface is explained in detail.

2.4.2 Flip Angle Optimisation

Having described the rest of the optimisation setup, it remains to choose exactly which parameters are subjected to optimisation. The simplest option is to only optimise one parameter, namely, the flip angle of the (double-)saltire pulse. The flip angle dependence of PSYCHE spectra is well-understood, which crucially allows us to evaluate the cost functions outlined above and determine whether they are functioning correctly.

I first sought to measure how the cost functions described above varied with the flip angle. To this end, JRSE spectra using a *single* saltire as the PSE and various flip angles (from 10° to 50°) were acquired (Fig. 2.12). Since both cost functions penalise both low sensitivity and low purity, we might expect that there is be an intermediate value where neither sensitivity nor purity was penalised too much: this would be the 'optimum' flip angle. In the event, it was found that only f_{diff} yielded a useful—albeit shallow—minimum at around 25° (recall that the PSE here is a single saltire, so this corresponds roughly to a 15°–20° double saltire). The f_{phase} cost function, on the other hand, was strictly decreasing within the range of flip angles tested: it is possible that there is an optimum at an even larger flip angle, but this would have been a scientifically unsound conclusion given that the sample was decently concentrated (50 mM).

This naturally raises the question of why f_{diff} yields an optimum which falls within what we would consider a 'sensible' region. It is tempting to believe that the form of f_{diff} (Eq. (2.33)), which is at first glance quite intuitive, naturally leads to a better result. I argue here, however, that this is mostly down to *coincidence*. This is difficult to explain quantitatively, but in a broader sense, we may imagine that the cost function separately penalises low sensitivity and low purity, i.e. it can be decomposed into something of the form

$$f = g(s) + \lambda h(p), \tag{2.34}$$

[13] Of course, heavily optimised code in low-level languages such as C and Fortran—or perhaps even MATLAB—would run faster. However, speeding up the code has virtually no impact on the optimisation, since its rate is limited by spectrum acquisition. In this situation, it makes far more sense to save *developer time*.

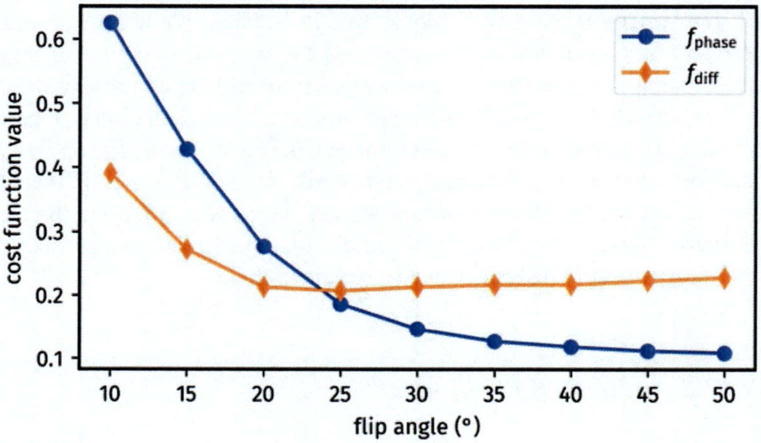

Fig. 2.12 Behaviour of the two cost functions, f_{phase} and f_{diff}, on experimental J-refocused spin echo spectra. *Data code:* `5C-190809`

where s and p are respectively the sensitivity and purity, and g and h are some unknown functions which *decrease* with increasing s and p (since we want to minimise the cost function). This is clearly a simplification, because the plots in Fig. 2.10 show that the effects of sensitivity and purity on the cost functions are not additive; however, it is sufficient to make the point here. The parameter λ represents the relative weighting of purity to sensitivity: if λ is large than the purity is more strongly emphasised, and vice versa if λ is small.

When we say that an optimum is 'sensible' or 'sound', this is with respect to the sensitivity/purity balance in the parent pseudo-2D homodecoupled spectrum. In other words, what we *really* seek is a cost function which measures the sensitivity and purity of that spectrum:

$$f' = g'(s') + \lambda' h'(p'), \tag{2.35}$$

where everything is marked with a prime symbol to indicate that it is with respect to the decoupled spectrum, and the parameter λ' is chosen to fit our judgement of the required balance, or in other words, yield an optimum of around 20°. Now, we may reasonably assume that s and s' are proportional, but p and p' are hardly the same thing: one is manifested in terms of artefact intensity and the other in terms of phase distortions. Furthermore, the λ provided to us by the cost function f may not necessarily be the same as our ideal choice of λ'; let alone the forms of the functions g and h. The fact that Eq. (2.34) *happens* to provide the same flip angle optimum as the idealised Eq. (2.35) cannot truly be attributed to design!

Of course, just because a cost function works mostly by serendipity does not mean that it cannot be used. So, I ran several *actual* optimisations of the flip angle using the cost function f_{diff}, which reliably converged to optima between 20° and 25° regardless of the initial point chosen. Typically, around 10 spectrum acquisitions were required, corresponding to a time of around 2–4 min. This is not surprising in light of our knowledge of f_{diff}; however, it provides us with a proof-of-principle that automated optimisation of NMR parameters is possible. Further optimisations of the PSYCHE flip angle are also discussed in Sect. 3.4.8.

2.4.3 *Waveform Parameterisation and Optimisation*

Given that a working optimisation setup, including cost functions, had been developed, it was a logical step to then test it out on a more challenging problem: namely, how the waveform used in the PSYCHE PSE could be modified. This goes beyond simply modifying the number of saltires, as was done in Sect. 2.3. There is no real reason why the pulse *must* be an integer number of saltires: in principle it can have *any* shape, although being symmetric about the centre of the pulse would likely still be beneficial in terms of preserving the mechanism of spatiotemporal averaging.

2.4.3.1 Sum-of-Chirp Parameterisation

A naive attempt at optimising the pulse would simply involve modifying every pulse point in the double-saltire waveform used in the PSYCHE element. As described in Sect. 1.2, each pulse point consists of a pair of x- and y-amplitudes (c_x, c_y); therefore, for a pulse with m points, we would have a parameter vector $\mathbf{x} \in \mathbb{R}^{2m}$. Unfortunately, for PSYCHE, m is on the order of 10000, and an optimisation with 20000 points is totally unfeasible.[14]

As a result of this, we must consider other ways of parameterising the waveform which use fewer degrees of freedom. Several approaches to this issue have surfaced in the literature, such as the use of Fourier series [37, 70–73], Gaussian cascades [74], or spline interpolation between a subset of pulse points [75]. In this instance, we can use the knowledge that the PSYCHE PSE is composed of saltire pulses to our advantage. Each saltire pulse is a linear combination of two chirps, defined by

$$\phi(t) = \phi_0 + 2\pi f_0 t + \pi\tau_{\mathrm{p}}(\Delta F)\left(\frac{t}{\tau_{\mathrm{p}}} - \frac{1}{2}\right)^2; \tag{2.36}$$

$$c_x(t) = A\cos[\phi(t)]; \tag{2.37}$$

$$c_y(t) = A\sin[\phi(t)], \tag{2.38}$$

[14] Although problems of this size have been tackled using optimal control theory [66–69], it is not really feasible to use it in cases where the pulse is applied *together* with a gradient, as-is the case in PSYCHE. On top of that, the coupling networks and spin systems of interest are rather more complicated than in typical applications of optimal control.

for $t \in [0, \tau_p]$. Here, τ_p is the duration of the chirp, A is the amplitude of the chirp (which is time-independent), ϕ_0 the phase of the chirp, ΔF the bandwidth, and f_0 its frequency offset (i.e. where the centre of the bandwidth lies). The latter two parameters are assumed to be given in units of ordinary frequencies (Hz), not angular frequencies. (Note that Eq. (2.37) and (2.38), defining the x- and y-coefficients of the pulse, are just a restatement of Eq. (1.30).)

Given these expressions, we see that there are five parameters of the chirp which can be modified: A, τ_p, ϕ_0, ΔF, and f_0. The two chirps which form one saltire pulse simultaneously sweep in opposite directions, which mean that ΔF for one chirp is the negative of the other; however, their parameters are otherwise equal. We may, however, also envision a case where the pulse is constructed from two chirps which are applied at a different point in time. This adds one more parameter to each chirp, namely, t_0, the starting time of the pulse. Each chirp, therefore, sweeps from the frequency $f_0 - (\Delta F)/2$ at a time t_0, to the frequency $f_0 + (\Delta F)/2$ at a time $t_0 + \tau_p$. In total, this gives us 12 parameters to optimise [36, 76].

After constructing the sum of two chirps, 'empty' regions of no RF application ($c_x = c_y = 0$) at the beginning and the end were trimmed off. Since a sum of two chirps is not necessarily symmetric (with respect to reflection in time), the waveform was then reflected about its end, thus doubling the length of the pulse. The entire waveform was then multiplied by a *smoothing function* $f_{\mathrm{sm}}(t)$ (Fig. 2.13b), which prevents large jumps in RF amplitude at the beginning and end of the pulse. f_{sm} depends on a smoothing parameter s_{sm}, which is typically 0.1–0.2:

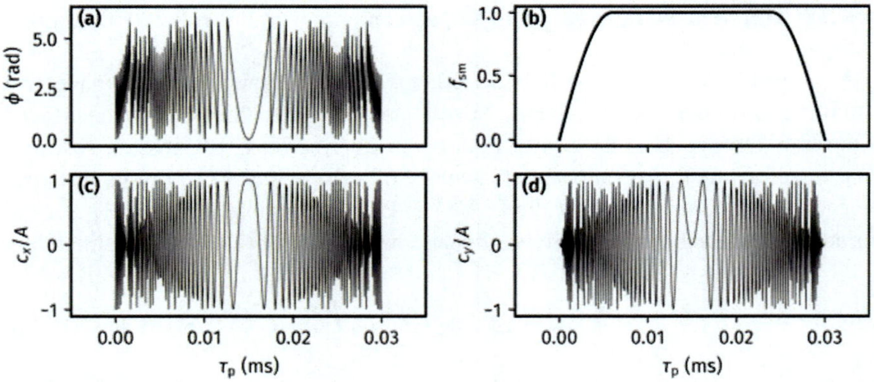

Fig. 2.13 Plots of the quantities in Eqs. (2.36) to (2.37), (2.38) for a typical chirp pulse ($\tau_p = 30$ ms, $\Delta F = 10$ kHz). **a** $\phi(t)$ wrapped to the range $[0, 2\pi)$. **b** The quarter-sine smoothing profile which is later applied to the entire waveform (Eq. (2.39)); here $s_{\mathrm{sm}} = 0.2$. **c** $c_x(t)/A$ prior to smoothing. **(d)** $c_y(t)/A$ prior to smoothing. The amplitude A is simply a constant, which is related to the flip angle of the chirp [35]

$$f_{\text{sm}}(t) = \begin{cases} \sin\left(\dfrac{\pi t'}{2s_{\text{sm}}}\right) & 0 \leq t' < s_{\text{sm}}; \\ 1 & s_{\text{sm}} \leq t' < 1 - s_{\text{sm}}; \\ \sin\left[\dfrac{\pi(1-t')}{2s_{\text{sm}}}\right] & s_{\text{sm}} \leq t' \leq 1, \end{cases} \tag{2.39}$$

where $t' = t/\tau_{\text{p}}$ (and here τ_{p} refers to the duration of the *entire* waveform, after reflection).

2.4.3.2 Optimisations

The initial point chosen was:

- Chirp 1: $\tau_{\text{p}} = 15\,\text{ms}$; $\Delta F = -5\,\text{kHz}$; $\phi_0 = 0$; $A = 36\,\text{Hz}$; $t_0 = 0$; $f_0 = 5\,\text{kHz}$;
- Chirp 2: $\tau_{\text{p}} = 15\,\text{ms}$; $\Delta F = 5\,\text{kHz}$; $\phi_0 = 0$; $A = 36\,\text{Hz}$; $t_0 = 0$; $f_0 = -5\,\text{kHz}$.

After the sum of these two pulses is reflected about its end, we obtain a single saltire with bandwidth 10 kHz, duration 30 ms, and an amplitude of 72 Hz, corresponding to a flip angle of approximately 32°.

Using this setup, several optimisations of the 12 parameters above were conducted using the f_{diff} cost function, which had performed well enough in the flip angle optimisations. However, it was quickly noticed that this led to spurious optima being located, such as the one in Fig. 2.14c. Although the initial point (a saltire) yielded a much larger SNR (Fig. 2.14b), the value of f_{diff} was still larger (i.e. worse) than

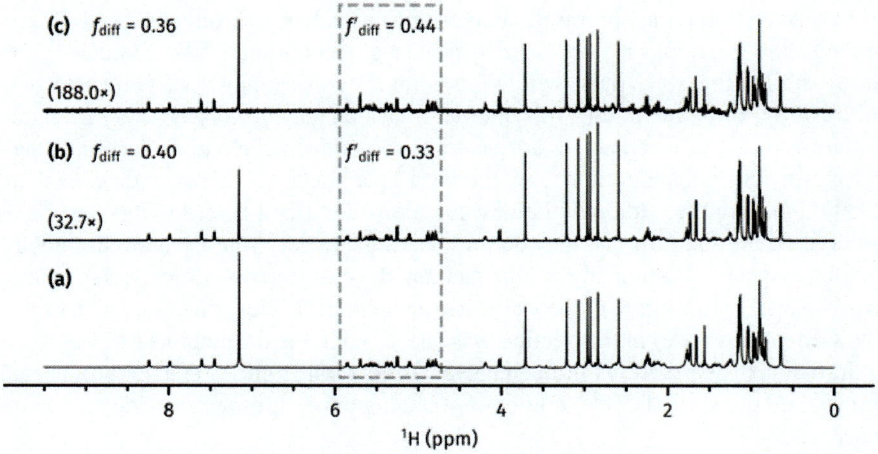

Fig. 2.14 **a** Target spectrum (pulse–acquire). **b** JRSE spectrum obtained using the initial guess (a saltire). **c** JRSE spectrum obtained with a spurious optimum point. The grey dotted box shows the restricted region over which f_{diff} was subsequently applied to, yielding a formally different cost function f'_{diff}; this yielded more sensible results where the cost function for the saltire pulse was smaller. *Data code:* `6C-190823`

for this false optimum. The reason for this is almost certainly because the saltire pulse distorts the relative intensity of the strong singlets in the spectrum, notably the C6D5H peak at 7.15 ppm, and the *N*-methyl groups between 2.5 and 3.8 ppm, whereas the false optimum does not (most likely by coincidence).

Of course, singlets are completely unimportant when devising a pure shift experiment. Unfortunately, singlets are also typically more intense than the rest of the spectrum and thus contribute disproportionately to the cost function. A simple and effective way to circumvent this is to restrict the region of the spectrum being evaluated. In this case, I chose to use the cyclosporin Ha region between 4.72 and 5.94 ppm (grey dotted box in Fig. 2.14). This yields a formally different cost function, which I label as f'_{diff}. With this, much more logical behaviour was observed: in particular, the saltire pulse performed better than the spurious optimum previously found.

While this new cost function could be successfully used to run optimisations, most of these unfortunately failed to find anything performing better than the original saltire pulse. On the rare occasion where something 'better' was found (as judged by the new cost function f'_{diff}), these 'optimised' pulses were fairly close to a saltire, and the corresponding decreases in the cost function extremely small—suggesting that the 'better' result may simply just have been due to noise in the cost function. Nevertheless, these new 'optima' *did* function as perfectly serviceable PSEs: for example, Fig. 2.15 compares a triple spin echo (TSE) PSYCHE spectrum obtained with an 'optimised' pulse to one obtained with the single saltire pulse. There is virtually no difference. This is a meaningful result, as it demonstrates that f'_{diff} is actually an accurate metric to determine the quality of a PSE (it is noisy, but this is to be expected of an experimentally measured cost function). Unfortunately, although the pulse shape is shown in Fig. 2.15b, the exact parameters which led to this pulse shape have been lost to time.

One issue is that the optimisation algorithms used here can only perform a *local* optimisation, i.e. it may not necessarily locate a global optimum. We cannot rule out the possibility that *some* pulse within this parameter space may in fact outperform a saltire pulse. However, finding this optimum—which may lie very far away from the initial guess of a saltire pulse—is not generally something which can be accomplished in a reasonable amount of time, even though 12 parameters is far more tractable than 20000. Furthermore, although the cost functions described here do work, they are generally quite 'flat', in that they do not discriminate very sharply between 'good' and 'bad' spectra. Combined with the fact that the cost function is noisy, this makes experimental optimisation of the waveform an uphill task. Nevertheless, much of the knowledge (and code) in this section was later used in the development of POISE.

In the next two sections in this chapter, I move on from PSYCHE and instead discuss completely different methods of obtaining pure shift spectra using only hard pulses in the PSE.

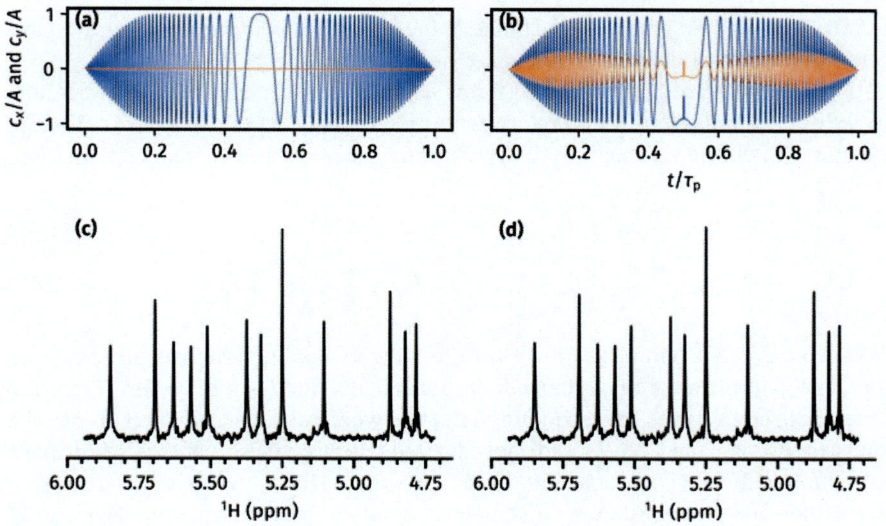

Fig. 2.15 **a** x- and y-coefficients of the initial saltire pulse (as a fraction of the maximum amplitude A). **b** x- and y-coefficients of the 'optimised' pulse (as a fraction of the maximum amplitude A). **c** TSE-PSYCHE spectrum obtained with the initial guess (a saltire pulse). **d** TSE-PSYCHE spectrum obtained with the 'optimised' pulse, with virtually equivalent performance. *Data code:* `6C-190831`

2.5 Time-Reversal Method

This section focuses on the time-reversal element briefly introduced in Sect. 2.2.2: the aim is to determine whether a pure shift method based on this can have better performance than PSYCHE. Before that, I will first provide a more detailed theoretical analysis of it. The time-reversal element consists of just a single hard pulse with a flip angle β, and we first analyse its potential use as a JRE (instead of a PSE), using Eq. (2.27) to (2.28), (2.29).

Recall that an ideal JRE should invert passive spins only, not active spins: thus, we seek a transformation of the form $I_{1+}I_{2\alpha} \rightarrow I_{1+}I_{2\beta}$. However, the hard pulse does this:

$$I_{1+}I_{2\alpha} \rightarrow \underbrace{c^2s^2I_{1+}I_{2\beta}}_{\text{term 1}} + \underbrace{c^4I_{1+}I_{2\alpha}}_{\text{term 2}} \underbrace{- \frac{S^2}{4}I_{1\alpha}I_{2+} + \frac{S^2}{4}I_{1\beta}I_{2+}}_{\text{terms 3 and 4}}, \tag{2.40}$$

where (as before) $S = \sin\beta$, $s = \sin(\beta/2)$, and $c = \cos(\beta/2)$. As in the analysis for PSYCHE (Sect. 2.2.3), term 1 here represents the desired signal; term 2 recoupling artefacts; and terms 3 and 4 COSY-type artefacts arising from coherence transfer. Terms with difference coherence orders have been neglected as they can be removed using CTP gradients.[15] Unlike in the PSYCHE analysis, however, we have not neglected any other terms of smaller order in s.

Since the desired and undesired terms have different coefficients, it is possible to *fully cancel out the recoupling artefacts* by recording (in this case) two different spectra with different values of β and performing an appropriate linear combination:

$$I_{1+}I_{2\alpha} \xrightarrow{\beta=0^\circ} I_{1+}I_{2\alpha}; \tag{2.41}$$

$$I_{1+}I_{2\alpha} \xrightarrow{\beta=90^\circ} \frac{1}{4}I_{1+}I_{2\beta} + \frac{1}{4}I_{1+}I_{2\alpha} - \frac{1}{4}I_{1\alpha}I_{2+} + \frac{1}{4}I_{1\beta}I_{2+}. \tag{2.42}$$

If we take Eq. (2.42) minus 1/4 of Eq. (2.41), the recoupling artefacts (arising from the $I_{1+}I_{2\alpha}$ term) are fully removed. In general, for an N-spin system, there are several different 'types' of recoupling artefacts where different numbers of passive spins (between 1 and $N - 1$) are not inverted. Each of these pathways will have different coefficients, as each spin that is flipped contributes s^2, whereas each spin that is not flipped contributes c^2. Suppressing all of these requires the acquisition and summation of N spectra with different flip angles and appropriate weights.

Before we go on further, notice even in the two-spin system that the COSY-type artefacts are *not* suppressed! In the original work [13], this time-reversal element was used in the middle of the t_1 period in a NOESY experiment. The coherence transfer peaks were not deemed to be problematic in this context: they gave rise to artefacts which had F_1 frequencies of $(\Omega_1 + \Omega_2)/2$, but different phase properties to genuine crosspeaks, allowing them to be easily identified. Of course, these artefacts are not acceptable in an actual pure shift spectrum.

To remove these peaks, I adopted the strategy first reported by Thrippleton *et al.* for suppression of COSY-type transfer pathways in 2DJ spectra [4]. In a 2DJ experiment, the central 180° pulse should in principle not cause coherence transfers between different spins; however, in strongly coupled systems this can happen. The solution was to bracket the 180° pulse, as well as half of the t_1 period, with a pair of opposing chirps and gradients. The same idea was also used in the TSE-PSYCHE experiment [11] to (further) suppress strong coupling responses in the parent PSYCHE experiment. This works because the unwanted CTPs have coherences on different spins during each of the two chirps; consequently, the coherences are inverted at different times by the chirp pulses and are ultimately dephased by gradients. The resulting time-reversal experiment was thus simply the same as the parent TSE-PSYCHE experiment, except that the central PSYCHE element was replaced by a β hard pulse (Fig. 2.16).

In such a sequence, the JRE is in fact not only the β pulse itself but also the second chirp (which takes the place of a 180° pulse). So, the β pulse here fulfils the role of a PSE, not a JRE. To accomplish this for a system containing N mutually coupled spins, N different experiments must be acquired using the following flip angles β_j and summed with the corresponding weights W_j $(j = 1, 2, \ldots, N)$:

[15] In the original work [13] which predated the widespread use of field gradients, other terms were removed through phase cycling, which is essentially equivalent.

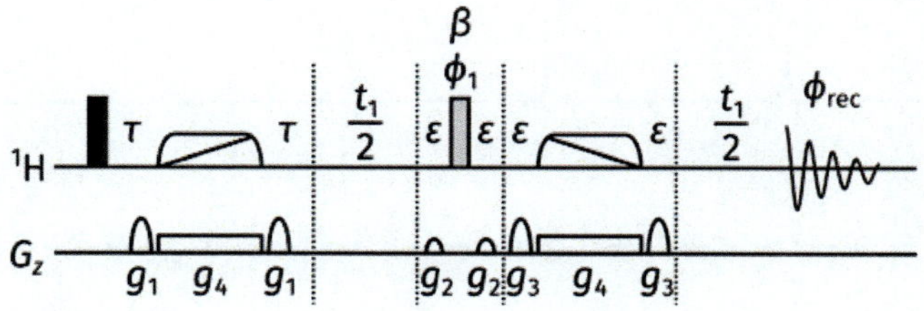

Fig. 2.16 Time-reversal pure shift pulse sequence. The flip angle β is varied as described in Eq. (2.43). Pulse phases are: $\phi_1 = (x, y, -x, -y)$; $\phi_{\text{rec}} = (x, -x, x, -x)$. The delay τ is set to $1/(4 \cdot T_{\text{chunk}})$ and allows for J-coupling to be refocused in the middle of the chunk. Gradient amplitudes are $(g_1, g_2, g_3, g_4) = (35\%, 49\%, 77\%, 1\%)$ (note that, in principle, g_4 should be calibrated according to the bandwidth of the chirp used)

$$\beta_j = \frac{j\pi}{N}; \tag{2.43}$$

$$W_j = \frac{N}{8} \cdot \frac{(-1)^j}{\sin^2(\beta_j/2)}. \tag{2.44}$$

For the sake of completeness, the values of β and W for a JRE are given here as well:

$$\beta_k = \frac{k\pi}{N}; \tag{2.45}$$

$$W_k = \frac{N}{8} \cdot \frac{(-1)^{k+N}}{\cos^2(\beta_j/2)}, \tag{2.46}$$

for $k = 0, 1, \ldots, N-1$. The derivation of these expressions is discussed more thoroughly in a paper by Griesinger et al. [5] In the context of this specific paper, note that the weights for the PSE correspond to that used for the ECOSY experiment, and the weights for the JRE correspond to that used for the complementary ECOSY experiment.

In practice, $N = 5$ is likely to cover most realistic spin systems. Explicitly evaluating Eqs. (2.43) and (2.44) yields $\beta = \{36°, 72°, 108°, 144°, 180°\}$ and $W = \{6.545, 1.809, 0.955, 0.691, 0.625\}$. Figure 2.17 shows insets of the five subspectra acquired with the above values of β and scaled by their respective weights W. The weighted sum (i.e. the pure shift spectrum) was phased, and the resulting phase correction values were propagated back to the individual subspectra.

Although the experiment seems to work, in that the weighted sum *is* indeed a pure shift spectrum, the fact that it is obtained through summation of N different experiments brings some immediate drawbacks. Firstly, the minimum duration of the experiment is lengthened by a factor of N: this is essentially the same as an N-step phase cycle. However, and perhaps more importantly in the context of *pure shift* NMR, the artefacts surrounding the main peaks are not perfectly cancelled through the process of summation. As a result, random distortions are observed around the

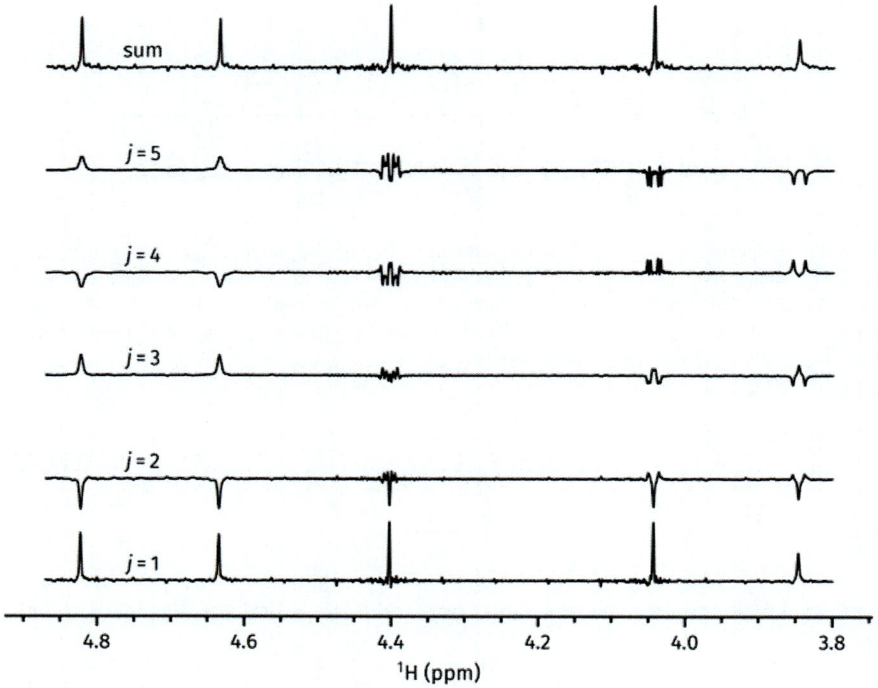

Fig. 2.17 Insets of weighted time-reversal subspectra (with $j = 1$ through 5), as well as their sum (the pure shift spectrum). *Data code:* `7A-201020`

desired peaks in the pure shift spectrum: this is noticeable in the 4.04 ppm peak in Fig. 2.17 and is even worse for more intense signals.

In terms of sensitivity, the time-reversal spectrum is not particularly exceptional either. Each of the five subspectra above was acquired with 2 scans; when compared against a typical TSE-PSYCHE experiment acquired with only 4 scans (i.e. 40% of the experiment duration), the TSE-PSYCHE experiment had comparable or perhaps even slightly better SNR (Fig. 2.18). The likely reason for this is because in the time-reversal experiment, signal is actually being *cancelled out* through the process of summation, as-is quite clearly shown in Fig. 2.17. In principle, the sensitivity of the time-reversal experiment could be optimised by acquiring the more heavily weighted spectra with more scans. However, I consider this unlikely to make a substantial difference to the conclusions drawn here. The idea of re-optimising the weights to better suppress artefacts was also briefly considered. However, given that Eq. (2.44) already yields *theoretically* complete suppression, it was deemed unlikely that anything substantially better could be obtained, considering that the artefacts arise from the summation process itself and are likely to appear regardless of what weights are chosen. Performing such an optimisation would also require a pure shift spectrum to have been acquired beforehand (for comparison), thus defeating the purpose of optimising the weights anyway.

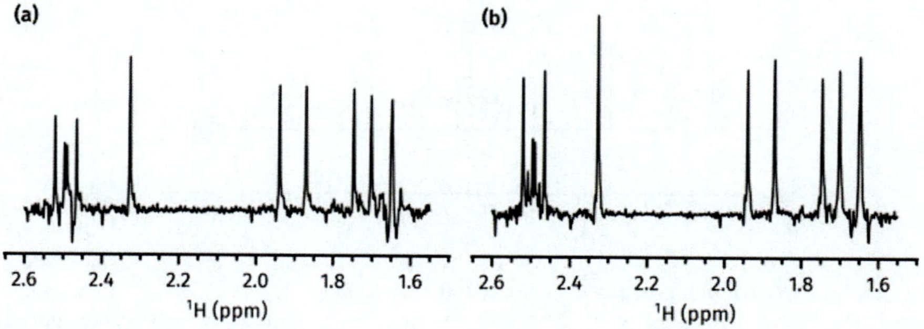

Fig. 2.18 Comparison of time-reversal and TSE-PSYCHE sensitivity. **a** Time-reversal pure shift spectrum (the same as the sum in Fig. 2.17) acquired with 2 scans for each subspectrum. **b** TSE-PSYCHE (double saltire, flip angle 15°) acquired with 4 scans. The spectra have been scaled so that their noise levels are similar: the signal intensity is comparable, or perhaps slightly better in the TSE-PSYCHE. *Data code:* `7A-201020`

Interestingly, Fig. 2.17 suggests that the $j = 1$ spectrum *on its own* already provides as good a result as the summed pure shift spectrum. This is not surprising, as the use of a hard pulse as the PSE yields a conceptually very similar result to PSYCHE in that the signal-to-artefact ratio depends on $\tan^2(\beta/2)$ (of course, the COSY-type artefacts must still be suppressed through the TSE scheme). This suggests that even without summation of multiple subspectra, the TSE time-reversal pulse sequence in Fig. 2.16 is a viable pure shift experiment—albeit one which does not have any significant advantage over PSYCHE.

2.6 'Discrete PSYCHE'

The last pure shift method in this chapter is completely original and represents perhaps the most fruitful attempt so far at optimising pure shift experiments. It relies on what is essentially a 'temporal discretisation' of the PSYCHE waveform and gradient combination: instead of applying a shaped pulse and a gradient simultaneously, hard pulses and gradients are interleaved in the PSE (Fig. 2.19).

For this reason I have dubbed this experiment the 'discrete PSYCHE', or dPSYCHE for short. There are two major reasons why this is more amenable towards optimisation than many of the previous experiments:

1. Pulses and gradients are no longer applied simultaneously, which makes simulation of the experiment *extremely* fast compared to the original PSYCHE. This opens up the possibility of entirely theoretical optimisations, as the noise can be completely eliminated from the cost function.

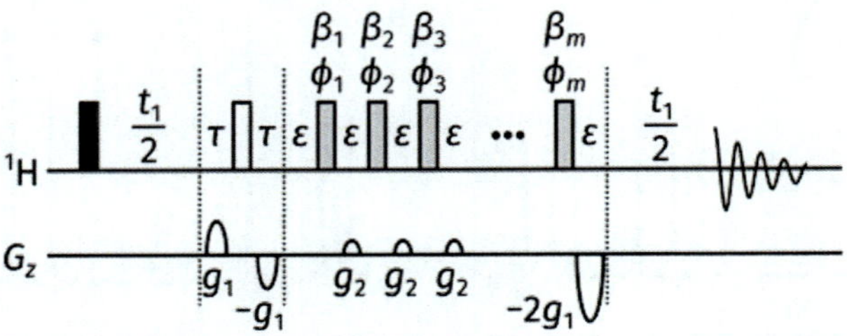

Fig. 2.19 dPSYCHE pulse sequence. Gradient amplitudes are $(g_1, g_2) = (35\%, 41\%)$; the gradients in the PSE g_2 are applied with a duration of 500 μs. The hard pulses in the PSE are applied with an RF amplitude of 18 kHz. The delay τ is set to $1/(4 \cdot T_{\text{chunk}})$ and allows for J-coupling to be refocused in the middle of the chunk

2. There are much fewer 'pulse points' than in the original PSYCHE: effectively, the phase and flip angle of every hard pulse has to be optimised, leading to $2m$ parameters. Even for $m \sim 10$, this is quite tractable if the optimisation is noiseless.

One downside of this is that it is difficult, or perhaps even impossible, to explain how the PSE works.[16] For a symmetric PSE where $\beta_1 = \beta_m$ (and so on) it is probably possible to reuse an explanation based on PSYCHE-style CTP selection, but this is clearly inapplicable if the flip angles and phases are scrambled.

2.6.1 Speeding up dPSYCHE Simulations

To begin, I first explain how the exact simulation of dPSYCHE experiments can be greatly accelerated through efficient propagator calculations. Although Spinach [65] is the leading simulation package for NMR and covers an extremely impressive range of experiments, this generality also prevents it from providing optimal performance for any single experiment. As it turns out, handwritten, specialised MATLAB code can outperform Spinach by orders of magnitude.

The NMR simulations developed here simply use the density operator formalism in Hilbert space, as outlined in Sect. 1.3. The Zeeman basis is used, and non-unitary transformations such as relaxation are neglected. Propagation under the Liouville–von Neumann equation (Eq. (1.38)) requires the calculation of matrix exponentials $\exp(-iH\tau)$. For an $N \times N$ matrix, the matrix exponential requires $O(N^3)$ time to calculate (and for a system containing p spins, we have $N = 2^p$); it is often this which

[16] Of course, we could simulate it and say that it works because the maths says it does; but that isn't very illuminating. Some of the optimisations done in this thesis are somewhat like a scaled-down version of machine learning, in that they produce better results at the cost of interpretability.

is the bottleneck in NMR simulations. Minimising the number of matrix exponentials, and/or their computational cost, is the key to achieving speedups, as will be shown in the following text.[17]

Generally, the accurate simulation of pulsed field gradients requires the sample to be divided up into n discrete slices, each simulated with a different H_{grad}.[18] Thus, a very naive implementation of the dPSYCHE experiment would require mn matrix exponentials, one per pulse per slice. The overall structure of the code would resemble listing 2.2.[19] Note that the term `H_free`, and the corresponding propagator `U_free`, technically refer to the interaction-picture free Hamiltonian, $H_{\text{free},I}$.

Listing 2.2 Rough structure of a naive dPSYCHE implementation. Note that I use the variable name `slce` as `slice` is an existing builtin Matlab function.

```
% loop over slices
for slce=1:n
    H_grad = I_z * G * z(slce);

    % loop over pulse points
    for pulse_point=1:m
        H_pulse = (c_x(pulse_point) * I_x) + (c_y(pulse_point) * I_y);

        % calculate propagators; m*n total matrix exponentials
        U_pulse = expm(-1j * (H_free + H_pulse) * t_pulse);
        U_grad = expm(-1j * (H_free + H_grad) * t_grad);

        rho = U_grad * U_pulse * rho * U_pulse' * U_grad';
    end
end
```

It is not difficult to come up with a more sensible approach which cuts this down by a factor of m: since the pulses are not applied together with the gradients, the pulse propagators U_{pulse} can be pre-calculated outside of the loop. Furthermore, all of the gradients within the PSE are the same, so U_{grad} can be moved out of the inner loop (Listing 2.3).

---

[17] Note that in my simulations, I simply used the built-in `expm` MATLAB function, which implements the matrix exponential using a combination of the scaling-and-squaring method and Padé approximation [77]. This is in contrast to Spinach, which primarily uses a scaled-and-squared Taylor series (according to the `propagator.m` file, various other methods supposedly did not 'live up to their marketing'). An in-depth discussion of matrix exponential methods is outside the scope of this thesis but can be found in a classic paper by Moler and Van Loan [78].

[18] In simple cases this can be avoided by simply removing all terms with the wrong coherence orders as we know they will be dephased (Eq. (1.110)), but this is too naive an approach for pure shift techniques.

[19] Strictly speaking, there is a slight inaccuracy in this code: the final gradient should have strength `-2G` and not `G`, but that is a minor detail which I leave out for clarity in the code.

Listing 2.3 Rough structure of a slightly faster implementation of dPSYCHE.

```
% precalculate pulse propagators; m total matrix exponentials
for pulse_point=1:m
    H_pulse = (c_x(pulse_point) * I_x) + (c_y(pulse_point) * I_y);
    U_pulse(m) = expm(-1j * (H_free + H_pulse) * t_pulse);
end

% loop over slices
for slce=1:n
    % calculate gradient propagators; n total matrix exponentials
    H_grad = I_z * G * z(slce);
    U_grad = expm(-1j * (H_free + H_grad) * t_grad);

    % loop over pulse points
    for point=1:m
        rho = U_pulse(m) * rho * U_pulse(m)';
        rho = U_grad * rho * U_grad';
    end
end
```

Spinach, which is designed to be general, has no idea that these optimisations are possible, so is naturally rather slower. However, even this is relatively inefficient. It can be shown that the two components of the gradient propagator, $H_{\text{free},I}$ and H_{grad}, actually commute with one another (even in the strong coupling case). Thus, we can write:

$$\exp[-\mathrm{i}(H_{\text{free},I} + H_{\text{grad}})\tau] = \exp(-\mathrm{i}H_{\text{free},I}\tau)\exp(-\mathrm{i}H_{\text{grad}}\tau) \tag{2.47}$$

(in general, for matrices A and B, $\exp(A + B) = \exp(A)\exp(B)$ if and only if $[A, B] = 0$). This on its own does not reduce the number of matrix exponentials required, but notice now that H_{grad} is a sum of I_{iz} terms and is therefore *diagonal* in the Zeeman basis. The exponential of a diagonal matrix is almost trivial to calculate, as we only need to exponentiate the diagonal *elements:*

$$\exp\begin{pmatrix} \lambda_1 & 0 & \dots & 0 \\ 0 & \lambda_2 & \dots & 0 \\ \vdots & \vdots & \ddots & \vdots \\ 0 & 0 & \dots & \lambda_n \end{pmatrix} = \begin{pmatrix} \exp(\lambda_1) & 0 & \dots & 0 \\ 0 & \exp(\lambda_2) & \dots & 0 \\ \vdots & \vdots & \ddots & \vdots \\ 0 & 0 & \dots & \exp(\lambda_n) \end{pmatrix}. \tag{2.48}$$

Instead of using the $O(N^3)$ `expm(M)` function, this can instead be done in $O(N)$ time using `diag(exp(diag(M)))` (the `diag` MATLAB function converts a diagonal matrix to a vector of its diagonal entries, and vice versa). So, we now require only $m + 1$ 'true' matrix exponentials. At this point, our matrix exponentials have almost been eliminated and the largest remaining bottleneck is almost certainly the matrix *multiplications* required for the propagation. We can cut these down by a factor of two simply by calculating the overall propagator

$$U_{\text{total}} = U_n \cdots U_2 U_1, \tag{2.49}$$

and then performing the propagation only at the very end:

$$\rho = U_{\text{total}} \rho_0 U_{\text{total}}^\dagger, \tag{2.50}$$

instead of performing every individual step $\rho \rightarrow U_1 \rho_0 U_1^\dagger$. The final optimised code therefore resembles that in Listing 2.4.

Listing 2.4 Rough structure of a fast dPSYCHE implementation.

```
% precalculate propagator due to free evolution during gradient
% only 1 matrix exponential
U_free = expm(-1i * H_free * t_grad);

% precalculate pulse propagators; m total matrix exponentials
for pulse_point=1:m
    H_pulse = (c_x(pulse_point) * I_x) + (c_y(pulse_point) * I_y)
    U_pulse(m) = expm(-1j * (H_free + H_pulse) * t_pulse)
end

% loop over slices
for slce=1:n
    % initialise propagator for this slice
    U_slce = eye(2 ^ p);

    % calculate gradient propagators; no matrix exponentials required
    H_grad = I_z * G * z(slce);
    U_grad = U_free * diag(exp(diag(-1j * H_grad * t_grad)));

    % loop over pulse points
    for point=1:m
        U_slce = U_pulse(m) * U_slce;
        U_slce = U_grad * U_slce;
    end

    % perform propagation only at the end
    rho = U_slce * rho * U_slce';
end
```

The performance of this handwritten code as compared to Spinach is summarised in Table 2.1. In all cases, the spectra produced by the two methods were entirely equivalent. It should be noted that the handwritten code does not even utilise CPU parallelisation, whereas Spinach does. I investigated the possibility of parallelising the loop over slices (replacing the outer `for` in Listing 2.4 with `parfor`): however,

Table 2.1 Comparison of wall-clock times for dPSYCHE simulations. The dPSYCHE sequence used contained 15 pulses, each applied with a flip angle of 15° and a phase of 0°. Spin systems were generated pseudo-randomly. All timings were measured on a 2019 MacBook Pro with a 2.6GHz 6-core Intel i7 CPU (although CPU parallelisation was not used)

Number of spins	Number of couplings	Execution time (s)	
		Spinach	Handwritten
1	0	3.33	0.35
2	0	4.59	0.32
	1	6.22	0.32
3	0	9.90	0.43
	1	12.95	0.45
	2	31.86	0.47
	3	35.01	0.48
4	0	30.59	1.02
	1	38.63	0.99
	2	44.61	1.04
	3	365.04	1.52
	4	446.03	1.57
	5	521.72	1.57
	6	588.31	1.69

this in fact made the code *slower*, presumably due to overhead. This is a good thing: it means that `parfor` can be used in an external loop, e.g. for the parallel simulation of the dPSYCHE experiment on different spin systems.

2.6.2 *Optimisations and Experimental Evaluation*

2.6.2.1 Spectrum-Comparison Cost Function

As previously discussed, the fact that the dPSYCHE experiment can be very quickly simulated opens up the possibility of entirely computational optimisation of the pulse sequence. For any arbitrary spin system, it is trivial to remove all the couplings *in silico* and simulate a pulse–acquire spectrum: this gives us a theoretically perfect pure shift spectrum. The simulated dPSYCHE experiment (on a system with couplings) can then be compared against this. The entire process is repeated using s different spin systems, and the cost function is defined as

$$f_{\text{diff},2} = \frac{1}{s}\sum_{s}\left\|\frac{\mathbf{S}_{\text{re}} - c\mathbf{T}_{\text{re}}}{c\|\mathbf{T}_{\text{re}}\|}\right\|^2, \quad (2.51)$$

where $\mathbf{S}$ is the dPSYCHE spectrum, and $\mathbf{T}$ is the target spectrum. The prefactor c will be discussed later; for now I treat it as 1. This cost function appears superficially very similar to f_{diff} (discussed in Sect. 2.4.1) and is based on the same principle that we want the spectra $\mathbf{S}$ and $\mathbf{T}$ to match one another, but there are several points of note:

- The spectrum $\mathbf{S}_{\text{re}}$ is not scaled down by its norm. This means that the sensitivity penalty no longer comes from the difference in the noise (as was previously the case), but rather directly from the difference in peak intensity. Since simulated spectra are noiseless, the original f_{diff} would not work here.
- The division of the entire cost function by $\|\mathbf{T}_{\text{re}}\|$ is not important if only one spin system is being simulated as it is simply a constant factor. However, if more than one spin system is being simulated, $\|\mathbf{T}_{\text{re}}\|$ differs from system to system and this factor helps to essentially normalise the contributions from each spin system.
- The norm in the cost function here is squared. Again, this makes no difference to the optimum if only one spin system is being investigated, because x^2 is strictly increasing for $x > 0$.[20] However, for multiple spin systems it makes sense to square the norm, as the largest deviations will be penalised more greatly: this means that a pure shift spectrum which works reasonably well across a wide range of spin systems will be prioritised over one which works perfectly well for some and fails badly for others.

I began by first checking how many t_1 increments (i.e. chunks) were required in the simulation to obtain reliable cost function values. If too few chunks are simulated, the resulting pure shift spectrum will have truncation artefacts, which are likely to mask artefacts from unwanted CTPs. The value of $f_{\text{diff},2}$ was thus tested with a wide variety of randomly chosen phases and angles, with the number of chunks set to 4, 8 and 16 (Fig. 2.20).

The 16-chunk and 8-chunk $f_{\text{diff},2}$ (Fig. 2.20a) do in fact line up quite well. Notably, as the inset shows, they both agree on the 'best' candidate (note that this is not necessarily anywhere near a *perfect* candidate, since these inputs were randomly generated). The 4-chunk $f_{\text{diff},2}$ also has the correct overall behaviour (Fig. 2.20b). However, its ranking of the 'best' candidates is not very accurate: the 4-chunk cost function rates the red cross in the inset as the optimum, but that is only the 13th-best candidate when using the 16-chunk cost function. Ultimately, I decided to use the 4-chunk cost function for 'quick and dirty' optimisations, where only an approximate optimum was required. However, for anything requiring more accuracy, the 8-chunk cost function was used.

An optimisation was then carried out with a (rather arbitrarily chosen) setting of $m = 9$, i.e. nine hard pulses in the PSE. A total of $s = 20$ randomly generated spin systems were used, matching the number of CPU cores on the computer used for the optimisations: these were further subdivided into four two-spin systems, eight three-spin systems and eight four-spin systems. The derivative-based BFGS algorithm was used to carry out the optimisation: this is a popular line search algorithm which uses

[20] It may affect the rate of convergence, but this is not something I tested.

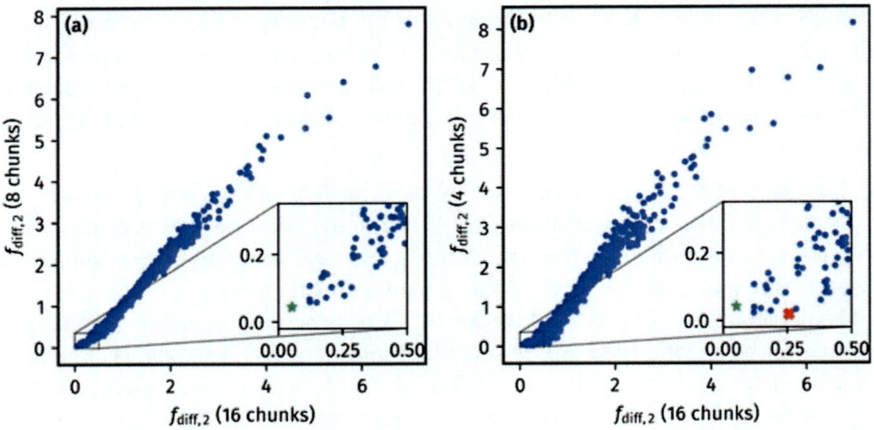

Fig. 2.20 Comparison of $f_{\text{diff},2}$ values when simulated with different numbers of chunks. 16 chunks is assumed to be the 'gold standard'. **a** Correlation between 16-chunk and 8-chunk cost functions. The 'optimum' identified by both cost functions is plotted as a green star in the inset. **b** Correlation between 16-chunk and 4-chunk cost functions. The 'optima' identified by the 16- and 4-chunk cost functions are respectively plotted as a green star and red cross in the inset

an approximate Hessian to calculate the search direction at each iteration [79, 80]. No lower or upper bounds were placed on the flip angles or phases: instead, phases can simply be wrapped to the region $[0, 2\pi)$, and in the simulations, the hard pulses were modelled as being instantaneous rotations, so their flip angles can also just be wrapped to $[0, 2\pi)$. (This would not be completely valid if realistic, finite pulses were used, since changing the flip angle would also change their duration.) This first optimisation yielded the following optimised parameters for the nine pulses:

$$\{\beta_i\} = \{118.6972^\circ, 29.8400^\circ, 107.4850^\circ, 190.4788^\circ, 138.4710^\circ, 144.5939^\circ, 18.9674^\circ, 73.8900^\circ, 130.6071^\circ\}; \tag{2.52}$$

$$\{\phi_i\} = \{144.5641^\circ, 173.3596^\circ, 38.9878^\circ, 146.3121^\circ, 127.7346^\circ, 7.5104^\circ, 36.9805^\circ, 110.2791^\circ, 182.3894^\circ\}. \tag{2.53}$$

To evaluate the quality of the decoupling on a real sample, the dPSYCHE experiment was performed experimentally, and the results compared against a PSYCHE experiment (Fig. 2.21). Firstly, it is worth pointing out that *a* pure shift spectrum (if not a very good one) was obtained: this validates the form of the PSE and the optimisation approach used here, especially considering that the optimisation was not tailored towards this particular sample. The dPSYCHE experiment clearly has much greater sensitivity than the PSYCHE experiment; however, the decoupling quality is extremely poor. The reason for this is likely because the optimisation is prioritising sensitivity too highly over purity, and ultimately stems from the factor c in Eq. (2.51) which we have ignored until now. Since $c = 1$, the optimisation is

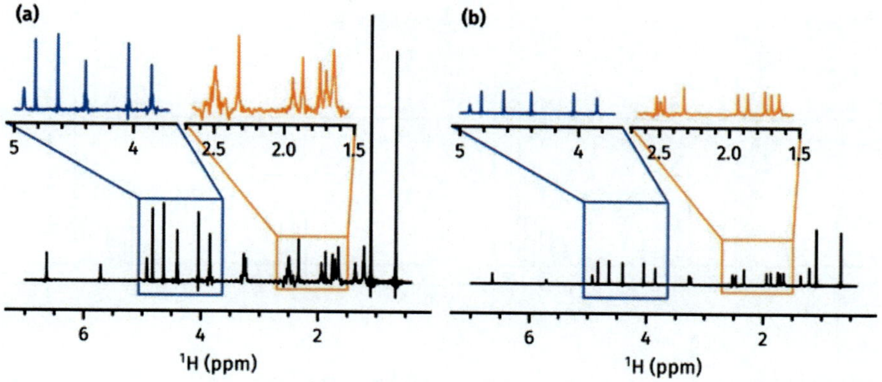

Fig. 2.21 **a** dPSYCHE experiment acquired using the flip angles and phases in Eq. (2.52) and (2.53). **b** (Double-saltire) PSYCHE experiment acquired with $\beta = 20°$. Spectra and insets with the same colour share the same scaling, allowing for direct sensitivity comparisons. *Data code:* `6A-211027`

essentially guiding the dPSYCHE experiment towards having *exactly the same sensitivity as a pulse–acquire experiment.*

Although this might be a nice idea in principle, it is not physically sensible: no pure shift method has sensitivity which is even close to that of a pulse–acquire experiment. It makes more sense to scale down the target spectrum by introducing a factor of $c < 1$ into the cost function (Eq. (2.51)); the parameter c therefore represents the 'target sensitivity'. By changing the parameter c, we can control whether the optimisation emphasises sensitivity or purity more. (Note that this option was *not* available to us in the experimental JRSE-based optimisations in Sect. 2.4.)

A series of new optimisations were therefore run, with c ranging from 0.2 to 1 (Table 2.2). The 4-chunk cost function was used, and the maximum number of function evaluations capped at 5000. Each optimisation was run 10 times with a different starting seed, and the best of the 10 results chosen for further evaluation (Table 2.2).

The resulting spectra (Fig. 2.22) show that adjusting c has the desired effect: larger values lead to greater sensitivity and lower purity, and vice versa. In this spectrum of andrographolide, the 3.5–5 ppm region (blue inset) is an 'easier' region to decouple: there are fewer couplings here and all spin systems are firmly within the weak coupling regime. The 1.5–2.8 ppm region (orange inset) is 'more difficult' to decouple, especially the two peaks at 2.5 ppm, which are mutually strongly coupled. The dPSYCHE spectrum with $c = 0.4$ appeared to be a good compromise between purity and sensitivity: its sensitivity was higher than that of the original ($\beta = 20°$) PSYCHE experiment. Furthermore, it provided good decoupling in the 'easier' deshielded region and somewhat acceptable performance in the shielded region, with the notable exception of the strongly coupled peaks at 2.5 ppm.

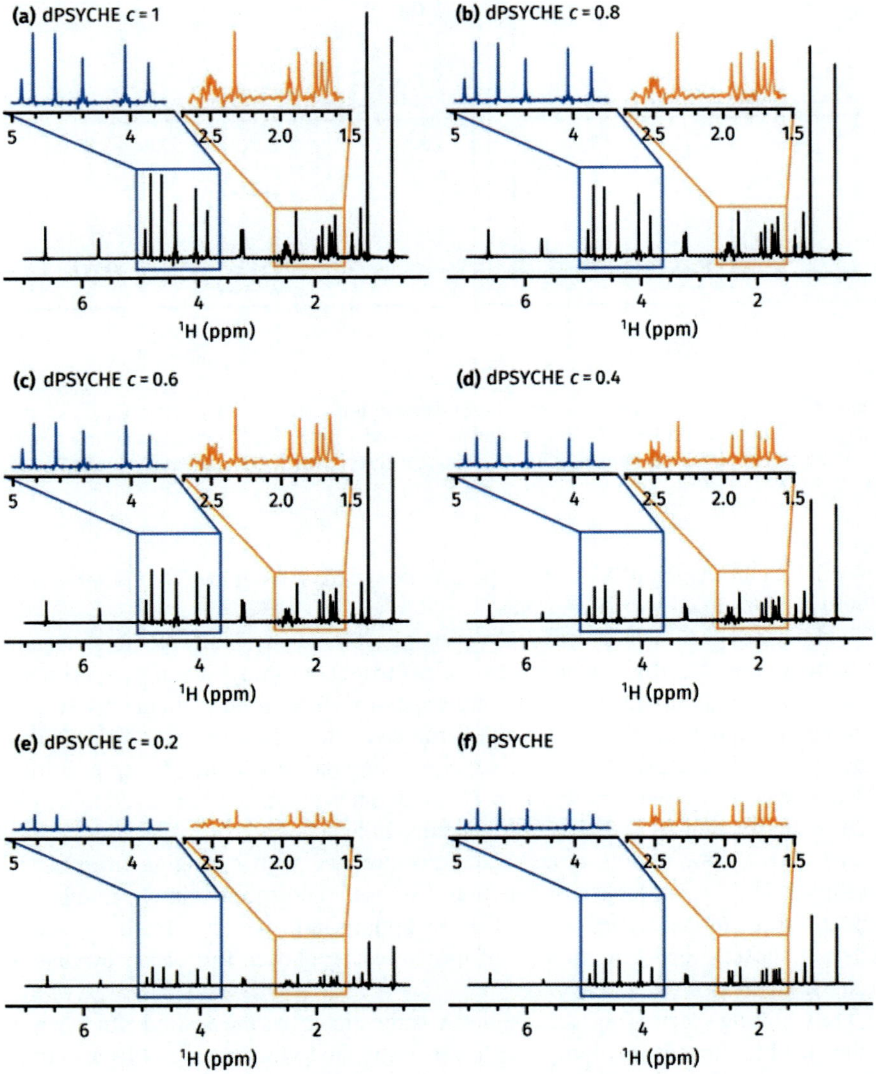

Fig. 2.22 **a** dPSYCHE, $c = 1$. **b** dPSYCHE, $c = 0.8$. **c** dPSYCHE, $c = 0.6$. **d** dPSYCHE, $c = 0.4$. **e** dPSYCHE, $c = 0.2$. **f** Double-saltire PSYCHE with $\beta = 20°$. Spectra and insets with the same colour share the same scaling, allowing for direct sensitivity comparisons. *Data code:* `6A-211231`

I therefore ran a longer optimisation using the same starting point and with no limit on the number of function evaluations, which successfully lowered the cost function value by twofold (the last entry in Table 2.2). The pure shift spectrum obtained using these optimised pulses is shown in Fig. 2.23. The decoupling quality in both regions is comparable to that in the PSYCHE spectrum, again with the exception of the strongly coupled peaks at 2.5 ppm.

Table 2.2 Results of dPSYCHE optimisations for different sensitivities. Note that since the cost function depends on the value of c, the exact value of $f_{\mathrm{diff},2}$ for these optimisations cannot be compared to one another; they are only presented here for completeness. The final entry, labelled with an asterisk, was run with no limit on the number of function evaluations (this is discussed further in the text)

c	Flip angles (°)	Phases (°)	$f_{\mathrm{diff},2}$
1	59.8813, 96.0748, 111.5862, 118.7285, 144.701, 176.9773, 29.2866, 40.0658, 63.8237	355.773, 81.741, 99.8752, 81.6675, 337.0056, 150.7004, 105.9666, 317.9979, 27.7897	0.0404
0.8	81.9461, 65.311, 90.3488, 106.6388, 73.3462, 196.7857, 51.791, 34.8188, 53.9506	329.8571, 60.3564, 137.3929, 68.5286, 340.74, 126.29, 64.4601, 145.4714, 27.3019	0.0321
0.6	53.1313, 82.2547, 88.6093, 161.9291, 83.462, 161.3548, 14.2976, 66.2199, 53.9709	347.2835, 59.5518, 68.072, 113.5358, 34.5537, 156.191, 268.7641, 302.7427, 16.29	0.0250
0.4	77.2998, 127.7274, 87.6663, 104.6371, 99.6474, 168.8171, 8.8569, 52.7865, 58.1408	342.2787, 47.7526, 76.4114, 97.3304, 14.2178, 163.9513, 94.2575, 280.0273, 16.0971	0.0271
0.2	120.6613, 107.6712, 84.0427, 133.5377, 82.1999, 223.122, 41.1617, 41.9398, 12.016	347.2051, 79.4503, 109.8796, 83.5017, 345.4284, 133.8355, 66.4118, 201.0041, 4.4419	0.0261
0.4*	92.4395, 133.2136, 38.9704, 34.9182, 56.0256, 57.4097, 31.5916, 140.1088, 128.7125	62.036, 16.411, 319.5634, 128.4443, 49.7357, 328.5103, 210.6498, 44.0123, 327.4517	0.0108

To make the dPSYCHE experiment more robust towards strong coupling, a TSE version of the sequence was also written and tested (Fig. 2.24). This did indeed improve the decoupling at 2.5 ppm, as expected. However, although the TSE-dPSYCHE version still retains its *overall* sensitivity advantage over TSE-PSYCHE (especially evident in the deshielded region), this is not true of every peak: the two peaks at 1.65 and 1.70 ppm seem to have decreased intensities in the TSE-dPSYCHE experiment. The reason for this is not fully clear. One could say that the optimised pulses are unlikely to work equally well on *all* combinations of chemical shifts and couplings, and these peaks may simply fall into one of the less effective regions; but in the (non-TSE) dPSYCHE experiment (Fig. 2.23a), these distortions in intensity are not observed. One possible conclusion is that some of the intensity in the non-TSE experiment in fact stems from artefacts, which are not fully suppressed: this may also explain the slight broadening at the base of the peaks in Fig. 2.23a.

2.6.2.2 State-Transfer Cost Function

The optimisations conducted so far have only used the simulated pure shift spectrum as an input to the cost function. However, since these optimisations are conducted in a purely theoretical manner, it is possible to directly calculate the effect of the PSE

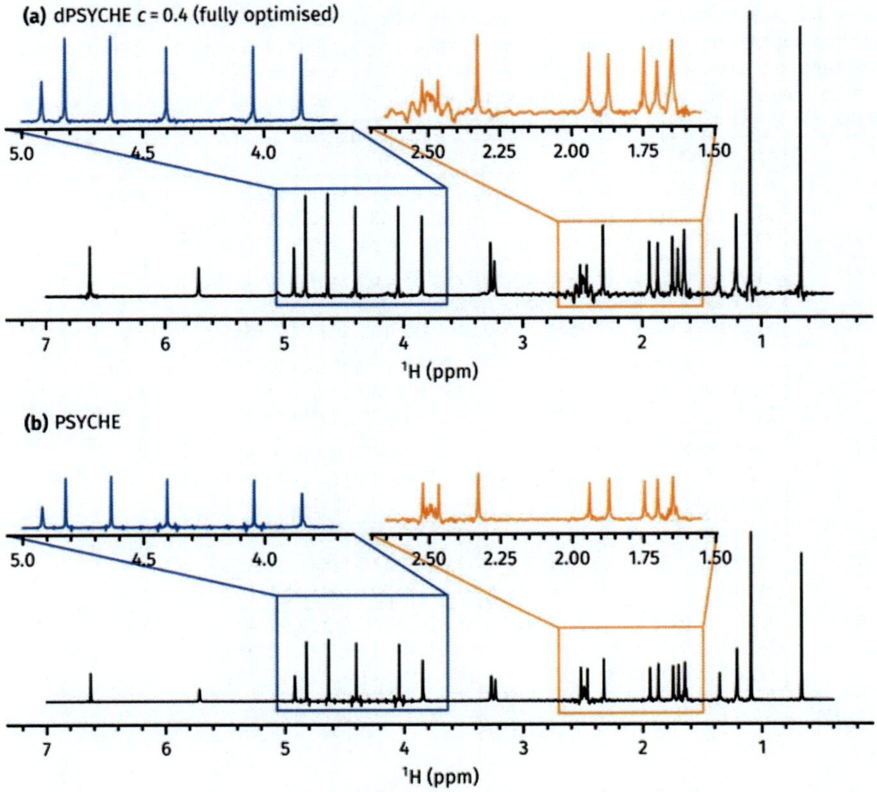

Fig. 2.23 **a** Fully optimised dPSYCHE experiment with $c = 0.4$. **b** Double-saltire PSYCHE with $\beta = 20°$. Spectra and insets with the same colour share the same scaling, allowing for direct sensitivity comparisons. *Data code:* `6A-211231`

on each input state to determine its quality. The possible state transfers caused by the PSE can be categorised as follows:

1. desired (e.g. $I_{1+}I_{2\alpha} \rightarrow I_{1-}I_{2\alpha}$);
2. undesired, but can be dephased by CTP gradients (e.g. $I_{1+}I_{2\alpha} \rightarrow I_{1+}I_{2\alpha}$);
3. undesired, but can be suppressed by the TSE element (e.g. $I_{1+}I_{2\alpha} \rightarrow I_{1\alpha}I_{2-}$);
4. undesired, but survive even with the TSE element (e.g. $I_{1+}I_{2\alpha} \rightarrow I_{1-}I_{2\beta}$).

We want to maximise the degree of desired state transfer, item (1), which leads to the pure shift signal; and minimise the degree of unwanted state transfers (which lead to artefacts). For a TSE-dPSYCHE experiment, the only relevant artefacts are contained in item (4); without the TSE addition, we would additionally need to take item (3) into account. These state-transfer coefficients would also have to be summed (or averaged) over all possible single-quantum input states: for an n-spin system, there are $n \cdot 2^{n-1}$ possible input states (n possible spins which could be active, and 2^{n-1} permutations of passive spin states). The cost function would then be

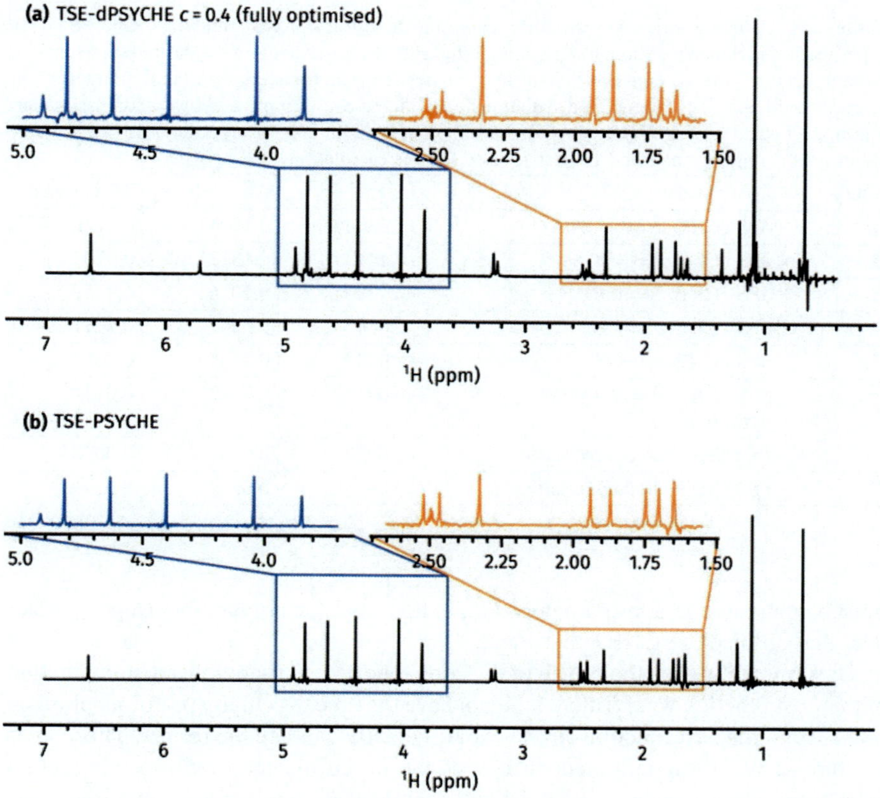

Fig. 2.24 **a** Fully optimised TSE-dPSYCHE experiment with $c = 0.4$. **b** Double-saltire TSE-PSYCHE with $\beta = 20°$. Spectra and insets with the same colour share the same scaling, allowing for direct sensitivity comparisons. *Data code:* `7A-220129`

constructed as a linear combination of the relevant terms, where the weightings serve a similar function to the parameter c in the spectrum-comparison cost function $f_{\text{diff},2}$ (Eq. (2.51)), in that they can be tweaked to control the balance between sensitivity and purity.

Table 2.3 shows the values of each of these components for all of the optima obtained so far. Entries 1 through 5 show the effect of varying c: as expected, decreasing c leads to less signal and also fewer artefacts. These are consistent with the spectra shown in Fig. 2.22. Unsurprisingly, the 'long' optimisation (entry 6) also performs better in all respects than the corresponding 'short' optimisation with $c = 0.4$ (entry 4). Finally, it is worth comparing the best dPSYCHE result (the $c = 0.4$ long optimisation, entry 6) with the standard PSYCHE method (entries 7 and 8). When PSYCHE is run with a 'normal' flip angle of 20°, the purity is far superior to any of the dPSYCHE experiments. However, the sensitivity is only around half that of the dPSYCHE. In order to accomplish the same sensitivity as the dPSYCHE, the PSYCHE flip angle

Table 2.3 Evaluation of previous optima using the state-transfer cost functions. One PSYCHE experiment is included for comparison. 'Signal' refers to the coefficient for the desired state transfer; 'all artefacts' are the average coefficient for all undesired state transfers; and 'TSE artefact' is the average coefficient for all undesired state transfers which are not suppressed by the TSE sequence. These are further averaged over all possible input states with coherence order +1. A three-spin system with couplings of 7 and 10 Hz was used for this calculation

Entry	Description	Signal	All artefacts	TSE artefacts
1	dPSYCHE, $c = 1$	0.1544	0.1906	0.1126
2	dPSYCHE, $c = 0.8$	0.1417	0.1313	0.0707
3	dPSYCHE, $c = 0.6$	0.1226	0.1016	0.0359
4	dPSYCHE, $c = 0.4$	0.0869	0.0561	0.0171
5	dPSYCHE, $c = 0.2$	0.0443	0.0315	0.0087
6	dPSYCHE, $c = 0.4$ (long optimisation)	0.0915	0.0282	0.0119
7	PSYCHE, 20° double saltire	0.0420	0.0083	0.0040
7	PSYCHE, 34° double saltire	0.0918	0.0342	0.0249

must be increased to approximately 34°; in this case, the artefact intensity is greater than that of the dPSYCHE experiment.

Unfortunately, even though all of the code for efficient (parallelised) computation of this cost function was written, I did not have the time to actually perform optimisations using this cost function and to experimentally validate the results. This comes as somewhat a disappointment to me, as the initial results shown in this section have been quite promising, and it is likely that the state-transfer-based optimisations would have improved this even further. However, even from just the results obtained so far, it is clear that the interleaved pulses and gradients used in the dPSYCHE PSE form a reasonable *ansatz* which can yield pure shift spectra which are comparable to the PSYCHE method.

2.7 Ultrafast PSYCHE-iDOSY

In the final section of this chapter, I turn to something entirely different: the combination of pure shift diffusion NMR with ultrafast acquisition. This work was done in collaboration with Jean-Nicolas Dumez (University of Nantes): the original project idea was first conceived and implemented there. Unfortunately, the results we obtained at Oxford were not promising enough to drive the project further, especially in light of time constraints. However, for the sake of completeness, I describe the overall concept as well as these results in this section.

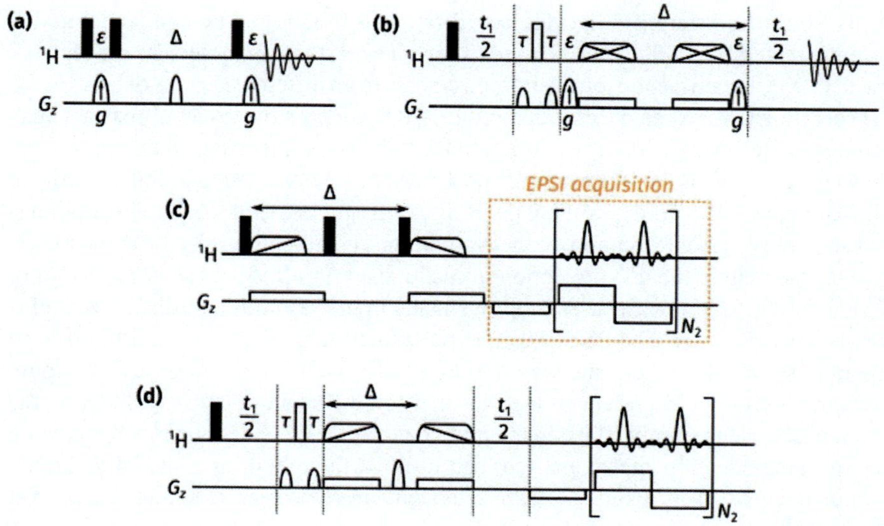

Fig. 2.25 **a** Basic stimulated echo DOSY experiment. To acquire the diffusion dimension, the amplitudes of gradients with arrows inside are incremented. Δ indicates the diffusion delay. **b** PSYCHE-iDOSY experiment. τ is set to $1/(4 \cdot T_{\mathrm{chunk}})$, as before. **c** Ultrafast DOSY experiment. The EPSI acquisition block is highlighted in orange, consisting of a prephasing gradient followed by detection in the presence of alternating gradients. **d** Ultrafast PSYCHE-iDOSY experiment. Note the use of unidirectional chirps rather than saltire pulses

I begin with a brief overview of diffusion NMR [81]; this topic will also be covered in some detail in Sect. 3.4.10. Figure 2.25a shows a typical 2D stimulated echo DOSY experiment, where the indirect (diffusion) dimension involves the incrementation of gradient amplitudes rather than an evolution delay. For each gradient amplitude, one time-domain signal is recorded. Molecular diffusion during a delay placed between a pair of gradients leads to attenuation of the signal, as the spatially dependent phase imparted by the first gradient is not perfectly refocused by the second. This attenuation is described by the Stejskal–Tanner equation [82, 83], which in its simplest form is

$$I(G) = I(0)\exp\left[-(\gamma\delta G)^2 D\Delta'\right]. \tag{2.54}$$

Here, $I(G)$ represents the signal intensity when measured with a gradient amplitude of G; γ is the magnetogyric ratio, δ is the duration of the bracketing gradients, D the diffusion coefficient, and Δ' a 'corrected' diffusion delay. The exact correction required depends on the shapes of the gradients used. By measuring the signal intensity I for at least two different values of G, the diffusion coefficient D can be estimated.[21]

---

[21] As will be discussed in Sect. 3.4.10, this should really be an *apparent* diffusion coefficient as there are various other factors which can affect the intensity, notably convection.

The fact that different molecules have different diffusion coefficients, and thus different attenuation profiles, can be used to separate mixtures of molecules through the basic DOSY experiment. However, when signals from different species overlap in the ^{1}H spectrum (as-is often the case in mixtures), it is not possible to cleanly extract peak intensities, resulting in poor resolution in the diffusion dimension. There are several ways to solve this issue: for example, peaks can be resolved using another chemical shift dimension in a 3D experiment [84–88]. A slightly less time-consuming option is to use pure shift NMR to improve the resolution. The first pure shift diffusion experiments used either the 45° projection of a 2DJ spectrum [89], or the Zangger–Sterk PSE [29, 90, 91]: in both cases, improvements in the resolution of diffusion coefficients were reported. The PSYCHE-iDOSY experiment [57] (Fig. 2.25b) improved upon these, much in the same way that PSYCHE itself improved on existing pure shift methods. The 'i' refers to the fact that the diffusion delay is *internal* to the sequence, i.e. it is simply added in the middle of the PSYCHE element, which (using the instantaneous flip approximation) can itself be thought of as a spatially parallel stimulated echo. This avoids the need to add an entirely separate stimulated echo at the beginning or end of the sequence.

While the addition of a pure shift dimension is less expensive compared to a full chemical shift dimension in terms of time, the fact remains that the PSYCHE-iDOSY is a (pseudo)-3D experiment. One way to shorten this is to use spatial encoding techniques to collapse one dimension, specifically the diffusion dimension [92]. Two separate steps are required for this: firstly, a gradient whose amplitude varies across the sample must be applied (the 'encoding' step), and then an imaging acquisition technique must be used to read out the signal in each slice of the sample (the 'detection' step). In the context of 'standard' 2D DOSY, this was first done by Loening et al. [93], using the field generated by a z^2 shim coil for spatial encoding. The readout was then performed by simply acquiring the FID in the presence of a weak gradient, such that the peak shapes directly reflect the distribution of signal intensities across the sample. Later, Thrippleton et al. [94] introduced the (by now familiar) chirp–gradient combination for spatial encoding: since spins in different slices are inverted at different times, the total gradient area applied in each slice is different. In later work by the groups of Frydman [95] and Dumez [96, 97], the spatial encoding of the diffusion attenuation is still accomplished using the very same chirp–gradient combination. However, the detection step is accomplished using the echo-planar spectroscopic imaging (EPSI) acquisition technique [98, 99], much like in ultrafast 2D NMR (Fig. 2.25c) [58–60, 62, 100–102].

Unsurprisingly, the aim in this section is to unite the PSYCHE-iDOSY and ultrafast DOSY techniques to form an ultrafast PSYCHE-iDOSY experiment. This would yield PSYCHE-iDOSY spectra in a much reduced time, at the cost of sensitivity. The resulting pulse sequence is shown in Fig. 2.25d. The EPSI acquisition is used to collapse the diffusion dimension, not the PSYCHE chunking dimension.[22] One benefit of this is that only one PSYCHE chunk needs to be acquired at a time, circumventing

[22] The other option would require an ultrafast PSYCHE experiment, which—to date—has not been developed.

the need for long EPSI detection periods (which can cause spectrometer damage). Note also that it is mandatory to use unidirectional chirps in the PSYCHE PSE: in the original PSYCHE-iDOSY, these pulses were used only for the purpose of homonuclear decoupling, and thus saltire pulses were acceptable (and indeed, superior). However, in the ultrafast version, the PSE is also responsible for creating the requisite spatial encoding of gradient areas. Using saltire pulses here would lead to an undesired 'double' encoding, yielding data which cannot be processed correctly.

The sequence itself was written and evaluated by Corentin Jacquemmoz and Jean-Nicolas Dumez at the University of Nantes, using a concentrated (1M) sample of ethanol in D2O. Their results are shown in Fig. 2.26a to b, c. When I later tried to reproduce these results, I additionally used the POISE routine described in Sect. 3.4.6 to optimise the amplitude ratio between the positive and negative gradients during the EPSI acquisition (without optimisation, the results obtained were even poorer); my results are shown in Figs. 2.26d, e, f. The data processing used to generate these figures was written by me,[23] and broadly consists of the following steps: reshaping of the EPSI raw data, concatenation of PSYCHE chunks, and Fourier transformation in both dimensions.[24] For simplicity, the data points acquired during the negative gradients are discarded.

In principle, the Fourier transformed (z, δ) plots should reveal a diffusion profile similar to that in Guduff et al. [96] This is the case for the Nantes data; in fact, it was possible to process this further to obtain a conventional 2D DOSY spectrum, where the diffusion coefficient for each component is extracted and plotted. For the Oxford data, although the general *form* of the curve is correct (in that there is a generally decreasing pattern), some unusual 'spikes' can clearly be seen in the (z, δ) diffusion profiles.

My current best hypothesis is that these spikes are artefacts from unwanted CTPs which are ordinarily suppressed through spatial averaging but are instead being refocused by the gradients during the EPSI acquisition. This is supported by the fact that the Nantes data were acquired on a microimaging probe equipped with triaxial gradients, which allowed CTP gradients to be applied along the x- and y-axes (note that the PSYCHE gradient must still be applied along the z-axis). Naturally, any artefacts dephased by these cannot be refocused by the z-gradients in the EPSI block. On the other hand, the Oxford data were acquired on a Prodigy cryoprobe with only z-gradients.

The intensity of these 'spike' artefacts can, in fact, be controlled by varying other acquisition parameters. For example, increasing the EPSI gradient amplitude leads to increased artefacts (Fig. 2.27a and b): this lends credence to the hypothesis above that the artefacts arise from refocused CTPs. Notice also, in Fig. 2.27a, the appearance of

[23] The Nantes team have a more complete MATLAB package for this, which goes several steps further than that shown here and includes features such as phase correction and fitting of the diffusion profiles to various modified Stejskal–Tanner equations. However, these were not necessary here as the project never got to that stage, and I preferred to use Python.

[24] As discussed in Frydman et al. [60], the k-domain in ultrafast NMR is directly proportional to the indirect-dimension frequency domain. So, to obtain the diffusion profile—which varies with space, i.e. z—a Fourier transform along this axis is required.

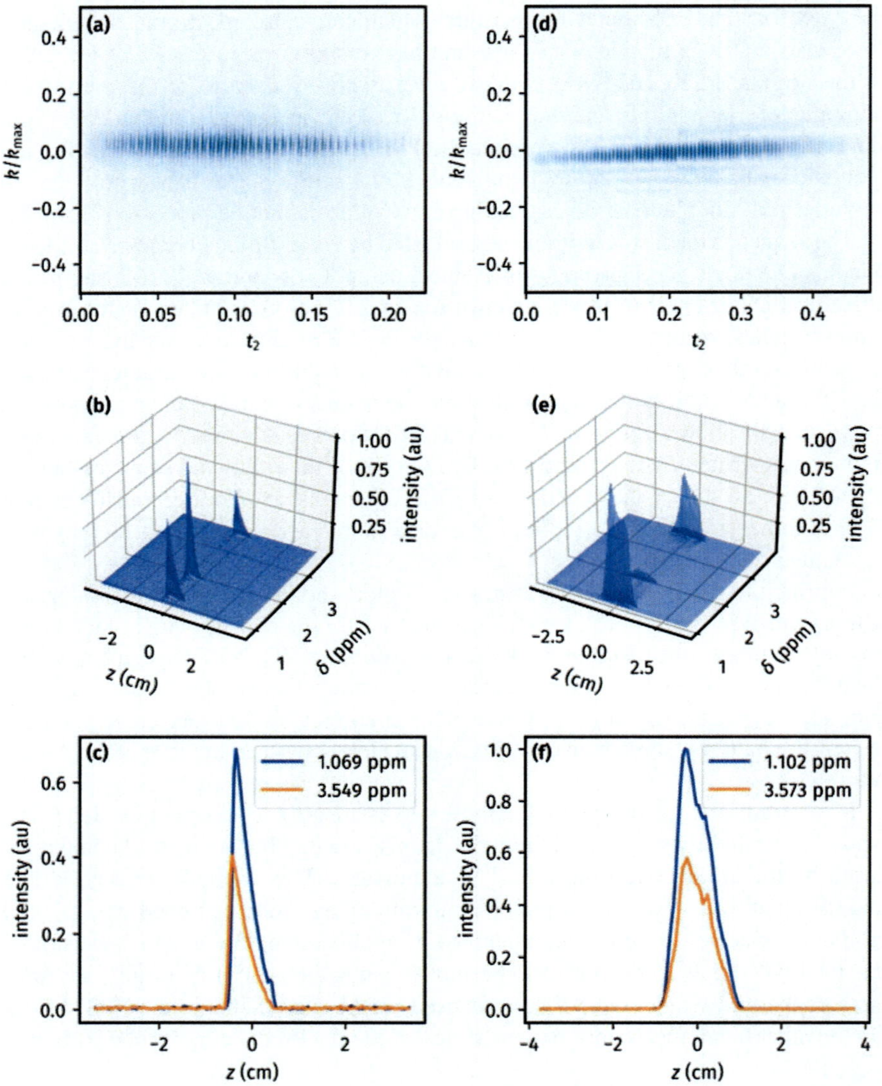

Fig. 2.26 **a–c** EPSI PSYCHE-iDOSY data obtained by Corentin Jacquemmoz and Jean-Nicolas Dumez at the University of Nantes. The three plots respectively show the 2D raw (k, t_2) data (after PSYCHE processing has been carried out); the Fourier transformed (z, δ) data; and the individual diffusion profiles for each peak. **d–f** The equivalent data which I recorded in Oxford. The central peak corresponds to HDO; the Oxford sample used was drier and thus it is less intense in (**e**) compared to (**b**). *Data code:* `6E-210426`

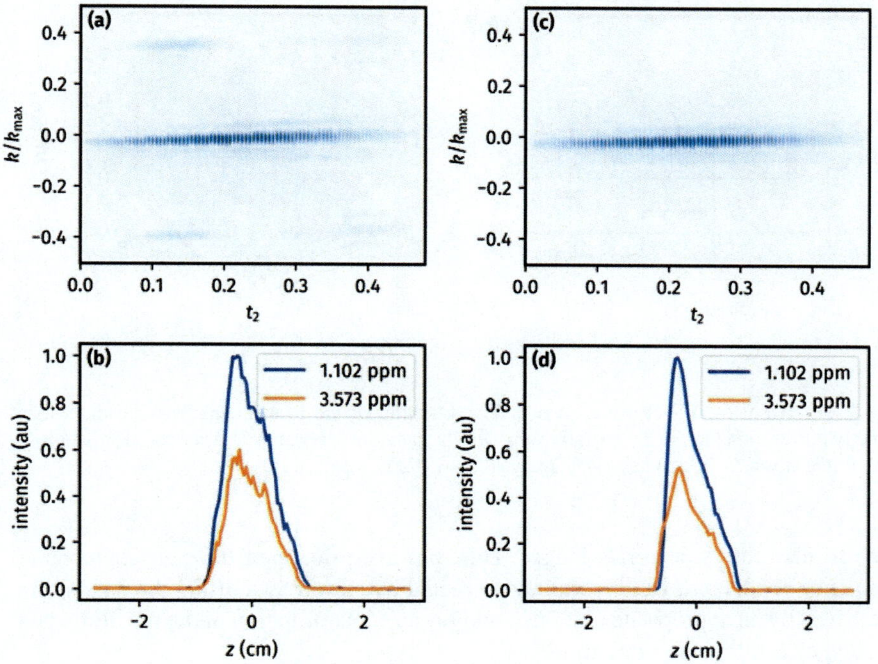

Fig. 2.27 **a–b** The same as in Fig. 2.26d and f, but with the EPSI acquisition gradient increased to 32% (from 24% in the original). *Data code:* `6E-210426` **c–d** The same as in Fig. 2.26d, f, but with the PSYCHE gradient increased to 5% (from 3% in the original). *Data code:* `6E-210505`

signals at $k \neq 0$: this should not happen because in ultrafast experiments, the

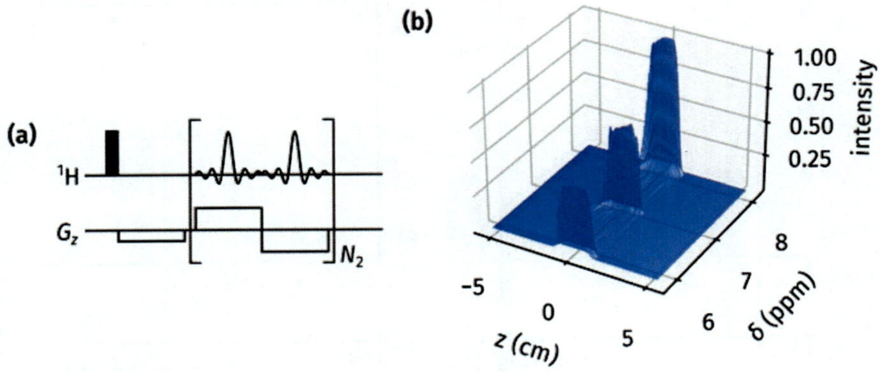

Fig. 2.28 **a** Pulse–EPSI pulse sequence. The variation of the peak intensity along the z-axis directly corresponds to the B_1 spatial profile. **b** The resulting data after Fourier transformation in both dimensions ($k \leftrightarrow z$ and $t_2 \leftrightarrow \delta$). *Data code:* `6E-210426`

time to dissipate when t_1 is longer. This, however, does not have an appreciable impact on the data: it can be crudely corrected by circularly shifting the data along the k-axis by an appropriate amount, and Fourier transformation of the resulting data yielded effectively the same results.

All in all, it appears that the fundamental idea behind the ultrafast PSYCHE-iDOSY experiment is sound: the Nantes data form an excellent proof-of-principle. However, the implementation of the sequence almost certainly needs to be optimised in order for high-quality data to be extracted. An important subsequent step would then be to derive the appropriate form of the Stejskal–Tanner equation for extracting diffusion coefficients from the data. Unfortunately, I simply did not have the time to pursue this in any further detail.

2.8 Conclusion

This chapter revolves around the technique of pure shift NMR, which yields spectra where couplings are suppressed and only chemical shift information is shown. Pure shift spectra provide significantly improved resolution, but with poorer sensitivity; the resulting spectra also suffer from artefacts due to detection of unwanted CTPs, especially in the case of strong coupling. Various forms of the pure shift element were studied, with the aim of improving on the sensitivity and purity of the PSYCHE method [34]. Section 2.3 describes PSYCHE with either one or four-saltire pulses, as opposed to the usual two; in Sect. 2.4, these saltire pulses are entirely replaced with an arbitrary shaped pulse obtained through optimisation; Sect. 2.5 looks at the time-reversal element, where different spectra are co-added to produce a pure shift spectrum; and Sect. 2.6 is devoted to the dPSYCHE method, where pulses and gradients are interleaved instead of being applied simultaneously.

Many of these did not yield any improvement over PSYCHE itself and were thus not considered further. In particular, the summation process of the time-reversal element introduced further spectral artefacts arising from incomplete cancellation of unwanted signals. The most promising of these appears to be the dPSYCHE experiment: although I did not complete my investigation into this, the results obtained to date suggest that the experiment may be optimised to provide comparable decoupling quality to PSYCHE but with better sensitivity (see, e.g. Table 2.3). There is a clear path towards this end, starting with investigating the use of a different (state transfer based) cost function; it is also worth increasing the flexibility of the optimisation, for example, by increasing the number of pulses or by optimising the gradient amplitudes as well, to see if better results can be attained.

Assuming that this can be accomplished, though, one may rightly still ask whether a slight 10–20% improvement in signal-to-artefact ratio is *useful*. PSYCHE already allows the experimentalist to control this quantity by adjusting the flip angle of the saltire pulses, and thus the same outcome may in principle be obtained by lowering the flip angle (and increasing the measurement time to compensate for the drop in signal-to-noise). Pure shift NMR is admittedly not a particularly widely used technique, and in my view, it is not for the lack of good pure shift elements. It would be very much worthwhile exploring other potential areas in which it can be applied with greater regularity (or at all): an obvious candidate would be natural product assignment, where spectra are often highly complex, and incorrect assignments more abundant than one might like [103].

In the final section (Sect. 2.7), I also briefly discussed some preliminary efforts towards an ultrafast version of the PSYCHE-iDOSY experiment [57]. Although the pulse sequence cannot be considered to be ready for use, the data collected so far show that the underlying concept is sound. The use of ultrafast techniques to record a PSYCHE-iDOSY is mainly attractive due to the potential time savings, as a pseudo-3D experiment is reduced to a pseudo-2D experiment. However, this comes with a significant reduction in sensitivity. The use of pure shift already incurs a sensitivity penalty, and the addition of the ultrafast component further diminishes the available signal. Thus, such a method would likely be applicable only to very concentrated samples.

While the work described in this chapter did not lead to published output, it did provide a motivating context for the idea of *optimising NMR experiments*—particularly for pulse sequences which are expensive to simulate, such as PSYCHE. In this chapter, the aim was purely to search for the best-performing pure shift sequence. However, the dual considerations of sensitivity and purity are relevant to almost all NMR experiments; it is, therefore, natural to ask how experiments can be set up to maximise these criteria. To answer this, I designed a software package called POISE: this is precisely the subject of the next chapter, which we now turn to.

References

1. Zangger K (2015) Pure shift NMR. Prog Nucl Magn Reson Spectrosc 86–87:1–20. https://doi.org/10.1016/j.pnmrs.2015.02.002
2. Castañar L (2017) Pure shift 1H NMR: what is next? Magn Reson Chem 55:47–53. https://doi.org/10.1002/mrc.4545
3. Keeler J (2010) Understanding NMR spectroscopy, 2nd edn. Wiley, Chichester, U.K
4. Thrippleton MJ, Edden RAE, Keeler J (2005) Suppression of strong coupling artefacts in J-spectra. J Magn Reson 174:97–109. https://doi.org/10.1016/j.jmr.2005.01.012
5. Griesinger C, Sørensen OW, Ernst RR (1986) Correlation of connected transitions by two-dimensional NMR spectroscopy. J Chem Phys 85:6837–6852. https://doi.org/10.1063/1.451421
6. Aue WP, Karhan J, Ernst RR (1976) Homonuclear broad band decoupling and two-dimensional J-resolved NMR spectroscopy. J Chem Phys 64:4226–4227. https://doi.org/10.1063/1.431994
7. Xu P, Wu X-L, Freeman R (1991) Broadband-decoupled proton spectroscopy. J Magn Reson 95:132–148. https://doi.org/10.1016/0022-2364(91)90329-r
8. Nuzillard J-M (1996) Time-reversal of nmr signals by linear prediction. Application to phase-sensitive homonuclear J-resolved spectroscopy. J Magn Reson Ser A 118:132–135. https://doi.org/10.1006/jmra.1996.0020
9. Simova S, Sengstschmid H, Freeman R (1997) Proton chemical-shift spectra. J Magn Reson 124(104):121. https://doi.org/10.1006/jmre.1996.1001
10. Pell AJ, Keeler J (2007) Two-dimensional J-spectra with absorption-mode lineshapes. J Magn Reson 189:293–299. https://doi.org/10.1016/j.jmr.2007.09.002
11. Foroozandeh M, Adams RW, Kiraly P, Nilsson M, Morris GA (2015) Measuring couplings in crowded NMR spectra: pure shift NMR with multiplet analysis. Chem Commun 51:15410–15413. https://doi.org/10.1039/c5cc06293d
12. Griesinger C, Soerensen OW, Ernst RR (1985) Two-dimensional correlation of connected NMR transitions. J Am Chem Soc 107:6394–6396. https://doi.org/10.1021/ja00308a042
13. Sorensen OW, Griesinger C, Ernst RR (1985) Time reversal of the evolution under scalar spin-spin interactions in NMR. Application for ..1 decoupling in two-dimensional NOE spectroscopy. J Am Chem Soc 107:7778–7779. https://doi.org/10.1021/ja00311a101
14. Oschkinat H, Pastore A, Pfandler P, Bodenhausen G (1986) Two-dimensional correlation of directly and remotely connected transitions by z-filtered COSY. J Magn Reson 69:559–566. https://doi.org/10.1016/0022-2364(86)90176-9
15. Pell AJ, Edden RAE, Keeler J (2007) Broadband proton-decoupled proton spectra. Magn Reson Chem 45:296–316. https://doi.org/10.1002/mrc.1966
16. Moutzouri P, Paruzzo FM, Simoes de Almeida B, Stevanato G, Emsley L (2020) Homonuclear decoupling in 1H NMR of solids by remote correlation. Angew Chem Int Ed 59:6235–6238. https://doi.org/10.1002/anie.201916335
17. Bruschweiler R, Zhang F (2004) Covariance nuclear magnetic resonance spectroscopy. J Chem Phys 120:5253–5260. https://doi.org/10.1063/1.1647054
18. Zhang F, Bruschweiler R (2004) Indirect covariance NMR spectroscopy. J Am Chem Soc 126:13180–13181. https://doi.org/10.1021/ja047241h
19. Jaeger M, Aspers RLEG (2014) Covariance NMR and small molecule applications. Annu Rep NMR Spectrosc 83:271–349. https://doi.org/10.1016/B978-0-12-800183-7.00005-8
20. Morris GA, Aguilar JA, Evans R, Haiber S, Nilsson M (2010) True chemical shift correlation maps: a TOCSY experiment with pure shifts in both dimensions. J Am Chem Soc 132:12770–12772. https://doi.org/10.1021/ja1039715
21. Aguilar JA, Colbourne AA, Cassani J, Nilsson M, Morris GA (2012) Decoupling two- dimensional NMR spectroscopy in both dimensions: pure shift NOESY and COSY. Angew Chem Int Ed 51:6460–6463. https://doi.org/10.1002/anie.201108888

22. Foroozandeh M, Adams RW, Nilsson M, Morris GA (2014) Ultrahigh-resolution total correlation NMR spectroscopy. J Am Chem Soc 136:11867–11869. https://doi.org/10.1021/ja507201t
23. Lupulescu A, Olsen GL, Frydman L (2012) Toward single-shot pure-shift solution 1H NMR by trains of BIRD-based homonuclear decoupling. J Magn Reson 218:141–146. https://doi.org/10.1016/j.jmr.2012.02.018
24. Meyer NH, Zangger K (2013) Simplifying proton NMR spectra by instant homonuclear broadband decoupling. Angew Chem Int Ed 52:7143–7146. https://doi.org/10.1002/anie.201300129
25. Mauhart J, Glanzer S, Sakhaii P, Bermel W, Zangger K (2015) Faster and cleaner real-time pure shift NMR experiments. J Magn Reson 259:207–215. https://doi.org/10.1016/j.jmr.2015.08.011
26. Kiraly P, Nilsson M, Morris GA (2018) Practical aspects of real-time pure shift HSQC experiments. Magn Reson Chem 56:993–1005. https://doi.org/10.1002/mrc.4704
27. Donovan KJ, Frydman L (2014) HyperBIRD: a sensitivity-enhanced approach to collecting homonuclear-decoupled proton NMR spectra. Angew Chem Int Ed n/a–n/a. https://doi.org/10.1002/anie.201407869
28. Taylor DA, Natrajan LS, Nilsson M, Adams RW (2021) SABRE-enhanced real-time pure shift NMR spectroscopy. Magn Reson Chem 59(1244–1252):5206. https://doi.org/10.1002/mrc
29. Aguilar JA, Faulkner S, Nilsson M, Morris GA (2010) Pure Shift 1H NMR: a resolution of the resolution problem? Angew Chem Int Ed 49:3901–3903. https://doi.org/10.1002/anie.201001107
30. Moutzouri P, Chen Y, Foroozandeh M, Kiraly P, Phillips AR, Coombes SR, Nilsson M, Morris GA (2017) Ultraclean pure shift NMR. Chem Commun 53:10188–10191. https://doi.org/10.1039/c7cc04423b
31. Zangger K, Sterk H (1997) Homonuclear broadband-decoupled NMR spectra. J Magn Reson 124:486–489. https://doi.org/10.1006/jmre.1996.1063
32. Garbow JR, Weitekamp DP, Pines A (1982) Bilinear rotation decoupling of homonuclear scalar interactions. Chem Phys Lett 93:504–509. https://doi.org/10.1016/0009-2614(82)83229-6
33. Aguilar JA, Nilsson M, Morris GA (2011) Simple proton spectra from complex spin systems: pure shift NMR spectroscopy using BIRD. Angew Chem Int Ed 50:9716–9717. https://doi.org/10.1002/anie.201103789
34. Foroozandeh M, Adams RW, Meharry NJ, Jeannerat D, Nilsson M, Morris GA (2014) Ultrahigh-resolution NMR spectroscopy. Angew Chem Int Ed 53:6990–6992. https://doi.org/10.1002/anie.201404111
35. Foroozandeh M, Morris GA, Nilsson M (2018) PSYCHE pure shift NMR spectroscopy. Chem Eur J 24:13988–14000. https://doi.org/10.1002/chem.201800524
36. Foroozandeh M (2020) Spin dynamics during chirped pulses: applications to homonuclear decoupling and broadband excitation. J Magn Reson 318:106768. https://doi.org/10.1016/j.jmr.2020.106768
37. Kupce E, Boyd J, Campbell ID (1995) Short selective pulses for biochemical applications. J Magn Reson Ser B 106:300–303. https://doi.org/10.1006/jmrb.1995.1049
38. Giraud N, Béguin L, Courtieu J, Merlet D (2010) Nuclear magnetic resonance using a spatial frequency encoding: application to J-edited spectroscopy along the sample. Angew Chem Int Ed 49:3481–3484. https://doi.org/10.1002/anie.200907103
39. Gubensäk N, Fabian WMF, Zangger K (2014) Disentangling scalar coupling patterns by real-time SERF NMR. Chem Commun 50(12254–12257):892e. https://doi.org/10.1039/c4cc05
40. Mishra SK, Suryaprakash N (2017) Pure shift edited ultra high resolution NMR spectrum with complete eradication of axial peaks and unwanted couplings. J Magn Reson 279:74–80. https://doi.org/10.1016/j.jmr.2017.04.014
41. Buchberger K, Walenta M, Zangger K (2018) Extracting unresolved coupling constants from complex multiplets by a real-time J-upscaled SERF experiment. Magn Reson Chem 56:934–940. https://doi.org/10.1002/mrc.4699

42. Uhrin D, Liptaj T, Kover KE (1993) Modified BIRD pulses and design of heteronuclear pulse sequences. J Magn Reson Ser A 101:41–46. https://doi.org/10.1006/jmra.1993.1005
43. Sakhaii P, Haase B, Bermel W (2009) Experimental access to HSQC spectra decoupled in all frequency dimensions. J Magn Reson 199:192–198. https://doi.org/10.1016/j.jmr.2009.04.016
44. Paudel L, Adams RW, Király P, Aguilar JA, Foroozandeh M, Cliff MJ, Nilsson M, Sándor P, Waltho JP, Morris GA (2013) Simultaneously enhancing spectral resolution and sensitivity in heteronuclear correlation NMR spectroscopy. Angew Chem Int Ed 52:11616–11619. https://doi.org/10.1002/anie.201305709
45. Reinsperger T, Luy B (2014) Homonuclear BIRD-decoupled spectra for measuring one-bond couplings with highest resolution: CLIP/CLAP-RESET and constant-time-CLIP/CLAPRESET. J Magn Reson 239:110–120. https://doi.org/10.1016/j.jmr.2013.11.015
46. Nolis P, Motiram-Corral K, Perez-Trujillo M, Parella T (2019) Broadband homodecoupled time-shared 1H.13C and 1H.15N HSQC experiments. J Magn Reson 298:23–30. https://doi.org/10.1016/j.jmr.2018.11.005
47. Singh U, Bhattacharya S, Baishya B (2020) Pure shift HMQC: resolution and sensitivity enhancement by bilinear rotation decoupling in the indirect and direct dimensions. J Magn Reson 311:106684. https://doi.org/10.1016/j.jmr.2020.106684
48. Zwahlen C, Legault P, Vincent SJF, Greenblatt J, Konrat R, Kay LE (1997) Methods for measurement of intermolecular NOEs by multinuclear NMR spectroscopy: application to a bacteriophage .. N-Peptide/boxBRNA complex. J Am Chem Soc 119:6711–6721. https://doi.org/10.1021/ja970224q
49. Kupe E, Freeman R (2007) Compensated adiabatic inversion pulses: broadband INEPT and HSQC. J Magn Reson 187:258–265. https://doi.org/10.1016/j.jmr.2007.05.009
50. Thrippleton MJ, Keeler J (2003) Elimination of zero-quantum interference in two-dimensional NMR spectra. Angew Chem Int Ed 42:3938–3941. https://doi.org/10.1002/anie.200351947
51. Timari I, Szilagyi L, Kover KE (2015) PSYCHE CPMG.HSQMBC: an NMR spectroscopic method for precise and simple measurement of long-range heteronuclear coupling constants. Chem Eur J 21:13939–13942. https://doi.org/10.1002/chem.201502641
52. Koos MRM, Kummerlowe G, Kaltschnee L, Thiele CM, Luy B (2016) CLIP-COSY: a clean in-phase experiment for the rapid acquisition of COSY-type correlations. Angew Chem Int Ed 55:7655–7659. https://doi.org/10.1002/anie.201510938
53. Sinnaeve D, Foroozandeh M, Nilsson M, Morris GA (2016) A general method for extracting individual coupling constants from crowded 1H NMR spectra. Angew Chem Int Ed 55:1090–1093. https://doi.org/10.1002/anie.201508691
54. Aguilar JA, Belda R, Gaunt BR, Kenwright AM, Kuprov I (2018) Separating the coherence transfer from chemical shift evolution in high-resolution pure shift COSY NMR. Magn Reson Chem 56:969–975. https://doi.org/10.1002/mrc.4727
55. Kaltschnee L, Knoll K, Schmidts V, Adams RW, Nilsson M, Morris GA, Thiele CM (2016) Extraction of distance restraints from pure shift NOE experiments. J Magn Reson 271:99–109. https://doi.org/10.1016/j.jmr.2016.08.007
56. Ilgen J, Nowag J, Kaltschnee L, Schmidts V, Thiele CM (2021) Gradient selected pure shift EASY-ROESY techniques facilitate the quantitative measurement of 1H,1H-distance restraints in congested spectral regions. J Magn Reson 324:106900. https://doi.org/10.1016/j.jmr.2020.106900
57. Foroozandeh M, Castanar L, Martins LG, Sinnaeve D, Poggetto GD, Tormena CF, Adams RW, Morris GA, Nilsson M (2016) Ultrahigh-Resolution Diffusion-Ordered Spectroscopy. Angew Chem Int Ed 55(15579–15582):09676. https://doi.org/10.1002/anie.2016
58. Frydman L, Scherf T, Lupulescu A (2002) The acquisition of multidimensional NMR spectra within a single scan. Proc Natl Acad Sci U S A 99:15858–15862. https://doi.org/10.1073/pnas.252644399
59. Pelupessy P (2003) Adiabatic single scan two-dimensional NMR spectrocopy. J Am Chem Soc 125:12345–12350. https://doi.org/10.1021/ja034958g

60. Frydman L, Lupulescu A, Scherf T (2003) Principles and features of single-scan two-dimensional NMR spectroscopy. J Am Chem Soc 125:9204–9217. https://doi.org/10.1021/ja030055b
61. Kiraly P, Kern N, Plesniak MP, Nilsson M, Procter DJ, Morris GA, Adams RW (2021) Single-scan selective excitation of individual NMR signals in overlapping multiplets. Angew Chem Int Ed 60:666–669. https://doi.org/10.1002/anie.202011642
62. Dumez J-N (2018) Spatial encoding and spatial selection methods in high-resolution NMR spectroscopy. Prog Nucl Magn Reson Spectrosc 109:101–134
63. Nelder JA, Mead R (1965) A simplex method for function minimization. Comput J 7:308–313. https://doi.org/10.1093/comjnl/7.4.308
64. McKinnon KIM (1998) Convergence of the Nelder-mead simplex method to a nonstationary point. SIAM J Optim 9:148–158. https://doi.org/10.1137/s1052623496303482
65. Hogben HJ, Krzystyniak M, Charnock GTP, Hore PJ, Kuprov I (2011) Spinach - A software library for simulation of spin dynamics in large spin systems. J Magn Reson 208:179–194. https://doi.org/10.1016/j.jmr.2010.11.008
66. Khaneja N, Reiss T, Kehlet C, Schulte-Herbrüggen T, Glaser SJ (2005) Optimal control of coupled spin dynamics: design of NMR pulse sequences by gradient ascent algorithms. J Magn Reson 172:296–305. https://doi.org/10.1016/j.jmr.2004.11.004
67. De Fouquieres P, Schirmer SG, Glaser SJ, Kuprov I (2011) Second order gradient ascent pulse engineering. J Magn Reson 212(412–417):023. https://doi.org/10.1016/j.jmr.2011.07
68. Glaser SJ, Boscain U, Calarco T, Koch CP, Köckenberger W, Kosloff R, Kuprov I, Luy B, Schirmer S, Schulte-Herbrüggen T, Sugny D, Wilhelm FK (2015) Training Schrödinger's cat: quantum optimal control. Eur Phys J D 69:279. https://doi.org/10.1140/epjd/e2015-60464-1
69. Goodwin DL, Kuprov I (2016) Modified Newton-Raphson GRAPE methods for optimal control of spin systems. J Chem Phys 144:204107. https://doi.org/10.1063/1.4949534
70. Geen H, Wimperis S, Freeman R (1989) Band-selective pulses without phase distortion. A simulated annealing approach. J Magn Reson 85:620–627. https://doi.org/10.1016/0022-2364(89)90254-0
71. Geen H, Freeman R (1991) Band-selective radiofrequency pulses. J Magn Reson 93:93–141. https://doi.org/10.1016/0022-2364(91)90034-q
72. Nuzillard JM, Freeman R (1994) Band-selective pulses designed to accommodate relaxation. J Magn Reson Ser A 107:113–118. https://doi.org/10.1006/jmra.1994.1056
73. Kupce E, Freeman R (1995) Band-selective correlation spectroscopy. J Magn Reson Ser A 112:134–137. https://doi.org/10.1006/jmra.1995.1023
74. Emsley L, Bodenhausen G (1990) Gaussian pulse cascades: new analytical functions for rectangular selective inversion and in-phase excitation in NMR. Chem Phys Lett 165:469–476. https://doi.org/10.1016/0009-2614(90)87025-m
75. Ewing B, Glaser SJ, Drobny GP (1990) Development and optimization of shaped NMR pulses for the study of coupled spin systems. Chem Phys 147:121–129. https://doi.org/10.1016/0301-0104(90)85028-u
76. Verstraete J-B, Foroozandeh M (2022) Improved design of frequency-swept pulse sequences. J Magn Reson 336:107146. https://doi.org/10.1016/j.jmr.2022.107146
77. Higham NJ (2005) The scaling and squaring method for the matrix exponential revisited. SIAM J Matrix Anal Appl 26:1179–1193. https://doi.org/10.1137/04061101x
78. Moler C, Van Loan C (2003) Nineteen dubious ways to compute the exponential of a matrix. Twenty-five years later. SIAM Rev 45:3–49. https://doi.org/10.1137/s00361445024180
79. Kelley CT (1999) Iterative methods for optimization. Society for Industrial and Applied Mathematics (1999). https://doi.org/10.1137/1.9781611970920.
80. Nocedal J, Wright SJ (2006) Numerical optimization, 2nd edn. Springer, New York
81. Johnson CS Jr (1999) Diffusion ordered nuclear magnetic resonance spectroscopy: principles and applications. Prog Nucl Magn Reson Spectrosc 34:203–256. https://doi.org/10.1016/s0079-6565(99)00003-5
82. Stejskal EO, Tanner JE (1965) Spin diffusion measurements: spin echoes in the presence of a time-dependent field gradient. J Chem Phys 42:288–292. https://doi.org/10.1063/1.1695690

83. Sinnaeve D (2012) The Stejskal-Tanner equation generalized for any gradient shape-an overview of most pulse sequences measuring free diffusion. Concepts Magn Reson 40A:39–65. https://doi.org/10.1002/cmr.a.21223
84. Wu D, Chen A, Johnson CS Jr (1996) Three-dimensional diffusion-ordered NMR spectroscopy: the homonuclear COSY-DOSY experiment. J Magn Reson Ser A 121:88–91. https://doi.org/10.1006/jmra.1996.0142
85. Jerschow A, Müller N (1996) 3D diffusion-ordered TOCSY for slowly diffusing molecules. J Magn Reson Ser A 123:222–225. https://doi.org/10.1006/jmra.1996.0241
86. Vitorge B, Jeannerat D (2006) NMR diffusion measurements in complex mixtures using constant-time-HSQC-IDOSY and computer-optimized spectral aliasing for high resolution in the carbon dimension. Anal Chem 78:5601–5606. https://doi.org/10.1021/ac060744g
87. McLachlan AS, Richards JJ, Bilia AR, Morris GA (2009) Constant time gradient HSQCi-DOSY: practical aspects. Magn Reson Chem 47(1081–1085):2518. https://doi.org/10.1002/mrc
88. Perea-Buceta J, Rico del Cerro D, Kilpeläinen I, Heikkinen S (2021) Incorporated diffusion ordered heteronuclear multiple bond correlation spectroscopy, 3D iDOSY-HMBC. Merging of diffusion delay with long polarization transfer delay of HMBC. J Magn Reson 323:106892. https://doi.org/10.1016/j.jmr.2020.106892
89. Cobas JC, Martín-Pastor M (2004) A homodecoupled diffusion experiment for the analysis of complex mixtures by NMR. J Magn Reson 171:20–24. https://doi.org/10.1016/j.jmr.2004.07.016
90. Nilsson M, Morris GA (2007) Pure shift proton DOSY: diffusion-ordered 1H spectra without multiplet structure. Chem Commun 933. https://doi.org/10.1039/b617761a
91. Glanzer S, Zangger K (2014) Directly decoupled diffusion-ordered NMR spectroscopy for the analysis of compound mixtures. Chem Eur J 20:11171–11175. https://doi.org/10.1002/chem.201402920
92. Telkki V-V, Urbańczyk M, Zhivonitko V (2021) Ultrafast methods for relaxation and diffusion. Prog Nucl Magn Reson Spectrosc 126–127:101–120. https://doi.org/10.1016/j.pnmrs.2021.07.001
93. Loening NM, Keeler J, Morris GA (2001) One-dimensional DOSY. J Magn Reson 153:103–112. https://doi.org/10.1006/jmre.2001.2423
94. Thrippleton MJ, Loening NM, Keeler J (2003) A fast method for the measurement of diffusion coefficients: one-dimensional DOSY. Magn Reson Chem 41:441–447. https://doi.org/10.1002/mrc.1195
95. Shrot Y, Frydman L (2008) Single-scan 2D DOSY NMR spectroscopy. J Magn Reson 195:226–231. https://doi.org/10.1016/j.jmr.2008.09.011
96. Guduff L, Kuprov I, van Heijenoort C, Dumez J-N (2017) Spatially encoded 2D and 3D diffusion-ordered NMR spectroscopy. Chem Commun 53:701–704. https://doi.org/10.1039/c6cc09028a
97. Jacquemmoz C, Mishra R, Guduff L, Heijenoort C, Dumez J-N (2022) Optimisation of spatially-encoded diffusion-ordered NMR spectroscopy for the analysis of mixtures. Magn Reson Chem 60:121–138. https://doi.org/10.1002/mrc.5194
98. Mansfield P (1977) Multi-planar image formation using NMR spin echoes. J Phys C: Solid State Phys 10:L55–L58. https://doi.org/10.1088/0022-3719/10/3/004
99. Stehling M, Turner R, Mansfield P (1991) Echo-planar imaging: magnetic resonance imaging in a fraction of a second. Science 254:43–50. https://doi.org/10.1126/science.1925560
100. Tal A, Frydman L (2010) Single-scan multidimensional magnetic resonance. Prog Nucl Magn Reson Spectrosc 57:241–292. https://doi.org/10.1016/j.pnmrs.2010.04.001
101. Giraudeau P, Frydman L (2014) Ultrafast 2D NMR: an emerging tool in analytical spectroscopy. Annu Rev Anal Chem 7(129–161):213–020208. https://doi.org/10.1146/annurev-anchem-071

102. Gouilleux B, Rouger L, Giraudeau P (2018) Ultrafast 2D NMR: methods and applications. Annu Rep NMR Spectrosc 75–144. https://doi.org/10.1016/bs.arnmr.2017.08.003
103. Nicolaou KC, Snyder SA (2005) Chasing molecules that were never there: misassigned natural products and the role of chemical synthesis in modern structure elucidation. Angew Chem Int Ed 44:1012–1044. https://doi.org/10.1002/anie.200460864

Chapter 3
POISE

'And have you never known the pleasure and triumph of a lucky guess?—I pity you.—I thought you cleverer—for, depend upon it a lucky guess is never merely luck. There is always some talent in it.'

— JANE AUSTEN, *Emma*

This chapter describes the development of a piece of software for on-the-fly optimisation of NMR experimental parameters, titled POISE (*Parameter Optimisation by Iterative Spectral Evaluation*). POISE is primarily written in Python 3 and includes an interface to Bruker's TopSpin software, allowing optimisations to be started directly from the TopSpin command line. Through this optimisation process, NMR acquisition parameters can be adjusted for individual spectrometers and samples, ultimately leading to improved spectral quality.

In this chapter, I first provide some details about the implementation of POISE. The bulk of the text which follows is devoted to a number of applications in solution-state NMR spectroscopy. At the end, the extension of the concept of on-the-fly optimisation to ESR spectroscopy is also briefly discussed: I contributed code for this, but the experimental ESR work and data analysis were carried out by Jean-Baptiste Verstraete and William Myers (University of Oxford).

The work in this chapter forms the subject of two publications [1, 2]:

- Yong, J. R. J.; Foroozandeh, M. On-the-Fly, Sample-Tailored Optimization of NMR Experiments. *Anal. Chem.* **2021**, 93, 10735–10739, https://doi.org/10.1021/acs.analchem.1c01767
- Verstraete, J.-B.; Yong, J. R. J.; Goodwin, D. L.; Myers, W. K.; Foroozandeh, M. Towards fully optimised and automated ESR spectroscopy. *Chem. Commun.* **2022**, 58, 10715–10718, https://doi.org/10.1039/d2cc02742a.

J. Yong, *Optimising NMR Spectroscopy Through Method and Software Development*, Springer Theses, https://doi.org/10.1007/978-3-031-46684-7_3

3.1 Introduction

In the previous chapter, I covered various approaches to improving pure shift NMR through the use of optimisation. Although the optimisation code written there was highly specialised and only designed to work on pure shift applications, it was envisioned that this optimisation approach could be applied to essentially *any* NMR experiment which required parameter optimisation. This qualification is technically true of *every* experiment: even the simplest pulse–acquire experiment can be optimised for maximal SNR per unit time through the use of Ernst angle excitation. More complex examples, such as 2D experiments, typically have parameters which should be chosen to optimally match coupling constants (e.g. the INEPT delay, Fig. 1.3) or relaxation rates (e.g. the NOE mixing time).

In routine NMR spectroscopy, this requirement for accurate parameters is often 'solved' through the use of compromise values, which fall in the middle of an expected range for typical molecules. For example, the INEPT delay may be optimised for a $^{1}J_{CH}$ value of 145 Hz, which is reasonably close to most actual $^{1}J_{CH}$ constants for sp^3 and sp^2 carbons. As an alternative, parameter values may be optimised 'by hand' prior to acquisition. Such a procedure is often recommended for more complicated experiments such as solvent suppression, diffusion, or saturation transfer difference experiments. However, compared to these, the use of experimental optimisation has several benefits. It is:

1. *sample-specific*, and as long as the default values are within the optimisation bounds, the optimisation will yield performance which is no worse than the defaults;
2. more *robust* towards unusual molecular structures, which have physical or chemical properties which fall outside of an expected range;
3. *instrument-specific*, so can compensate for spectrometer imperfections;
4. *automated*, so does not require an expert to adjust parameter values manually, or even any user intervention for that matter;
5. *objective*, in that the quality of a spectrum can (in principle) be mathematically measured through a cost function; and
6. *fast*, in that it uses an algorithm which is designed to achieve rapid decreases in the objective function: many 'manual' optimisations involve either trial-and-error or an exhaustive grid search (i.e. increasing a parameter value one step at a time), neither of which are efficient.

Despite these advantages, experimental optimisation of NMR parameters has seen only limited use. In fact, although there are several examples of such optimisations in laser [3], nuclear quadrupole resonance [4–6], and ESR [7] spectroscopies, the only direct parallel in NMR which I am aware of is that of the eDUMBO pulses for heteronuclear [8, 9] and homonuclear dipolar [10] decoupling in solid-state magic-angle spinning experiments. In this work, the Emsley group used 'direct spectral optimisation' to determine the best coefficients for a pulse expressed as a Fourier series. The performance of these pulses was measured by a cost function which primarily took into account the intensity of the detected peaks: a larger intensity corresponds to better decoupling performance. Interestingly, the aim of using an experimental optimisation here was not to obtain sample-specific pulses (point (1)), but rather to account for the 'spectrometer response', i.e. instrumental non-idealities (point (3)). It was assumed that the compound used for the optimisation was a suitably representative choice, so that the optimisation result could simply be applied to other samples with no change.

The likely reason for the low popularity of experimental optimisations is *time*. Each *function evaluation* (FE), i.e. each measurement of the cost function, corresponds to the acquisition of an NMR spectrum which may take seconds to minutes. Thus, where possible, it is easier to run NMR optimisations in a *theoretical* manner, which can be much faster and also circumvents the effect of noise. Examples of such optimisations include the design of shaped pulses through optimal control theory [11–17], by simple parameterisation [18–23], or the optimisation of entire pulse sequences [24–28] (the dPSYCHE experiment in Sect. 2.6 falls under this category). In fact, even the aforementioned eDUMBO pulses were not *originally* designed as an experimental optimisation: they are actually an enhancement of the DUMBO decoupling schemes, which were optimised using numerical simulations [29].

In this chapter, I aim to provide a convincing argument that experimental optimisation is not necessarily slow. In particular, I show that it is often possible to devise optimisation routines which yield improved results in a matter of minutes. All the optimisations here are performed using a software package written by me, called POISE (Parameter Optimisation by Iterative Spectral Evaluation). POISE is open-source (https://github.com/foroozandehgroup/nmrpoise) and can be installed in a single step through `pip install nmrpoise`. Furthermore, it comes with extensive user documentation, both in the form of a text guide (https://foroozandehgroup.github.io/nmrpoise) as well as video (https://www.youtube.com/watch?v=QTCeSCRZs4I).

In contrast to previous work, which typically feature optimisations targeted at one specific application, I have endeavoured to make POISE as customisable and as broad as possible. Although it is just a single piece of software, this generality is what allows POISE to perform all the optimisations described in this chapter. Furthermore, it also means that other users can devise specific cost functions and optimisation procedures for their own use which can also be run using POISE. Thus, *POISE is more than just the applications shown later in this chapter*: it is really a platform which makes it possible to carry out arbitrary optimisations on an NMR spectrometer.

3.2 Technical Overview

In this section, I first cover the general principles underlying, and the implementation of, POISE. The basic operation of POISE is summarised in the flowchart of Fig. 3.1: this is a generalised version of the pure shift optimisations carried out in Sect. 2.4. Almost all aspects of this can be customised by the user, as will be described.

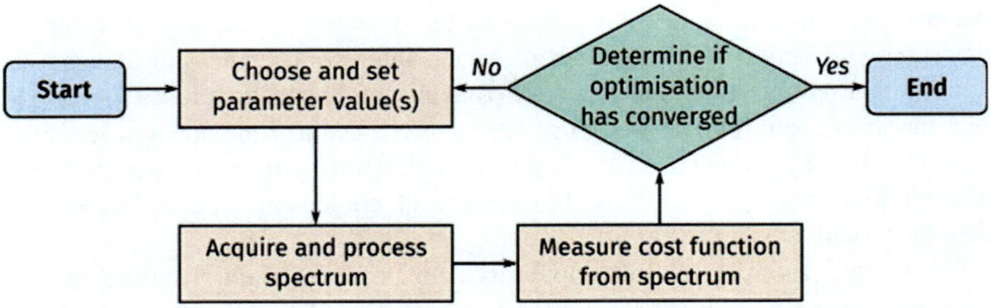

Fig. 3.1 Flowchart depicting the main steps in a POISE optimisation

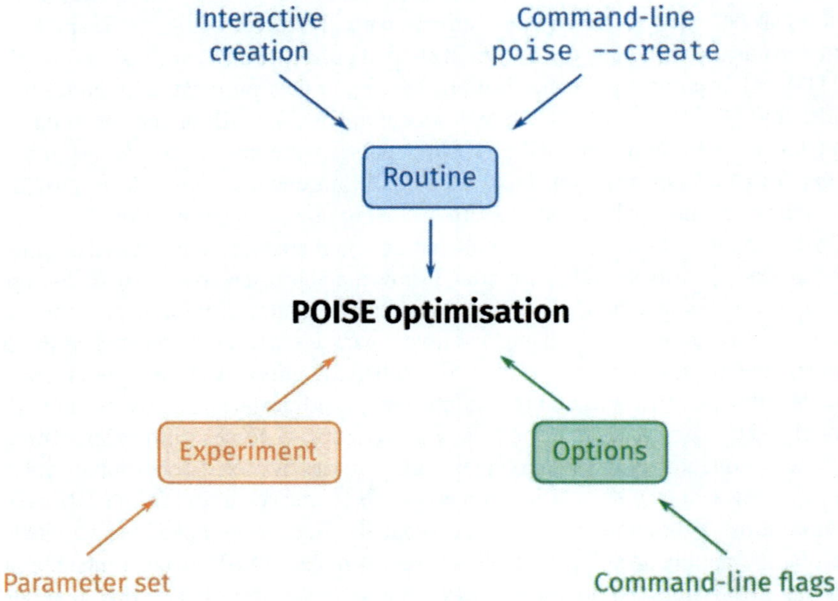

Fig. 3.2 The various 'ingredients' of a POISE optimisation and how they can be set up

I begin by drawing a distinction between an optimisation *routine*, as well as the *way* in which these routines can be run (Fig. 3.2). Routines consist of a series of predefined variables, such as the parameter(s) to be optimised: however, these may be optimised in different ways, which is where the experiment used to perform the optimisation, as well as other options, come in. These 'ingredients' of an optimisation will be repeatedly referred to in the discussion of individual applications in Sect. 3.4.

3.2.1 Routines

An *optimisation routine*, as defined in POISE, consists of the following components:

1. *Name*
   This is an identifier used to refer to the entire routine, which is arbitrary, but should ideally be descriptive.
2. *Parameters*
   The parameters to be optimised. These are given as strings and directly correspond to TopSpin parameter names, for example, `P1` for a pulse width.
3. *Initial guess* (one per parameter)
   The point at which the optimisation is started. Naturally, this should represent the user's best guess as to where the optimum lies. It is generally sensible to choose unoptimised, 'default' values for these.
4. *Lower* and *upper bounds* (one of each per parameter)
   These may be chosen based on a range of values which are 'chemically sensible'; alternatively, instrumental specifications may sometimes restrict the range of parameter values which can be explored.
5. *Tolerances* (one per parameter)
   The tolerance loosely corresponds to the level of accuracy required for the optimisation. Of course, setting too large a tolerance will simply yield an imprecise and meaningless result. However, there is no point in setting this to be too small (i.e. requesting an overly accurate optimum): this increases the time required for convergence, because the value of the cost function evaluated at two points too close together will differ only by noise. Furthermore, parameter values cannot be implemented with arbitrary precision on a spectrometer, and the available resolution sets a lower bound for the tolerance (see Sect. 3.4.6 for a more concrete discussion of this). These conditions make it sound as if there is little room for error in choosing the tolerance, but in practice the desired accuracy is often reasonably clear from the context, and it generally suffices to get the order of magnitude correct.
6. *AU program*
   The AU program defined in an optimisation routine is used to acquire and process the spectrum. The user may leave this empty, in which case POISE automatically detects the dimensionality of the experiment and performs standard processing steps (Fourier transformation, window multiplication, phase correction, and baseline correction). This basic setup is often sufficient for many applications. However, the AU program does allow for almost infinite customisation of the actual spectral measurement: for example, it may call other scripts in TopSpin which create shaped pulses (as shown in Sect. 3.4.8).
7. *Cost function*
   The cost function measures the 'badness' of the spectrum thus recorded. As in the previous chapter, the optimisation algorithm seeks to minimise this value. POISE cost functions must be written in Python 3: this design decision is considered later in Sect. 3.2.5. For ease of use, several cost functions which cover 'typical'

optimisation scenarios, such as maximising or minimising signal intensity, come pre-installed with POISE. This means that users do not necessarily need to write their own cost function if they are not familiar with Python.

Collectively, a routine is therefore a description of *what is to be optimised*. POISE allows users to create new routines interactively through a series of dialogue boxes. Alternatively, routines themselves can be created on-the-fly using the `poise --create` command: this is useful when some components are not known beforehand, such as if the optimum from a different optimisation is to be used as the initial point in a new one. However, this is limited to single-parameter routines.

After being created, routines are stored in the human-readable JSON format: they can therefore be modified using any text editor. Examples of these JSON files are presented in subsequent sections.

3.2.2 The Experiment

Once the user has defined a (named) routine, it can then be run from the TopSpin command line using the command `poise ROUTINE_NAME`. However, the routine itself does not specify what experiment is to be run (i.e. the pulse program), nor any of the other parameters in the experiment. These must be set by the user and can most conveniently be stored in a TopSpin parameter set which can simply be loaded before starting the optimisation.

This flexibility means that the same *type* of optimisation may be applied to different pulse sequences. For example, a routine for optimising the NOE mixing time (as described in Sect. 3.4.4) can be run with different versions of the NOESY sequence, as long as the parameter being optimised always has the same identifier in TopSpin (`D8` for the mixing time). Likewise, parameters such as the number of scans can be adjusted in order to run optimisations on samples with different concentrations, without having to redefine a new routine.

3.2.3 Optimisation Options

Finally, POISE provides a few command-line options which control how the optimisation is carried out:

- the `--maxfev` option allows the user to control the maximum number of FEs, or in other words, the maximum number of experiments run. If the optimisation has not converged after acquiring this many spectra, the best result so far is simply returned. This indirectly allows a cap to be placed on the time spent on optimisation.
- the `--quiet` option silences all output from POISE. This is useful when, for example, a optimisation is to be run under automation. (The optima found are

stored in the dataset itself after the optimisation ends and can therefore always still be retrieved.)

- the `--separate` option allows each FE to be run in a new experiment number, so that the optimisation trajectory can be analysed after its conclusion.
- perhaps most importantly, the `--algorithm` option allows the user to choose one of three optimisation algorithms, which I now describe in more detail.

3.2.4 Optimisation Algorithms

As was briefly mentioned in Sect. 2.4.1, derivative-based optimisation algorithms cannot be used for experimental optimisations. To be more precise, when analytical gradients are not available, derivative-based algorithms calculate gradients using a finite difference approximation:

$$\nabla f(x) \approx \frac{f(x + \varepsilon) - f(x - \varepsilon)}{2\varepsilon}, \tag{3.1}$$

where ε is the step size used for the finite difference calculation. In Nocedal and Wright [30], it is shown that the error in this finite difference gradient (as compared to the true gradient) has an upper bound of

$$\frac{\eta(x; \varepsilon)}{\varepsilon} + O(\varepsilon^2), \tag{3.2}$$

where $\eta(x; \varepsilon)$ is (roughly) the noise in the region $[x - \varepsilon, x + \varepsilon]$. The first term represents the error due to this noise, and if ε is small, this term is large; on the other hand, the second term represents the error due to the finite difference approximation, and if ε is large *this* term is large. This means that finite difference gradients, and any algorithms which use them, are entirely unreliable in the presence of noise. Instead, POISE provides a choice of three derivative-free optimisation algorithms: the Nelder–Mead (NM) method [31], the multidirectional search (MDS) method [32, 33], and the Py-BOBYQA trust-region method [34, 35].

3.2.4.1 Nelder–Mead

The NM method is a highly popular derivative-free optimisation algorithm, which maintains a set of points $\{y_1, y_2, \ldots, y_{n+1}\}$ during the optimisation, n being the number of parameters being optimised (or the *dimension* of the optimisation problem). The convex hull of these points, Y, is the smallest possible set of points which contains all the y_k and is convex, in that

$$\forall x_1, x_2 \in Y, \forall \alpha \in [0, 1], \alpha x_1 + (1 - \alpha)x_2 \in Y, \tag{3.3}$$

and in this context is called a *simplex*. If this convex hull is nonempty—or equivalently, if the n vectors $y_k - y_1$ $(2 \leq k \leq n+1)$ are linearly independent—then the simplex is called *nonsingular*. (To provide an analogy for $n = 2$, the convex hull is the shape obtained by stretching a rubber band around three pins placed at the points $\{y_1, y_2, y_3\}$. The simplex would be singular if the three points were collinear.)

The NM algorithm is in fact quite intuitive to understand. The initial simplex is first constructed using the supplied initial point: POISE specifically uses the method of Spendley et al. [36] to construct a *regular simplex*. The optimisation itself begins by measuring the cost function f at every point of the simplex, and sorting the points in ascending order of cost function values (i.e. from best to worst), such that $f(y_1) \leq f(y_2) \leq \cdots \leq f(y_{n+1})$. The centroid of the simplex is defined by the n best points:

$$\bar{y} = \sum_{i=1}^{n} y_i. \tag{3.4}$$

On each iteration of the NM algorithm, we attempt to replace the worst point y_{n+1} with a better point (Fig. 3.3). The search for the new point is performed in several steps: first, the worst point is *reflected* about the centroid of the simplex to obtain a new point y_r:

$$y_\mathrm{r} = \bar{y} - (y_{n+1} - \bar{y}). \tag{3.5}$$

The value of the cost function is evaluated at this point and is compared against the existing points in the simplex. If this reflected point falls in the middle of the pack, such that $f(y_1) \leq f(y_\mathrm{r}) < f(y_n)$, this represents a 'modest' improvement in the cost function: we simply replace the worst point with this and continue to the next iteration.

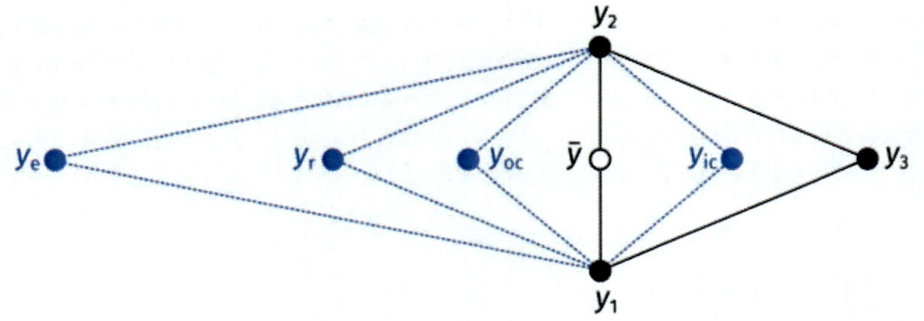

Fig. 3.3 Diagram showing various points evaluated in one iteration of the Nelder–Mead algorithm (for an optimisation of two parameters). The solid black lines indicate the boundaries of the simplex, which is assumed to be ordered such that y_1 is the best point (has the lowest cost function value) and y_3 the worst. The current points in the simplex are shown as filled black circles, and the centroid $\bar{y}$ as an empty black circle. The blue circles indicate the trial points which the algorithm attempts to replace y_3 with and are further discussed in the text. Blue dashed lines indicate the simplex which would result if the corresponding trial point were accepted

On the other hand, if the reflected point is better than all the other points (i.e. $f(y_r) < f(y_1)$), then we ambitiously attempt to *expand* the simplex even further in that direction by testing the point

$$y_e = \bar{y} - 2(y_{n+1} - \bar{y}). \tag{3.6}$$

Of course, there is no guarantee that this is necessarily better than y_r; therefore, we choose whichever point of y_r or y_e had a lower value of f, and replace the worst point with this and continue to the next iteration.

If the reflected point is an improvement on the worst point but is no better than the remaining points, in that $f(y_n) \leq f(y_r) < f(y_{n+1})$, then the algorithm performs an *outside contraction*, which resembles a half-hearted reflection:

$$y_{oc} = \bar{y} - (1/2)(y_{n+1} - \bar{y}). \tag{3.7}$$

Conversely, if the reflected point is even worse than the worst point, in that $f(y_{n+1}) \leq f(y_r)$, then this suggests that search direction is very poor: we thus perform an *inside contraction*, which uses a point halfway between the worst point and the centroid:

$$y_{ic} = \bar{y} + (1/2)(y_{n+1} - \bar{y}). \tag{3.8}$$

If either of these contracted points are any better than y_r, then we replace the worst point in the simplex and continue to the next iteration; otherwise, we conclude that no search direction was good, and simply shrink the simplex towards the current best point by replacing each point y_k with $(y_k + y_1)/2$. In practice, these 'last-resort' shrink steps occur very rarely.

Finally, convergence is signalled when in every dimension of the optimisation, the width of the simplex is smaller than the chosen optimisation tolerance.[1] For a multiple-parameter optimisation, this can potentially mean that extra accuracy is obtained in one of the parameters (because the simplex may have shrunk along that dimension more quickly). However, it does guarantee that *at least* the specified level of accuracy in every dimension is achieved.

3.2.4.2 Multidirectional Search

In the preceding discussion, we noted that the simplex Y was nonsingular if the n vectors $y_k - y_1$ were linearly independent. Equivalently, the matrix M formed by concatenating these vectors

[1] The implementation of the NM algorithm in the `scipy` library only accepts a single value for the 'tolerance', which is then used in all dimensions. This is designed to be used by scaling the parameters beforehand such that the tolerance in each dimension is equal (and in fact, POISE was later updated to do exactly this). However, during initial development I chose instead to re-implement the NM algorithm with a convergence check which allowed for different tolerances to be specified in each dimension.

$$M = \big(y_2 - y_1, y_3 - y_1, \ldots, y_{n+1} - y_1\big) \tag{3.9}$$

must be nonsingular, i.e. have a nonzero determinant. We can quantify how 'close' the simplex Y is to being singular, using the l^2 condition number of the matrix M, which in this context is usually referred to as the *simplex condition*:

$$\kappa(Y) = \|M\| \|M^{-1}\|, \tag{3.10}$$

where $\|M\|$ is the matrix norm induced by the Euclidean norm,

$$\|M\| = \max_{x \neq 0} \frac{\|Mx\|}{\|x\|}. \tag{3.11}$$

A singular simplex Y of course does not have a well-defined condition, since M^{-1} does not exist. However, the larger the condition of a simplex is, the closer it is to being singular. Very loosely speaking, a long and thin simplex has a large condition number and would be singular if its width were to go to zero.

The simplex updates made in the process of the NM algorithm mean that the simplex condition changes throughout the course of the optimisation. This is good for achieving rapid decreases in the cost function, since the simplex shape adapts to the cost function being optimised. However, if the simplex condition gets too large, it is possible that the optimisation will stall at a nonstationary point, since the search directions of the simplex are severely limited. The MDS algorithm was proposed partially for the purpose of avoiding this ill-conditioning.[2]

The MDS method is also simplex-based and uses similar reflection/expansion/contraction steps to the NM algorithm. However, instead of (e.g.) reflecting a single worst point y_{n+1} about the other points, it reflects *all* of the n worst points $\{y_2, y_3, \ldots, y_{n+1}\}$ about the best point y_1. This means that the shape of the simplex Y, and thus its condition $\kappa(Y)$, is always preserved: this provides it with much better convergence properties, as was shown by Torczon et al. [32, 37][3] The increased reliability of the MDS algorithm over the NM algorithm was demonstrated on a variety of example optimisation problems: even in the very simple case where the cost function was simply the norm of a vector,

$$f(y) = \|y\|, \tag{3.12}$$

it was shown that the NM algorithm stalled when the dimension of the problem, n, was sufficiently large. The value of n needed to precipitate this failure depended on the problem being solved and generally ranged from 8 to 40. On the other hand, the

[2] The main reason was in fact to better exploit computer parallelism, but it was also noticed that the MDS method proved to be generally more robust than NM.

[3] Specifically, it can be concluded that at least one of the search directions was bounded away from being orthogonal to the gradient; or in simpler (and less precise) terms, at least one of the search directions is close enough to a direction in which the cost function f decreases.

MDS method proved to be robust under the same conditions, eventually converging to the optimum—although in the cases where NM *did* work, the MDS method generally required more FEs.

It was this improved robustness of the MDS algorithm which prompted Goodwin et al. to use it in their (experimental) optimisation of ESR pulse shapes [7], and me to later include it in POISE. In the ESR work, the number of pulse points being optimised was 11 or 21, which fell into the regime where the MDS method would likely have better convergence properties than NM. However, in NMR, each experiment takes a substantially longer time than in ESR, and even optimisations with $n > 2$ become rather time-consuming due to the number of FEs required (the largest n explored in the present work is 4). Thus, it is unlikely that the NMR optimisations explored in POISE will lie in the regime where the NM algorithm is prone to failure. As will be shown later, I generally found that the NM and MDS methods were equally reliable in POISE optimisations, with NM generally being faster.

3.2.4.3 Py-BOBYQA

Unlike the NM and MDS algorithm, Py-BOBYQA is not simplex based but is a trust-region algorithm [34, 35]. The fundamental idea behind a (derivative-free) trust-region method is to sample the cost function at a set of points Y, and construct a model m through interpolation, which matches the cost function at these points:

$$\forall y \in Y, m(y) = f(y). \tag{3.13}$$

The model at iteration k is labelled m_k. Most trust-region methods, including Py-BOBYQA, use a quadratic model, which can be expressed as

$$m_k(x_k + p) = c + g^T p + p^T Gp, \tag{3.14}$$

where G is a symmetric matrix and x_k is the centre of the model at iteration k (x_0 being the user-specified initial point). For this model to be fully determined, the set Y must therefore contain $(n + 1)(n + 2)/2$ points in total.

The algorithm further maintains a *trust-region radius* Δ_k at each iteration, which is a measure of how reliable the model is. The initial trust-region radius, Δ_0, can be arbitrarily chosen: in the case of POISE, I elected to set Δ_0 to be 10 times the desired tolerance. The model m_k is then used to calculate the next step s_k, which is obtained by minimising m_k over all points within a radius of Δ_k from the centre x_k (the *trust-region subproblem*):

$$s_k = \underset{\|s\| \leq \Delta_k}{\arg\min}\, m_k(x_k + s). \tag{3.15}$$

Since m_k itself is noiseless, this subproblem can be solved using almost any algorithm: Py-BOBYQA uses a conjugate gradient method. The cost function is then evaluated at the trial point $x_k + s_k$ and compared against the value predicted by the model. If the ratio of 'actual improvement' to 'predicted improvement' is large enough, i.e.

$$r_k = \frac{f(x_k) - f(x_k + s_k)}{m_k(x_k) - m_k(x_k + s_k)} \geq \eta \tag{3.16}$$

for some threshold value η, then the step s_k is accepted and x_{k+1} is set to $x_k + s_k$, replacing the worst point in Y. Additionally, the trust-region radius Δ_k may be increased so that the next step(s) can be more ambitious. Conversely, if $r_k < \eta$, then there are one of two possibilities: either the model is poorly conditioned (in that the points in Y are very unevenly distributed), in which case one of the points is replaced and the model recalculated; or the model is sufficiently well-conditioned, in which case the step is rejected, and Δ_k is decreased.

Py-BOBYQA goes beyond a standard derivative-free trust-region algorithm in further limiting the rate at which the radius Δ_k can change (amongst others). Specifically, Py-BOBYQA also maintains a lower bound on the trust-region radius (denoted ρ_k), and on unsuccessful iterations Δ_k is not allowed to decrease below ρ_k. This prevents Δ_k from decreasing too quickly until the algorithm is certain that Y is sufficiently well-conditioned [38]. Another core feature of Py-BOBYQA is the implementation of multiple restarts, which endows it with greater robustness towards noise and also allows it to escape local minima [35, 39]. However, the multiple-restarts feature in Py-BOBYQA was disabled in POISE as this often led to overly long optimisations.[4]

Crucially, Py-BOBYQA differs from the simplex-based methods in that *it cares about the actual value of the cost function*. In the NM and MDS methods, only the relative ordering of the points in the simplex matters; it makes no difference to the algorithm whether the worst point has a cost function value of 10 or 1000. However, in Py-BOBYQA, the value of f is used in constructing the model, and thus directly influences the optimisation trajectory. Although this is beneficial in cases where the underlying cost function is relatively well-behaved,[5] and is reflected in faster convergence rates for such problems, it can be problematic for some cost functions. Py-BOBYQA is set as the default optimiser in POISE, but the user is strongly recommended to try the NM method as a first step when troubleshooting optimisations.

---

[4] Most mathematics papers on optimisation have no qualms in using hundreds or even thousands of FEs, and it is this context in which Py-BOBYQA outperforms other algorithms. Unfortunately for me, POISE works in an *extremely* restrictive regime where even 50 FEs would be considered very expensive.

[5] I *think* this means cases where the cost function is well described by a quadratic model. Of course, because of Taylor's theorem, every non-noisy cost function can be locally described by a quadratic model within a sufficiently small region. However, for meaningful progress to be made with noisy cost functions, the model must be built over a large enough region such that noise becomes less relevant.

3.2.5 Implementation Details

In this section, I discuss some behind-the-scenes details about how POISE is implemented and several design choices. This information is relevant for anybody looking to improve or otherwise modify the POISE codebase.

Firstly, POISE is written in Python 3, and since TopSpin does not have a built-in Python 3 interface,[6] this means that POISE is not entirely self-contained within TopSpin: in particular, an external installation of Python 3 is required, which may be a slight inconvenience. This choice was necessary because it would have been too time-consuming to implement numerical optimisation algorithms using the existing C or Python 2 application programming interfaces (APIs) in TopSpin: notably, the Python 2 API uses the Jython implementation of the language, which is incompatible with the `numpy` Python library. However, a benefit of this is that since the 'cost' of installing Python 3 is already paid, we can also allow users to define their own cost functions using libraries such as `numpy` and `scipy`. Without these, performing any kind of non-trivial data processing or analysis is significantly more awkward.

Like most Python 3 packages, POISE is available on the Python Package Index (PyPI), so can be installed using a single command: `pip install nmrpoise`. As usual, POISE is first installed in the Python `site-packages` directory. If the `nmrpoise` package is imported from a Python 3 script, this code in the `site-packages` directory is read. This may be required on occasion, as the `nmrpoise` package provides a few functions to analyse optimisation logs created by POISE. We might refer to this code as the 'library' component of `nmrpoise`.

This library component, however, is not used at all when running optimisations. When POISE is installed, it automatically searches for TopSpin installations in either `C:\` (on Windows), or `/opt/` (on Unix/Linux). The installation then creates the following components:

- a *frontend script* at `$TS/exp/stan/nmr/py/user/poise.py`, which allows POISE to be invoked by simply typing `poise` in the TopSpin command line and is responsible for controlling data acquisition; as well as
- a *backend directory* at `$TS/exp/stan/nmr/py/user/poise_backend`, within which all of the POISE data and logic is stored. For example, routines can be found in the `routines` subdirectory, and cost functions in the `costfunctions.py` and `costfunctions_user.py` files.

Here, `$TS` denotes the TopSpin installation directory detected during installation. As the name suggests, a non-standard TopSpin installation location can be specified if necessary using the `$TS` environment variable.

[6] Version 4.1.4 of TopSpin now comes with a Python 3 API; however, this was introduced too late for the work in this chapter.

All optimisations are run using the code *only* in the backend directory, and not anything in Python's `site-packages` folder. This is because the frontend script must know how to launch the backend (i.e. where to find the files), and it is simply easiest to predefine this location.[7]

Listing 3.1 Excerpt from the POISE frontend script, illustrating the two-way communication between frontend and backend.

```
try:
    # Launch backend
    backend = subprocess.Popen([p_python3, "-u", p_backend],
                               stdin=subprocess.PIPE,
                               stdout=subprocess.PIPE)

    # Pass information from frontend to backend
    for item in [args.algorithm, routine_id, p_spectrum, args.maxfev]:
        print >>backend.stdin, item
    backend.stdin.flush()

    while True:
        # Receive information from backend
        line = backend.stdout.readline()
```

Having files in two different places does mean that some form of communication between the two must be established. In POISE, this is accomplished through the use of anonymous pipes, one for each direction of communication (Listing 3.1). In this way, the backend can signal to the frontend what values of parameters should be evaluated; the frontend can then begin data acquisition, and signal to the backend when this is complete so that the cost function can be calculated. Although this setup works perfectly fine when left to run untouched, a frustrating number of 'tricks' are required to keep these synchronised if either the frontend or the backend are terminated unexpectedly, or if acquisition is prematurely stopped by the user (which usually suggests that they wish to stop the optimisation). This includes the backend creating a file with its process ID every time it is called and deleting it upon exit (Listing 3.2), meaning that the frontend can locate any stray backend processes which were not appropriately terminated.

[7] In fact, it is possible to dynamically determine the `site-packages` installation location at runtime, meaning that the entire backend does not need to be copied to TopSpin directories. However, that would mean the cost functions would be buried inside the `site-packages` directory, which is more difficult for the user to find.

Listing 3.2 Simplified excerpt from POISE backend script, showing a context manager used to keep track of backend process IDs. The context manager ensures that when the script is started, a file with the process ID is created; and when the script exits, this file is deleted. The `main()` function carries out the actual optimisation.

```
from contextlib import contextmanager

@contextmanager
def pidfile():
    # Create a file with the PID
    pid = os.getpid()
    pid_fname = Path(__file__).parent / f".pid{pid}"
    pid_fname.touch()
    # Run the code in the 'with' block
    try:
        yield
    # Delete the file after the 'with' block is exited
    finally:
        if pid_fname.exists():
            pid_fname.unlink()

if __name__ == "__main__":
    with pidfile() as _:
        main()
```

There is one other quirk of TopSpin surrounding data acquisition: it is possible to start the acquisition from a Python script (such as the frontend `poise.py` script), but it is not possible to block execution of the Python script while acquisition is running. Thus, it is not possible to trigger acquisition and wait until it is done before sending a signal to the backend.[8] The workaround is to call an AU program containing acquisition commands, which (somehow) blocks the Python script. A nice side effect of this is that the AU program can be extended by the user to include other commands (e.g. to perform special setup or processing routines).

Finally, during optimisations, the frontend must also be careful not to overwrite other experiments: this can easily happen if, for example, a user opens a new dataset in TopSpin just before acquisition is started. To ensure that this is the case, the frontend *always* brings the optimisation dataset to the foreground immediately before the acquisition AU program is executed. This creates a slight issue in that it can be difficult to view other spectra in TopSpin while an optimisation is proceeding. (Unfortunately, this is out of my control: TopSpin does not provide any documented way of running an acquisition AU program on a background dataset.)

[8] The TopSpin Python documentation claims that this *can* be accomplished using, for example, `XCMD("zg", wait=WAIT_TILL_DONE)`. However, none of the suggestions in the documentation worked as intended.

3.3 What POISE Is Not

Before moving on to cover applications of POISE, I want to make a note about several limitations of the approach chosen.

Firstly, *POISE is not specialised.* Generality is a strength in that POISE can be applied to a diverse range of NMR experiments. However, it can also be a weakness, because the capabilities of POISE must be restricted such that it can be adapted to different scenarios. POISE *always* follows the framework in Fig. 3.1: in particular, it simply seeks to find the optimum $\mathbf{x}^*$, defined by

$$\arg\min_{\mathbf{x}} f(\mathbf{x}). \tag{3.17}$$

Although this conforms to the mathematical notion of an optimisation, there are many other ways of searching for ideal NMR parameters, such as fitting data to a model or directly reading parameters off a spectrum.

This rigidity in the underlying logic means that it is very conceivable that in specific instances, specialised optimisation routines which use customised strategies for data acquisition and analysis *can* outperform POISE in terms of speed and/or accuracy. For example, we see this in Sect. 3.4.1: the TopSpin `pulsecal` routine for pulse width calibration can be much faster than POISE, because it only needs to perform one experiment to obtain an answer. I particularly want to distinguish POISE from other types of 'optimisations' reported in the literature, which typically *accumulate* data points until a given threshold is reached (in terms of, say, SNR or confidence in a parameter). Examples of such procedures can be found in the contexts of relaxation measurements [40, 41] and undersampling in multidimensional NMR [42–44].

Secondly, *POISE is not a global optimiser.* The optimisation algorithms provided within POISE are not designed to go beyond the first local minimum found (Py-BOBYQA technically can, but as described in Sect. 3.2.4, I disabled the multiple-restarts option responsible for this). In challenging optimisation cases where multiple local minima exist, it is not generally possible to predict which local minimum the algorithm will converge to. What *can* be guaranteed is that if the initial point is not already an optimum, then the optimisation will always provide a decrease in the cost function: in other words, it will always lead to an improvement in the spectrum (insofar as the cost function accurately represents the quality of the spectrum).

Finally, *POISE is not universally applicable.* It should be noted that there is always an inherent tradeoff against the time required for the optimisation itself. For example, it makes little sense to spend several minutes optimising the sensitivity of a pulse–acquire experiment: the time could simply be used to improve the SNR by collecting more scans. There is also the critical—though undeniably subjective—question of whether the optimisation is *necessary*: even if an optimisation can be run in a relatively short time, are the results really significantly better than a 'compromise' value in a default parameter set?[9] I do not profess to have a definitive answer to this, and I leave the reader to form their own conclusions in the specific contexts where they may consider using POISE. In any case, a 'meaningful' optimisation is likely to either be fast and/or solve a problem which cannot simply be tackled through signal averaging in the same amount of time. It is my hope that this is true of (most of) the examples shown.

3.4 Applications

In this section, I describe a number of scenarios in solution-state NMR where POISE can be used. These are generally ordered from simple to complex, and progressively show how the features in POISE can be used to customise optimisation procedures.

All POISE optimisations run in this chapter were performed five times to check for potential reproducibility issues. Due to noise in the cost function, these optimisations are not deterministic, and the optima obtained (as well as the time required to locate them) typically span a range. Where possible, this range is quoted in all the results shown in this chapter.

3.4.1 Pulse Width Calibration

The first of these applications is the calibration of a 90° ^{1}H pulse, which is applicable to virtually every NMR experiment. Essentially, we seek to determine τ_{p} for which

$$\tau_{\mathrm{p}}\omega_1 = \frac{\pi}{2}, \tag{3.18}$$

where the RF amplitude ω_1 is not known *a priori* (it is only indirectly controlled via the power level). This pulse width is conventionally specified as the `P1` parameter in TopSpin.

3.4.1.1 Optimisation Setup

In theory, performing a pulse–acquire experiment with a perfect 180° or 360° pulse would yield no detectable (i.e. transverse) magnetisation, i.e. a *null*. Generally, the 360° null is preferred as it minimises effects due to radiation damping and also allows a smaller recovery delay to be used. We can use POISE to search for this by acquiring

[9] Of course, a similar argument can be applied to *many* scientific advances. To use an example from the next chapter, is it really necessary to acquire NOAH spectra when one can just acquire the standalone 2D experiments? I have seen arguments on both sides—some people simply do not need the speedups provided and do not want to spend the time to set up or troubleshoot new experiments.

the spectrum, performing a magnitude-mode calculation, and using the intensity of the resulting spectrum as a cost function:

$$f_{\text{minabsint}} = \sum_i |S_i|, \tag{3.19}$$

where (reusing notation from Sect. 2.4.1) $\mathbf{S}$ is the spectrum under consideration represented as a complex-valued vector, and the i-th point of the spectrum is $S_i = \sqrt{\mathbf{S}_{\text{re},i}^2 + \mathbf{S}_{\text{im},i}^2}$. The label `minabsint` refers to how this cost function drives the optimisation to *min*imise the *abs*olute *int*ensity of the spectrum. To show how easily this can be coded in Python 3, an implementation of this is shown in Listing 3.3 (for all other cost functions in this chapter, the reader is directed to the POISE source code for their implementations). The `get1d_...` helper functions are provided by POISE.

Listing 3.3 The implementation of the `minabsint` cost function in POISE.

```
import numpy as np

def minabsint():
    r = get1d_real()
    i = get1d_imag()
    mag = np.abs(r + 1j * i)
    return np.sum(mag)
```

To check whether this cost function was sensible, I manually acquired a series of spectra with a range of pulse widths and calculated $f_{\text{minabsint}}$ for all of these (Fig. 3.4). In this chapter, I will refer to this process as a *reference grid search*. It should be noted that reference grid searches are a time-consuming procedure, and an end-user of POISE generally does *not* need to do this (it would defeat the purpose of the optimisation). I only perform one here to provide some insight into the nature of the optimisation. In any case, the reference grid search makes it clear that there is a well-defined minimum, located in this case at 48.3 μs; if the POISE optimisation process is accurate, it should converge to this point.

I chose the initial point to be four times the `prosol` value for `P1`, which evaluates to 48 μs for the spectrometer used here: this represents our 'best guess' and is derived from prior calibration of the pulse width on a standard sample. The tolerance is set to 0.2 μs, which corresponds to an accuracy of 0.05 μs for the 90° pulse width itself. The lower and upper bounds are set to be 8 μs away from the initial point, representing a 'sensible' region within which we expect the null to lie (this may need to be adjusted for samples with high ionic strength, but for typical organic samples this is more than enough). A standard `poise_1d` acquisition AU program is used, which simply acquires the spectrum and performs Fourier transformation, phase correction and baseline correction. The routine in JSON format is shown in the caption of Table 3.1. Finally, in order to reduce the time taken for the optimisation,

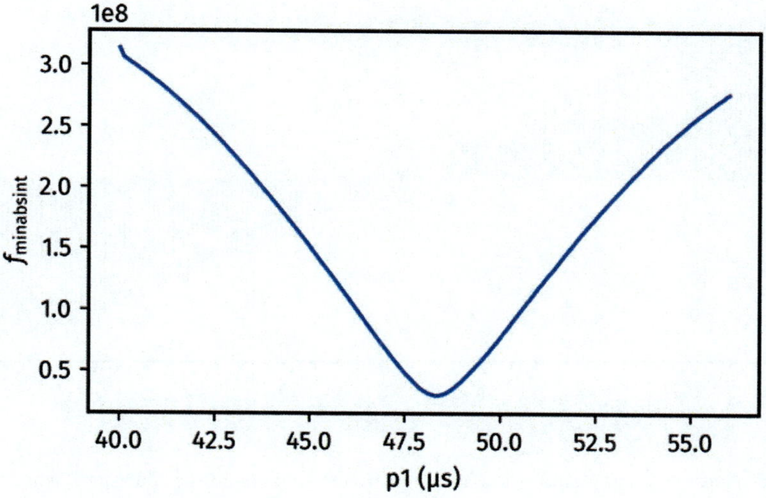

Fig. 3.4 Reference grid search, showing how the `minabsint` cost function varies with the pulse width `P1`. *Data code:* `6F-200826`

several tricks are used: each FE is run using no dummy scans and only one scan, and the acquisition time is set to just 1.1 μs (high resolution is not required for a reliable cost function value), and the recovery delay `D1` is set to 0. In practice, there is an extra gap of around 5 s between successive FEs due to spectrometer initialisation, so an extra recovery delay is not needed.

3.4.1.2 The Competition

The performance of POISE can be compared against two 'competitors' in this area. The traditional method of determining the 90° pulse width is to measure a pulse width array (colloquially, 'to array the pulse width') [45]. This entails measuring a series of pulse–acquire spectra, over which τ_p is evenly incremented: in optimisation parlance this would be called a *grid search.*[10] This leads to a sinusoidal pattern in the peak intensities, from which the 360° null can be directly read off. An example of this is shown in Fig. 3.5, where the 360° null at $\tau_p \approx 48$ μs is visible (it is never a *perfect* null because of off-resonance effects and ω_1, i.e. B_1, inhomogeneity).

TopSpin provides a built-in mechanism for measuring a pulse width array using the `popt` command. The pulse–acquire spectrum is measured for pulse widths between a lower and upper bound, and the user specifies a spectral region of interest which `popt`

[10] This differs from the *reference grid search* above only in spirit (hence the similar names). The process is exactly the same, but the grid search is used to actually locate the optimum, whereas in the reference grid search my aim is purely to verify whether the faster optimisation algorithms converge to the correct point.

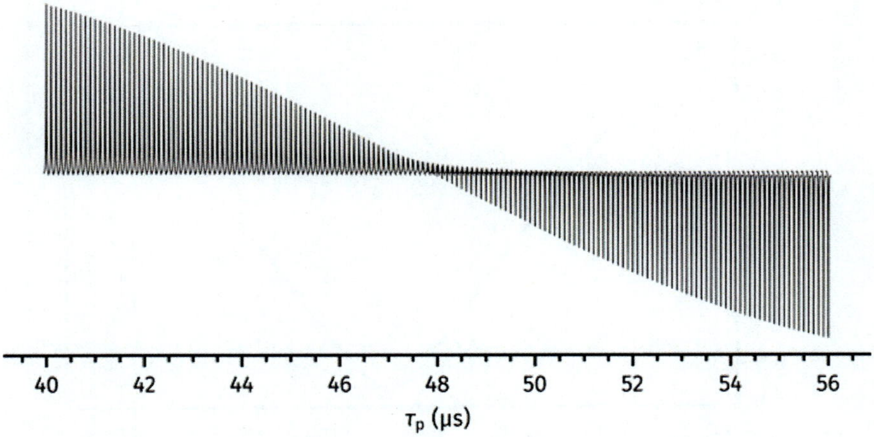

Fig. 3.5 An example of a pulse width array, where the variation of the water peak is monitored with changes in pulse width (this sample is in DMSO-d_6, but is slightly wet). These spectra were acquired manually before plotting; the TopSpin `popt` command would yield essentially identical results. *Data code:* `6F-200826`

uses to determine the null in spectrum intensity.[11] While this usually yields highly accurate results, the acquisition of so many spectra is relatively time-consuming and arguably unnecessary if the only purpose is to determine the null.

A more rapid method for pulse calibration is the nutation experiment of Wu and Otting [46], which allows the 90° pulse width to be determined in a single-scan experiment. In this experiment, an RF pulse with a given power level, corresponding to an (unknown) amplitude of ω_1', is applied during acquisition.[12] Assuming that the pulse is applied along the x-axis, this leads to the following product operators during the acquisition period:

$$I_z \xrightarrow{\omega_1' t_2} I_z \cos(\omega_1' t_2) - I_y \sin(\omega_1' t_2), \tag{3.20}$$

and (since we only detect I_-) an FID of

$$s(t_2) = \frac{1}{2\mathrm{i}} \sin(\omega_1' t_2) = -\frac{1}{4}[\exp(\mathrm{i}\omega_1' t_2) - \exp(-\mathrm{i}\omega_1' t_2)]. \tag{3.21}$$

[11] In fact, `popt` uses the notion of a cost function as well, in that it determines the point where the cost function is minimised. In this case, I set it to use the `MAGMIN` cost function, which seeks to minimise the intensity of the magnitude-mode spectrum; this is essentially identical to the `minabsint` cost function which was used for the POISE optimisations except that it only applies to the spectral region of interest.

[12] To be precise, it is applied for a proportion d of the dwell time between acquisition of successive points in the FID; d is called the *duty cycle* and must be accounted for when calculating the pulse width using this method.

When Fourier transformed, this yields an antiphase doublet where the two peaks are separated by the frequency $2\omega_1'$. Measuring the separation between the two peaks directly yields the unknown amplitude ω_1', from which $\tau_\mathrm{p}' = \pi/(2\omega_1')$ can be calculated. Typically, the RF amplitude ω_1' is rather smaller than the amplitude ω_1 which we would like to apply the hard pulse at, and thus $\tau_\mathrm{p}' > \tau_\mathrm{p}$. However, this can be adjusted for using the power levels applied (which are known to the user).

Although the nutation experiment can be performed extremely quickly using the TopSpin `pulsecal` command, often only requiring a few seconds, the pulse widths calculated are generally slightly shorter compared to the value obtained from a 360° null. This is a known effect which arises because the separation is calculated from the top of the peaks, which correspond to the most homogeneous region of the B_1 profile [46]. On the other hand, the 360° null (as measured through `popt` or POISE, for example) measures a signal which is averaged over the entire B_1 profile.

3.4.1.3 Optimisation Results

Compared to these two existing methods, we expect POISE to be faster than the `popt` grid search, and also more accurate than the nutation experiment in `pulsecal`. This is borne out in practice (Table 3.1). `popt` yields an optimum of 48.4 μs, which is closely matched by POISE. However, across all five optimisation runs performed for each algorithm, POISE locates this optimum using far fewer FEs (and far less time) because its algorithms are more efficient than a simple grid search. While the `pulsecal` routine is even faster than POISE,[13] it underestimates the 90° pulse width by about 4%. In this particular case, POISE is the only option which strikes a useful balance between speed and accuracy. These results also provide the first evidence that Py-BOBYQA is generally faster than the simplex-based methods: this observation is consistently reproduced in many of the other optimisations in this chapter.

3.4.1.4 Different Initial Points

One question we might reasonably ask is how robust POISE is towards poor initial guesses. In the case of the pulse width calibration, the answer is *very* robust. Tables 3.2 and 3.3 summarise the results obtained with an initial guess of 43 μs and 53 μs respectively. There is slightly decreased performance in that a few more FEs are required for convergence, but the accuracy of the result is unchanged.

It is tempting to use this example to draw the conclusion that the initial point does not matter in POISE optimisations. However, this is only really true for a simple optimisation like this. Looking again at the reference grid search in Fig. 3.4, it is

[13] `pulsecal` could be yet faster if it was instructed to not optimise the receiver gain prior to performing the nutation experiment.

Table 3.1 Comparison of methods for 360° pulse width determination. popt grid searches were run between values of 40 μs and 56 μs, with a linear increment of 0.4 μs (which, through interpolation, provides a precision of approximately 0.2 μs in the result, matching the tolerance used for POISE). pulsecal was run as normal and the reported pulse width multiplied by 4 to obtain the 360° pulse width for comparison. POISE optimisations were run according to the routine in Table 3.1. *Data code:* 6F-200826

Entry	Method	Optimum found (s)	FEs	Time taken (s)
1	popt	48.40	41	299
2	pulsecal	46.64	–	37
3	POISE (NM)	48.38	10	76–79
4	POISE (MDS)	48.38	10	77–80
5	POISE (BOBYQA)	48.29–48.41	6–7	46–54

Table 3.2 Pulse width optimisations with an initial point of 43 μs. The POISE routine is the same as in Table 3.1, except with "init": [43.0]. *Data code:* 6F-200826

Entry	Method	Optimum found (s)	FEs	Time taken (s)
1	POISE (NM)	48.38	14	109–114
2	POISE (MDS)	48.38	14	108–112
3	POISE (BOBYQA)	48.27–48.33	9	70

Table 3.3 Pulse width optimisations with an initial point of 53 μs. The POISE routine is the same as in Table 3.1, except with "init": [53.0]. *Data code:* 6F-200826

Entry	Method	Optimum found (s)	FEs	Time taken (s)
1	POISE (NM)	48.38	14	110–114
2	POISE (MDS)	48.25–48.38	16	123–126
3	POISE (BOBYQA)	48.26–48.33	9	69–70

clear that there is no other possible minimum that the optimiser could converge to. Furthermore, the noise in the cost function is almost indiscernible. These represent the *ideal* conditions for an experimental optimisation to work, and it is not surprising that extremely good performance is obtained with POISE. Some of the subsequent examples include more difficult or more noisy cost functions. We will see that POISE does indeed have *some* tolerance towards poor initial points, even in the presence of noise (after all, this is the entire purpose of using derivative-free algorithms). However, for very challenging optimisations (especially ones with many parameters) it is very likely that the optimisation will ultimately converge to a local minimum close to the initial point, which may not necessarily yield a substantial reduction in the cost function.

3.4.2 *Ernst Angle Optimisation*

Often, in a simple 1D pulse–acquire spectrum it is not hugely important to know the exact 90° pulse width: instead, it is more valuable to optimise the sensitivity per unit time of the spectrum. Before launching straight into how this may be obtained through optimisation, it is instructive to first consider which parameters are worth optimising. For a pulse–acquire experiment (Fig. 3.6), the repetition time τ_r is the sum of the acquisition time `AQ` plus the recovery delay `D1`; the flip angle θ is controlled via the pulse width `P1`. We assume that the experiment has been repeated enough times to reach a *steady state*, that is, the amount of z-magnetisation prior to the excitation pulse (point ①) is a constant, $M_{z,\mathrm{ss}}$. Application of the excitation pulse leads to a signal scaling as $M_{z,\mathrm{ss}} \sin\theta$, and residual (unexcited) longitudinal magnetisation of $M_{z,\mathrm{ss}} \cos\theta$ at point ②. After the repetition time τ_r (point ③), it can be shown using the Bloch equations [47] that the z-magnetisation recovers to

$$M_{z,0}(1-c) + cM_{z,\mathrm{ss}} \cos\theta, \tag{3.22}$$

where $c = \exp(-\tau_r/T_1)$ and $M_{z,0}$ is the initial, equilibrium z-magnetisation (before the experiment begins). Since the experiment has reached a steady state, points ① and ③ are equivalent: thus, we have that

$$M_{z,0}(1-c) + cM_{z,\mathrm{ss}} \cos\theta = M_{z,\mathrm{ss}}, \tag{3.23}$$

which can be rearranged to give

$$\frac{M_{z,\mathrm{ss}}}{M_{z,0}} = \frac{1-c}{1-c\cos\theta}. \tag{3.24}$$

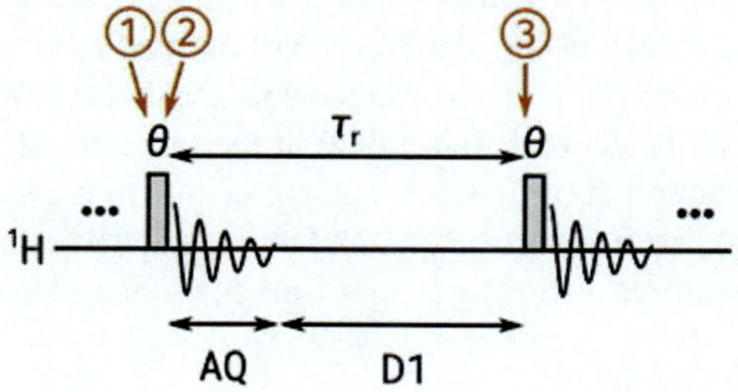

Fig. 3.6 Steady-state pulse–acquire experiment. The excitation flip angle is θ, and the repetition time between experiments is τ_r

The signal amplitude s therefore scales as

$$s = M_{z,\text{ss}} \sin\theta \propto \frac{1-c}{1-c\cos\theta} \cdot \sin\theta, \tag{3.25}$$

and is maximised when $\mathrm{d}s/\mathrm{d}\theta = 0$, the solution of which is the celebrated *Ernst angle* [48]:

$$\theta_\text{E} = \arccos c = \arccos\left[\exp\left(-\frac{\tau_\text{r}}{T_1}\right)\right]. \tag{3.26}$$

In general, T_1 and hence θ_E varies across the different spins in a given sample, so some degree of compromise is required in order to maximise sensitivity for all peaks.

To begin with, we may consider fixing τ_r and optimising `P1` to locate the Ernst angle (or to be precise, the pulse width which corresponds to the Ernst angle). This is by no means an illogical proposition. However, we can go one step further, because τ_r itself is comprised of two parameters, and the sensitivity *per unit time* may be affected by varying τ_r. Since the signal scales as $1/\tau_\text{r}$ (a shorter τ_r means more repetitions per unit time) but the noise scales only as $\sqrt{1/\tau_\text{r}}$, the sensitivity per unit time is

$$S = \frac{(1-c)\sin\theta}{(1-c\cos\theta)\sqrt{\tau_\text{r}}}. \tag{3.27}$$

Assuming that θ is always set to the respective Ernst angle for different τ_r, it can be shown that the best sensitivity per unit time is attained when $\tau_\text{r} \to 0$ [49, 50]. Of course, this limit is not physically possible: τ_r comprises the acquisition time which must be nonzero. However, it does imply that `AQ` should be kept as short as possible, and `D1` set to zero, as shown in Fig. 3.7.

3.4.2.1 Optimisation Setup

Knowing this, we can then set up a meaningful optimisation routine. We seek to optimise the pulse width `P1` such that the intensity of the real part of the spectrum is maximised (corresponding to a `maxrealint` cost function). In practice, I took the extra step of calibrating the 90° pulse width (as per the previous section) and modifying the pulse program such that the flip angle could be specified as the parameter `CNST20`: this is not generally necessary and is only useful for evaluating the accuracy of the results, as will be shown later. `AQ` and `D1` were set to be 1.2 s and 0 respectively, in accordance with the theory outlined above. The number of scans (`NS`) can be set to 1, but unlike in the pulse width calibration (Sect. 3.4.1), we must use enough dummy scans (`DS`) to ensure that a steady-state signal intensity is recorded: in practice I set `DS = 4`. This means that each FE, and thus the overall optimisation, requires a slightly longer time than the pulse width calibrations previously shown.

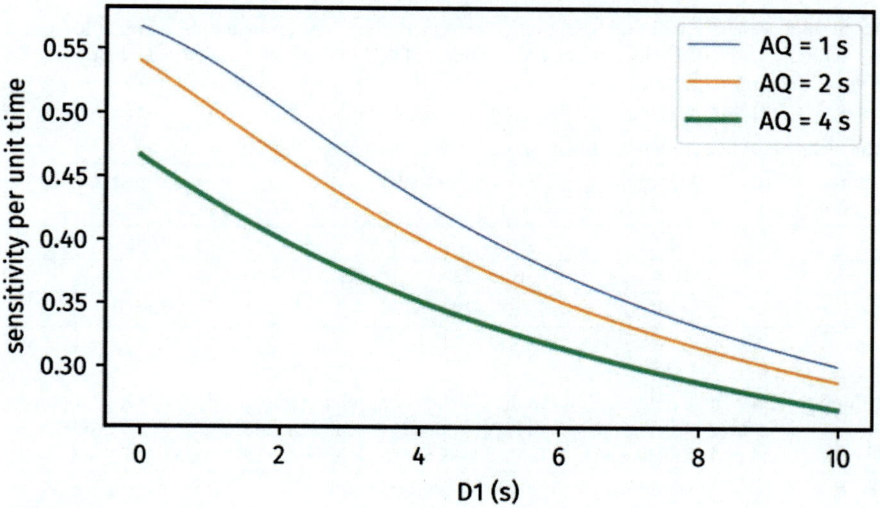

Fig. 3.7 Simulated sensitivity per unit time as a function of AQ and D1 (as given by Eq. (3.27)), assuming that an Ernst angle excitation pulse is used. T_1 was set to 1.5 s

3.4.2.2 Optimisation Results

The Ernst angle optimisation was run with two different spectral regions of interest: firstly, on all the aromatic and olefinic peaks in the sample of ferulic acid, and secondly, on only a single peak at 6.79 ppm. (This spectral region can be selected using the TopSpin `dpl` command, which stores the bounds using the parameters `F1P` and `F2P`: all built-in cost functions in POISE respect these two parameters.)

To assess the accuracy of the optima found, rather than performing a reference grid search as in Sect. 3.4.1, I measured T_1 values for each of these peaks using a typical gradient-enhanced inversion–recovery experiment. From these, the theoretical Ernst angles for each peak could be calculated (Table 3.4): they range from 60° to 73°.

Table 3.4 T_1 and corresponding Ernst angles for each peak in ferulic acid, calculated for a repetition time of 1.20 s (peak assignments can be found in table A.3). The values of $T_1 \ln 2$, measured via a full pseudo-2D inversion–recovery experiment, are also provided here in anticipation of the optimisations in Sect. 3.4.3. *Data code:* `5F-210619`

Peak	^{1}H chemical shift (ppm)	T_1 (s)	θ_E (°)	$T_1 \ln 2$ (s)
1	7.49	1.750	59.8	1.213
2	7.27	0.977	73.0	0.677
3	7.08	1.279	67.0	0.887
4	6.79	1.615	61.6	1.119
5	6.36	1.415	64.6	0.981

Table 3.5 Ernst angle optimisation, performed on all aromatic and olefinic peaks in ferulic acid (between 6 and 8 ppm). The POISE routine used here is: `{"name": "ernst","pars": ["cnst20"], "lb": [10.0], "ub": [90.0], "init": [30.0], "tol": [3.0],"cf": "maxrealint", "au": "poise_id"}`.
The mean of all five theoretical Ernst angles is 65.2. *Data code:* `5F-210619`

Entry	Algorithm	Optimum found (°)	FEs	Time taken (s)
1	NM	67.5–73.1	9–13	91–132
2	MDS	67.5–73.1	9	90–92
3	BOBYQA	70.1–70.7	7	70–71

Table 3.6 Ernst angle optimisations on the peak at 6.79 ppm in ferulic acid. The POISE routine is the same as in Table 3.5, but the spectral region under optimisation was set to be 6.71–6.87 ppm. The theoretical optimum, as given in Table 3.4, is 61.6°. *Data code:* `5F-210619`

Entry	Algorithm	Optimum found (°)	FEs	Time taken (s)
1	NM	60.0–67.5	9–11	91–111
2	MDS	65.6–67.5	11	110–111
3	BOBYQA	60.0–65.2	6–7	59–71

Generally, these optimisations could be completed in under two minutes. The optima found for these two optimisations are different: this is because the former searches for a compromise Ernst angle which balances T_1 of all peaks within the region (Table 3.5), whereas the latter optimises only for one T_1 (Table 3.6). At first glance, the results of the first optimisation appear to be biased towards the upper end of this range. However, this can be rationalised by the fact that a flip angle larger than θ_E is less detrimental to sensitivity compared to one that is smaller (this can be seen by plotting Eq. 3.25). On the other hand, the second optimisation yields an accurate value for the relevant peak (61.6°, peak number 4 in Table 3.4).

Although these results are in line with expectations, it is worth considering whether this optimisation is truly worth it. ^{1}H pulse–acquire spectra already have a very high intrinsic sensitivity, and in the two minutes taken to optimise the flip angle, one could easily just acquire (around) 64 more scans, at which point knowledge of the Ernst angle would cease to be useful. It *may* be more useful for nuclei which have lower sensitivity, such as ^{13}C: I did not evaluate this possibility. However, it must be borne in mind that a low-sensitivity experiment will also require more scans per FE, which leads to a corresponding increase in the optimisation time.

In my estimation, a more useful application of this optimisation routine would be to use it to determine an average T_1 value for a group of peaks. This could then be used to inform the choice of recovery delay for multidimensional experiments

[51, 52] or quantitative NMR experiments [53, 54]. It is, however, possible to more directly obtain T_1 values from an inversion–recovery experiment, which I describe next.

3.4.3 *Inversion–Recovery*

3.4.3.1 Optimisation Setup

If what we really want to measure is T_1 for a particular peak (or a set of peaks), a more direct way is to perform an inversion–recovery experiment (Fig. 3.8). This can either be recorded in a 2D form where the τ delay is incremented and the resulting intensities fit to an exponential curve, or in an iterative fashion by searching for the null in spectral intensity, which occurs at $\tau = T_1 \ln 2$: this latter option is particularly suited to optimisation. The delay τ was specified as the parameter `D27`, and the `zerorealint` cost function used here is just the absolute value of the integral over the region of interest: an ideal null would have a cost function value of 0. Table 3.4 provides the theoretical values of $T_1 \ln 2$, which were calculated using the more accurate 2D fitting procedure.

A slight drawback of using this method, compared to the Ernst angle optimisations in the previous section, is that the recovery delay must be sufficiently long to allow for complete relaxation between FEs: in this case, I used a `D1` of 5 s. This is compensated for by the fact that dummy scans are no longer needed to reach a steady state; I therefore set `DS=0` and `NS=1`.

3.4.3.2 Optimisation Results

Just as in the Ernst angle optimisations, two different types of optimisations were performed: one on the entire 6–8 ppm region, and one on just a single peak (note that here, the chosen peak is at 7.08 ppm, or peak 3 in Table 3.4). The results are shown in Tables 3.7 and 3.8. The optimisation over the entire aromatic region correctly converges to an 'averaged' value of $T_1 \ln 2$ over the five peaks (the mean value for the five peaks is 0.975 s). The second optimisation performed only on a single peak is no less accurate, yielding an optimum which closely matches the theoretical value for the peak in question (0.887 s).

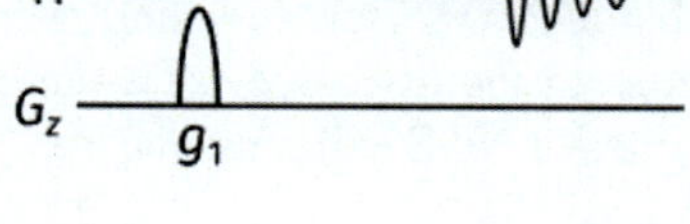

Fig. 3.8 Inversion–recovery pulse sequence

Table 3.7 Inversion–recovery optimisations on all aromatic and olefinic peaks in ferulic acid (6 ppm to 8 ppm). The POISE routine used here is: `{"name": "invrec","pars": ["d27"], "lb": [0.35], "ub": [1.75], "init": [0.6], "tol": [0.01],"cf": "zerorealint", "au": "poise_1d"}`. The mean of $T_1 \ln 2$ values for all five peaks is 0.975 s. *Data code:* `5F-210619`

Entry	Algorithm	Optimum found (s)	FEs	Time taken (s)
1	NM	0.938–0.969	14–16	204–235
2	MDS	0.956–0.975	16	233–235
3	BOBYQA	0.953–0.971	9–11	130–160

Table 3.8 Inversion–recovery optimisations on the peak at 7.08 ppm in ferulic acid. The POISE routine is the same as in Table 3.5, but the spectral region under optimisation was set to be 7.02–7.15 ppm. The theoretical optimum (from Table 3.4) is 0.887 s. *Data code:* `5F-210619`

Entry	Algorithm	Optimum found (s)	FEs	Time taken (s)
1	NM	0.863–0.875	14	202–205
2	MDS	0.863–0.869	14	203–204
3	BOBYQA	0.862–0.873	9–10	128–145

The only downside of these optimisations is the time required, which is on the order of 2–4 min. Although this is less time than required for a full pseudo-2D inversion–recovery experiment, POISE has the drawback that an optimisation can only be run on one peak at a time. Thus, if the aim is to determine T_1 for all peaks, then the 2D experiment may well end up being faster. On top of that, there are many other ways to measure T_1 which are faster than a full 2D inversion–recovery experiment and almost certainly also faster than POISE [55–59]. However, no explicit comparisons were performed in this work.

3.4.4 NOE Mixing Time

In all of the optimisations so far, the cost functions used are relatively simple, simply seeking to maximise or minimise some intensity. In fact, the Bruker `popt` interface does come with a number of cost functions itself, which can be used for grid search-based optimisations: so, in principle, all the previous examples could have been done with `popt` (albeit with a much longer time). However, POISE goes beyond this in that it allows users to define their own cost functions In this section, we exploit this customisability to devise a more complicated cost function for optimising mixing times in NOE experiments.

3.4.4.1 Optimisation Setup

The ideal NOE mixing time for a given compound depends on the rates of various relaxation processes: too short a mixing time does not allow for sufficient buildup of the NOE, but too long a mixing time leads to loss of signal through relaxation. In this section, I do not deal with this theoretically: the optimisation process is merely used to find the empirical best value (for the sample under study).

In the chosen sample, 3-fluorophenylboronic acid, there are four crosspeaks of interest in the 2D NOESY spectrum. I first performed a reference grid search to determine where the crosspeak intensities were maximised (Fig. 3.9). Generally, a broad minimum between 2.5 and 4 s is observed: any result within this range should be considered as 'correct'. The relatively large value for the optimal mixing time reflects the small size of the molecule used in this study (it has a molar mass of 139.92 g mol^{-1}), which corresponds to rapid tumbling and long T_1 values.[14] Although the broad minimum may at first glance seem imprecise, it merely reflects the underlying physical characteristics of the sample under study: an optimisation process cannot 'discover' extra precision where there is none to be found. We also see here the first example of a cost function where the noise is significant: this provides a good test of the derivative-free algorithms used in POISE.

While the 2D reference grid search offers first-hand insight into what our target optimum should be, it is unwise to run an optimisation using the full 2D experiment, simply because of the time required for each FE. A more sensible method is to use a selective 1D NOESY experiment. Although this is much faster, it does come with two caveats:

1. The frequency for selective irradiation must be first chosen, likely after acquisition of a 1D ^{1}H spectrum. Thus, the optimisation does require some *a priori* knowledge of the system being studied.
2. The peaks observed in the 1D NOESY *must* be sufficiently representative of those in the full 2D NOESY.

As for the cost function, it must be able to pick out only the peaks from the 1D NOESY spectrum which correspond to genuine NOE transfer, rejecting (for example) the selectively irradiated peak as well as other artefacts. The cost function used here (`noe_1d`) does this by reading the frequency used for the selective irradiation from the `SPOFFS2` parameter, excising a region of ca. 100 Hz around this irradiation frequency, and integrating the remainder of the spectrum. In order to account for the fact that the desired peaks may be either positive or negative, the absolute value of the integral is taken, and the negative of this is used as the cost function (since we

[14] There is a complicating factor in that the use of such long mixing times also leads to a noticeable increase in the experiment duration. As such, it may be more prudent to consider the sensitivity *per unit time*. In the optimisations which follow, I have neglected this issue: the 2× increase in sensitivity demonstrated below would only be cancelled out by a 4× increase in time, which is not the case here. However, to be more rigorous, this factor could be accounted for by modifying the form of the cost function.

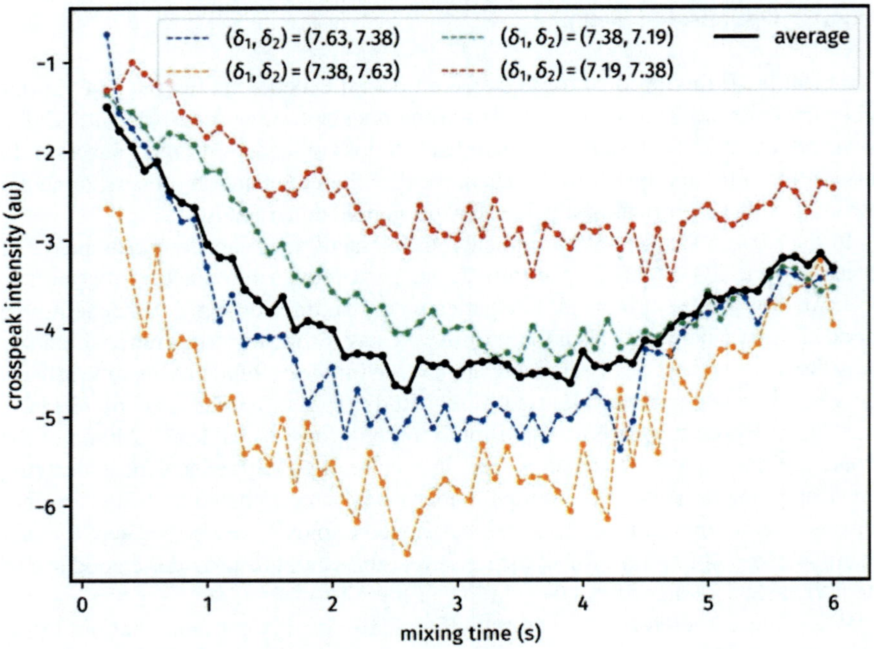

Fig. 3.9 Reference grid search of 2D NOE crosspeak intensities. The 2D NOESY spectra themselves are shown in Fig. 3.11. The individual crosspeak intensities are shown in dashed lines; the solid black line is the average of these four, which represents the quantity we seek to optimise. *Data code:* `7B-200725`

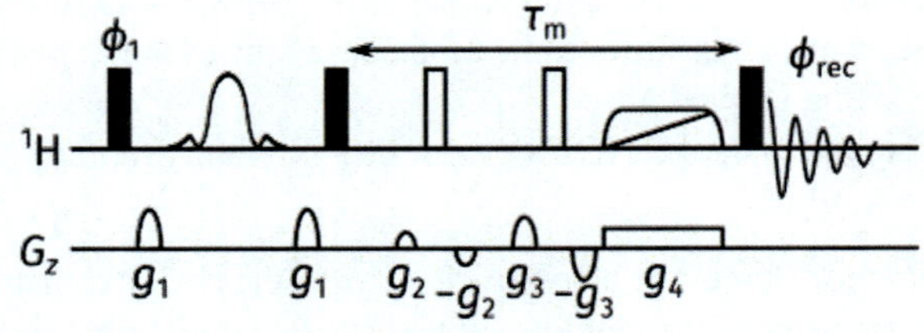

Fig. 3.10 Selective 1D NOESY pulse sequence used for POISE optimisations. τ_m denotes the NOE mixing time. Phase cycling was performed using $\phi_1 = \phi_{rec} = (x, -x)$. Gradient amplitudes were set as follows: $(g_1, g_2, g_3, g_4 = 17, 40, 25, 10\%)$

seek to maximise the intensity). Note that if a different 1D NOESY pulse program is used with different parameter definitions, then the cost function must be adjusted accordingly.

In the event, I used a modified 1D NOESY pulse program (Fig. 3.10), with two extra inversion pulses inserted during the mixing time: these minimise artefacts arising from relaxation during the mixing time, which can be especially problematic for long mixing times of several seconds. (In principle, these artefacts have equal positive and negative components and thus should not contribute to the cost function except in terms of noise; however, it is always a good idea to minimise the noise in the cost function as much as possible.) The mixing time is represented by the parameter `D8`. The initial value for the optimisation was set to 0.5 s, which is a reasonable compromise value for most 'small' organic molecules.

3.4.4.2 Optimisation Results

The results of this optimisation are shown in Table 3.9. The peak at 7.38 ppm was used for selective irradiation. In all cases, the optimisations converged to the correct region within 2.5 min (for BOBYQA) or 5 min (for the simplex-based algorithms). The resulting 2D NOESY spectra, with the initial and optimised mixing times of 0.5 s and 3.5 s respectively, are shown in Fig. 3.11, where the improvement in crosspeak sensitivity is clearly visible: in this case, an approximate 2× enhancement is obtained.

Table 3.9 NOE mixing time optimisations on a sample of 3-fluorophenylboronic acid. The POISE routine used here is: `{"name": "1dnoe", "pars": ["d8"], "lb": [0.2], "ub": [6.0], "init": [0.5], "tol": [0.1], "cf": "noe_1d", "au": "poise_1d"}`. *Data code:* `7B-200721`

Entry	Algorithm	Optimum found (s)	FEs	Time taken (s)
1	NM	3.25–3.88	16–18	268–312
2	MDS	3.63–3.75	16–18	269–305
3	BOBYQA	3.38–3.80	6–10	88–162

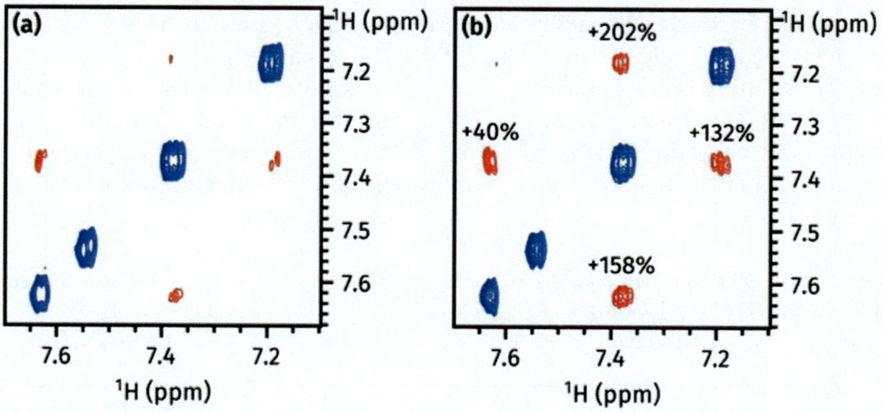

Fig. 3.11 2D NOESY spectra of 3-fluorophenylboronic acid, obtained **a** before and **b** after optimising the mixing time on the 1D NOESY sequence in Fig. 3.10. The mixing times used were 0.5 and 3.5 s respectively. Both spectra are plotted with the same contour levels; sensitivity increases for each peak are indicated as percentages relative to the unoptimised spectrum. *Data code:* `7B-200725`

It should be mentioned here that each FE was run using one dummy scan and two scans. This was only made possible by the high SNR afforded by a concentrated sample (120 mM), as well as a cryogenic probe. For more dilute samples where SNR is insufficient, the POISE optimisation will require more scans per FE and consequently will take longer. However, it can be argued that the *benefit* reaped from the optimisation is also larger, since the final (optimised) 2D NOESY would also take a correspondingly longer time.

3.4.4.3 A Different Sample

To more clearly illustrate the *sample-specific* nature of POISE optimisations, I also performed the same optimisation on a different sample: in this case, the decapeptide gramicidin S. Although I did not perform a reference grid search here, for this rather larger compound, we would expect the optimisation to converge to a shorter mixing time; this is indeed observed in practice (Table 3.10). The peak at 4.76 ppm was used for selective irradiation: this is the H^{α} proton of the ornithine residue. In this case, the sample is rather less concentrated, and I used `DS=2` and `NS=4`; this is still a relatively small number of scans, but was sufficient to yield reproducible results.

3.4.5 ASAP-HSQC Excitation Delay

The next example of a POISE optimisation is the ASAP-HSQC experiment [60, 61]: here, I use it to again illustrate the new possibilities which custom cost functions enable. In the ASAP-HSQC experiment (Fig. 3.12), an HSQC spectrum is recorded using only ^{13}C–bound proton magnetisation, and the ^{12}C–bound ('bulk') proton magnetisation is returned to the equilibrium $+z$ axis (i.e. the I_z state). In this way, instead of having a conventional recovery delay, the ^{13}C–^{1}H magnetisation can be directly replenished using isotropic mixing, which causes transfer of z-magnetisation from bulk protons. The elision of the recovery delay from the sequence thus leads to significantly shorter experiment durations: in one of the more extreme examples, an HSQC spectrum could be recorded within 7 s. This separation of different 'magnetisation

Table 3.10 NOE mixing time optimisations on a sample of gramicidin S. The POISE routine used here is identical to before. *Data code:* `7G-210815`

Entry	Algorithm	Optimum found (s)	FEs	Time taken (s)
1	NM	0.63–0.78	9–14	253–384
2	MDS	0.44–0.75	9–11	254–309
3	BOBYQA	0.58–0.77	5–8	136–223

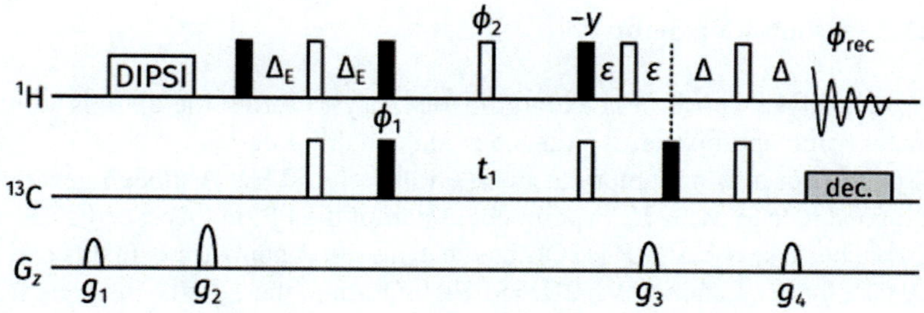

Fig. 3.12 ASAP-HSQC pulse sequence. Phase cycling is performed using $\phi_1 = (x, -x)$, $\phi_2 = (x, x, -x, -x)$, and $\phi_{\text{rec}} = (x, -x)$. Gradient amplitudes are as follows: $g_1 = 33\%$, $g_2 = 43\%$, and for echo–antiecho selection, $(g_3, g_4) = (80, 59.9\%)$ and $(63.9, 80\%)$ respectively. The delay Δ is set to $1/(4 \cdot {}^1J_{\text{CH}})$; the delay Δ_{E} was optimised as described in the text. The BIBOP, BEBOP and BURBOP optimal control pulses [12, 14, 15] were used for ^{1}H 180° pulses and all ^{13}C pulses; for more details, refer to the original paper by Luy and coworkers [60]

pools' is conceptually very similar to that in NOAH experiments (Sect. 4.1.2), and in anticipation of that, I use the notation $^1\text{H}^{\text{C}}$ and $^1\text{H}^{!\text{C}}$ to represent magnetisation belonging to protons coupled and not coupled to ^{13}C respectively.

By modifying the length of the delay Δ_{E} in the initial INEPT block, it is possible to adjust the proportion of the $^1\text{H}^{\text{C}}$ magnetisation excited during the sequence: the remainder is stored along z. The delay Δ_{E} can be related to an *effective flip angle*, θ_{eff}, through

$$\theta_{\text{eff}} = 2\pi J \Delta_{\text{E}}, \tag{3.28}$$

where J is here short for $^1J_{\text{CH}}$. This partial excitation means that the signal is decreased by a factor of $\sin\theta_{\text{eff}}$, but means that on the next scan or increment, there is a larger pool of $^1\text{H}^{\text{C}}$ magnetisation to start from. The combination of these factors means that there is an optimum Δ_{E} which yields the greatest steady-state signal. Equivalently, we can define an 'effective J-coupling', J_{eff}, for which the INEPT delay is optimised:

$$\Delta_{\text{E}} = \frac{1}{4J_{\text{eff}}} \iff J_{\text{eff}} = \frac{1}{4\Delta_{\text{E}}} = \frac{\pi J}{2\theta_{\text{eff}}}, \tag{3.29}$$

and we can search for the ideal J_{eff} to maximise the sensitivity of the ASAP-HSQC spectrum.

Notice that this is entirely analogous to the Ernst angle previously discussed in Sect. 3.4.2. However, here, the magnetisation is being recovered partly through isotropic mixing and partly through relaxation during the FID. This makes an analytic description of the ASAP-HSQC experiment rather more complicated (although it has been studied before [62]). Directly optimising the excitation delay Δ_{E} with POISE, which circumvents the theory entirely, is therefore an attractive option for practical day-to-day use.

3.4.5.1 Optimisation Setup

Given the analysis above, it is extraordinarily easy to set up the optimisation: the parameter being optimised is J_{eff} (encoded in the pulse program as `CNST3`), and the initial point chosen is a 'compromise' $^1J_{\text{CH}}$ value of 150 Hz. Although generally it is not a good idea to use a 2D experiment as part of the FE, in this specific case it is acceptable because the ASAP-HSQC has such a short running time. In this case, we simply take the projection of the 2D ASAP-HSQC onto the F_2 axis, integrate it, and take the negative of that to obtain the cost function.

A reference grid search was performed so that I could later verify the optimisation results (Fig. 3.13). The exact point where the crosspeak intensities are maximised is not obvious: generally, the signal increases up until $J_{\text{eff}} \approx 230$ Hz, after which it plateaus off. In a similar fashion to the NOE optimisations (Sect. 3.4.4), we may therefore consider any value above this to be 'correct'.

Optimisations were run using a 'low-quality' ASAP-HSQC spectrum with only 32 t_1 increments to minimise the time required. A recovery delay of 0.1 s was used.

3.4.5.2 Optimisation Results

The results from the POISE optimisations are collated in Table 3.11. As can be seen, even though the FE involves acquisition of a 2D spectrum, the optimisation

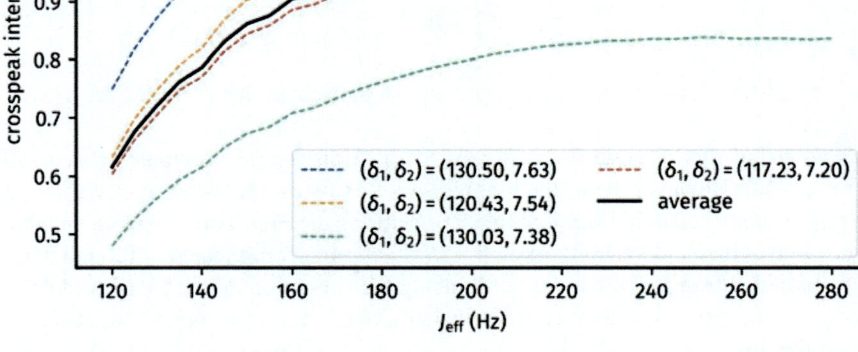

Fig. 3.13 Reference grid search showing how the ASAP-HSQC signal intensity for the 3-fluorophenylboronic acid sample varies with J_{eff}. *Data code:* `7B-200722`

Table 3.11 ASAP-HSQC INEPT delay optimisations. The POISE routine used was: `{"name": "asaphsqc", "pars": ["cnst3"], "lb": [120.0], "ub": [280.0], "init": [150.0], "tol": [10.0], "cf": "asaphsqc", "au": "poise_2d"}`.
Data code: `7B-200722`

Entry	Algorithm	Optimum found (Hz)	FEs	Time taken (s)
1	NM	250.0–256.3	8–9	169–195
2	MDS	237.5–243.8	8	171–179
3	BOBYQA	229.8–245.6	4–7	114–157

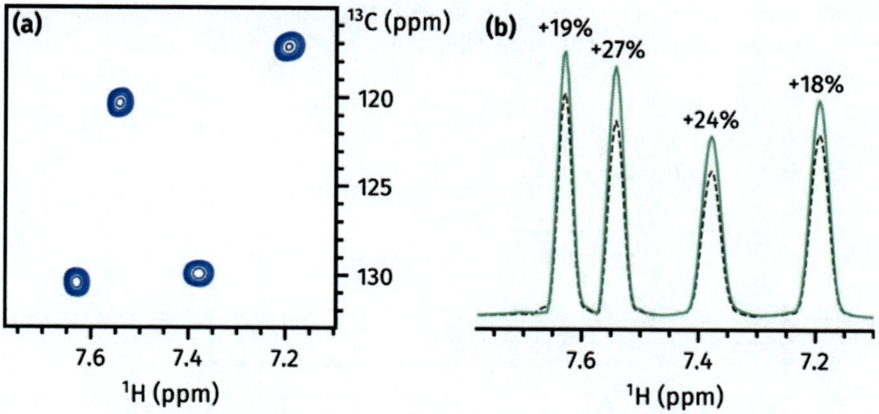

Fig. 3.14 **a** ASAP-HSQC spectrum of the 3-fluorophenylboronic acid sample. **b** Projections of the ASAP-HSQC spectra before (grey dotted line, using $J_{\text{eff}} = 150\,\text{Hz}$) and after (green solid line, using $J_{\text{eff}} = 245$ Hz) optimisation of the INEPT delay. Sensitivity increases for each peak are indicated as percentages relative to the unoptimised spectrum *Data code:* `7B-200722`

itself completes in just 2–3 min, correctly converging to J_{eff} values in the region of around 240 Hz. When re-evaluated on a 'full' ASAP-HSQC experiment with 64 t_1 increments (Fig. 3.14a),[15] the optimisation yielded an 18–27% improvement in the sensitivity of the ASAP-HSQC spectrum, as shown in Fig. 3.14b.

In fact, this specific example is not particularly impressive in itself. The ASAP-HSQC experiment shown in Fig. 3.14a takes only around 45 s to acquire, so simply repeating it two or three more times would yield a similar or even larger increase in SNR. However, this argument may easily be swung in the opposite direction if more t_1 increments were to be used, such as 256 for a more 'typical' ^{13}C spectral window, or even more if extremely high F_1 resolution is desired (the original paper [60] provides an example of a three-hour ASAP-HSQC experiment where 16384 t_1 increments were recorded to differentiate two peaks separated by 3 Hz in F_1).

[15] The ^{13}C resonances of interest in this compound fall within a very small window, so relatively few t_1 increments are needed.

3.4.6 Ultrafast NMR

The final example of a single-parameter POISE optimisation is also the most complicated: it pertains to the EPSI acquisition technique, which was previously introduced in Sect. 2.7. EPSI acquisition allows signals from different slices of the sample to be simultaneously detected in a single FID, and (in liquid-state NMR) has most famously been used in the context of *ultrafast* 2D experiments [63–69], where t_1 evolution is spatially encoded and read out using the EPSI technique.

During an EPSI acquisition period, alternating gradients of equal magnitude but opposite sign are applied, as shown in Fig. 3.15a and b. Each data point collected therefore depends not only on t_2 (the acquisition time domain) but also on k, a value which measures the extent of gradient dephasing caused by the acquisition gradients:

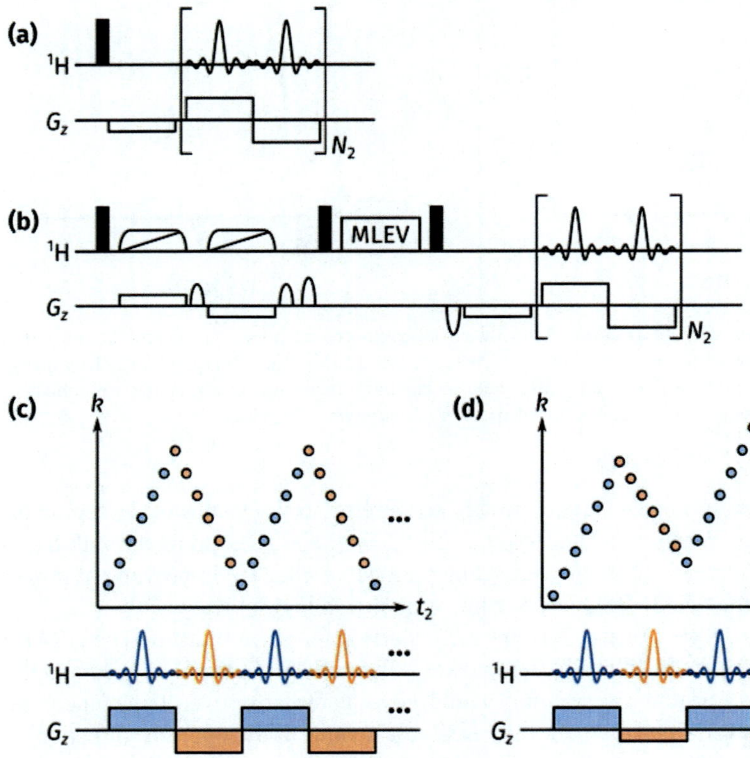

Fig. 3.15 Pulse sequences used for EPSI optimisation, and depiction of the (k, t_2) data matrices thus obtained. **a** Pulse–EPSI experiment. **b** Ultrafast TOCSY experiment; gradients were set up according to the protocol of Giraudeau et al. [68] **c** Illustration of the (k, t_2) data matrix obtained during an EPSI acquisition period. Blue dots indicate data acquired during a positive gradient, and orange dots data acquired during a negative gradient. **d** The (k, t_2) data matrix obtained when the positive and negative EPSI gradients (blue and orange in the pulse sequences) are not balanced. In this case, the negative gradients are slightly weaker, leading to a 'drift' towards larger values of k

$$k = \int_0^{t_2} \gamma G(t)\,\mathrm{d}t. \tag{3.30}$$

This value of k increases during positive gradients and decreases during negative gradients, leading to a zigzag pattern in k-space, all while t_2 is still increasing (Fig. 3.15c). The *prephasing gradient* immediately before acquisition has the same duration as the EPSI gradients, but has half the amplitude, meaning that k begins from a negative value, specifically, $-k_{\text{max}}/2$, where k_{max} is the change in k caused by one complete positive EPSI gradient. The indirect-dimension frequencies F_1 can be immediately read off from the k-domain as F_1 is directly proportional to k: this can also be understood as the EPSI gradients unwinding a phase-modulated 'helix' created by chemical shift evolution during the t_1 period, as has been beautifully illustrated in previous reviews [65, 66].

This rapid alternating of gradients is very demanding on spectrometers, and it is often the case that the positive and negative EPSI gradients—although *nominally* specified with the same amplitude—are not perfectly balanced. This causes a 'drift' in the k-domain over time, as illustrated in Fig. 3.15d. In this section, POISE is used to perform an *instrument-specific* optimisation in order to correct this effect.

3.4.6.1 Optimisation Setup

To measure the drift in k-space, it is easiest to use a pulse sequence for which there is no spatial encoding. The pulse–EPSI experiment (Fig. 3.15a) is perfectly suited for this. In the pulse program, I multiplied the amplitudes of the negative gradients by a factor α, represented as `CNST16` in TopSpin: the objective of POISE is therefore to determine the optimum value for this factor which minimises the drift. Calculating this drift requires fairly substantial processing, which is most easily done inside the cost function using the `numpy` library. The `epsi_gradient_drift` cost function therefore consists of the following steps:

1. reading in the 1D FID and reshaping it into a 2D (k, t_2) matrix;
2. discarding data obtained during negative gradients;
3. determining, for each point in t_2, the value of k for which the maximum signal is found. We denote this by $k' = k'(t_2)$, and for perfectly balanced gradients, we expect that $k' = 0$ for all values of t_2. However, if the gradients are imbalanced, then we generally have that $k' = mt_2 + c$ for some constants m and c;
4. performing linear regression to determine the value of m, which directly represents the extent of k-space drifting. The cost function is then simply given by $|m|$.

Since the cost function constructed here was relatively complicated, I carried out a reference grid search to confirm that it was working as intended (Fig. 3.16). This yielded an expected optimum of $\alpha = 1.0004$. (The sample used here was ferulic acid, but in principle this should not matter as the drift is due to the spectrometer and not the sample.)

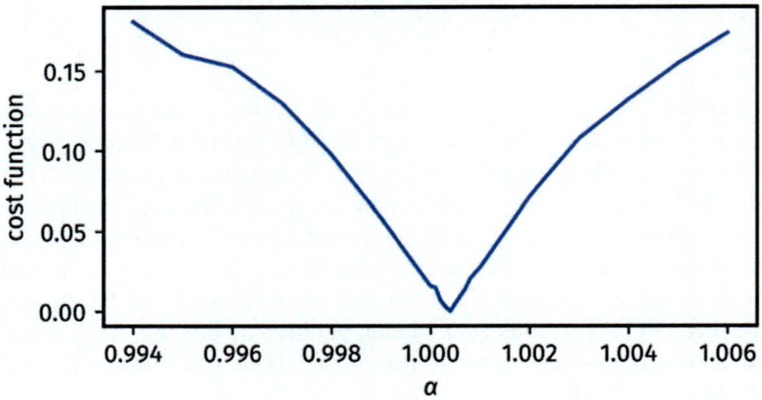

Fig. 3.16 Reference grid search for EPSI optimisation. *Data code:* `6F-210316`

To save time in the optimisation itself, I reused the strategy of setting `DS=0` and `NS=1`. As before, since each FE is separated by approximately five seconds of processing and spectrometer initialisation, this means that there is no need for another recovery delay; thus, `D1` was set to 0.5 s. However, note that such a short recovery delay should *not* be used if `DS` is larger than 0 or `NS` larger than 1, as EPSI acquisitions should not be carried out in rapid succession. Furthermore, `AQ` should be kept under 100 ms (I used 91.8 ms here).

3.4.6.2 Optimisation Results

In the event, all three optimisation algorithms successfully found the optimal value of α in under a minute (Table 3.12). This is *extremely* fast and the resulting value of α can be used for any subsequent ultrafast experiment, including the ultrafast PSYCHE-iDOSY experiment described in Sect. 2.7.

Table 3.12 Results of EPSI gradient imbalance optimisations. Note that in practice, all of these optimisations actually converge to $\alpha = 1.0004$; the extra decimal places are not used by the spectrometer (see text for further discussion). The POISE routine used here is: `{"name": "epsi", "pars": ["cnst16"], "lb": [0.99], "ub": [1.01], "init": [1.0], "tol": [0.0001], "cf": "epsi_gradient_drift", "au": "poise_1d"}`. . *Data code:* `6F-210316`

Entry	Algorithm	Optimum found	FEs	Time taken (s)
1	NM	1.000375	10	49–52
2	MDS	1.000375	10	49–50
3	BOBYQA	1.000383–1.000388	7	34–35

However, it is worth clarifying one subtlety of POISE here. In the table, the optima are quoted to a rather greater precision than one might expect at first: however, the actual precision of the optimisation result is not given by the number of significant figures,[16] but rather a combination of two factors:

- first, the *optimisation tolerance*, which is purely determined by the optimisation routine. Thus, in Table 3.12, the optimum of 1.000375 actually means 1.000375 ± 0.0001 (and not 1.000375 ± 0.0000005, as 'significant figures' would imply); and
- second, the available *spectrometer resolution*, which represents how precisely the spectrometer can implement a given parameter value. In this case, the available resolution is 0.0001, which means that even if we input the 'optimum' of $\alpha = 1.000375$, the experiment is executed with $\alpha = 1.0004$ (see also Fig. 3.17). The optimisation algorithm has no knowledge of this, so it still reports a value of 1.000375. **When reporting optima from POISE, users should be wary of any rounding performed by the spectrometer.**

This in fact further underlines the importance of choosing a sensible optimisation tolerance: it should be *at least* as large as the instrument resolution (larger values can be used when an extremely accurate result is not needed).

Returning to the current optimisation, the (k, t_2) data matrices obtained from the excitation–EPSI pulse sequence are shown in Fig. 3.17a and b. In the former, where the default value of $\alpha = 1$ is used, there is a noticeable drift in k-space. This is removed in the optimised experiment with $\alpha = 1.0004$. The optimised value of α can then be transferred to other ultrafast experiments, such as the ultrafast TOCSY (Fig. 3.15b). The improvements thus obtained are shown in Fig. 3.17c and d: in the optimised spectrum, substantial improvements in lineshapes are obtained, even allowing some couplings to be properly resolved in the indirect dimension. In principle, this k-drift could be compensated for by shearing of the (k, t_2) data matrix prior to Fourier transformation. However, similar improvements in lineshapes could not be accomplished using any existing processing scripts, including those in the protocol of Giraudeau et al. [68]; I did not attempt to manually shear the data myself either.

3.4.7 HMBC Low-Pass J-Filter

In this and the remaining examples, we look at optimisations involving *multiple parameters*, one of the key strengths of POISE. Grid searches are extremely ineffective for this purpose, as the number of FEs required scales exponentially with the number of parameters; human trial-and-error can also be very difficult as parameters may be tightly *coupled*, meaning that the optimum cannot simply be found by optimising one parameter at a time. In contrast, the optimisation algorithms used

[16] The entire concept of 'significant figures' is anyway a very crude method of expressing uncertainty.

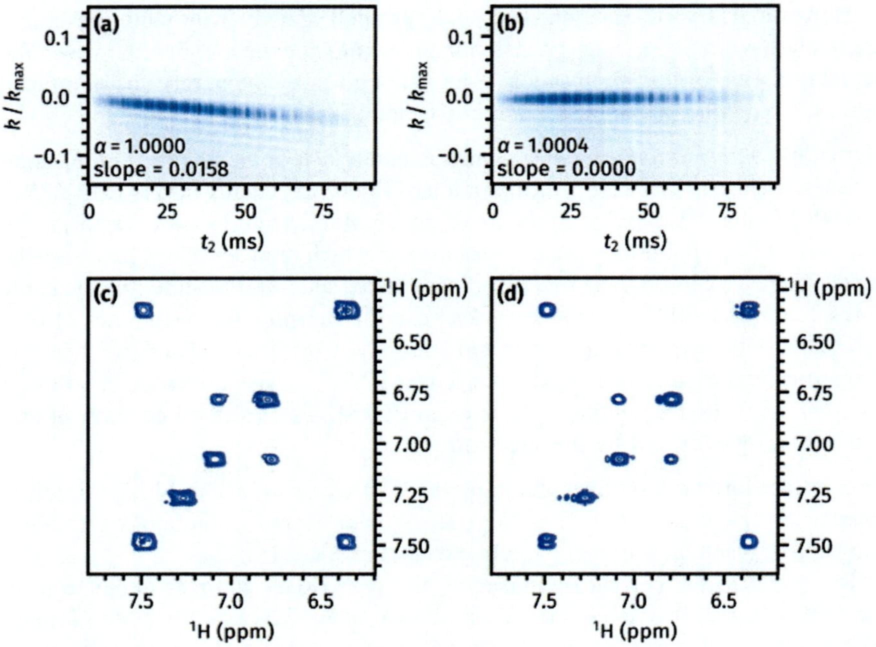

Fig. 3.17 **a, b** Magnitude-mode (k, t_2) data matrices obtained from the pulse–EPSI experiment, before and after optimisation of α respectively. **c, d** The corresponding 2D ultrafast TOCSY spectra, recorded with the unoptimised and optimised value of α respectively. The MLEV mixing period used was 10 ms.

in POISE make no assumption about the relationship between the two parameters. Furthermore, since only local minima are sought, the time required for convergence scales approximately linearly with the number of parameters, at least in the regimes which I have tested.[17]

In the simplest example of a multiple-parameter optimisation, I use POISE to search for the best delays in the low-pass J-filter (LPJF) of an HMBC experiment (Fig. 3.18a). Specifically, a second-order LPJF is used, which has the form τ_1–90°(^{13}C)–τ_2–90°(^{13}C). The role of the LPJF is to destroy ^{1}H^C magnetisation at the start of the HMBC, which would otherwise generate one-bond artefacts in the resulting spectrum. Assuming that the $^1J_{CH}$ values span a range of J_{min} to J_{max}, Sørensen and coworkers have provided 'theoretical' best values for the delays: [71, 72]

[17] I am not actually aware of any *theoretical* results which show how the number of function evaluations scales with the number of parameters in an optimisation; I suspect the answer depends on the exact problem being solved.

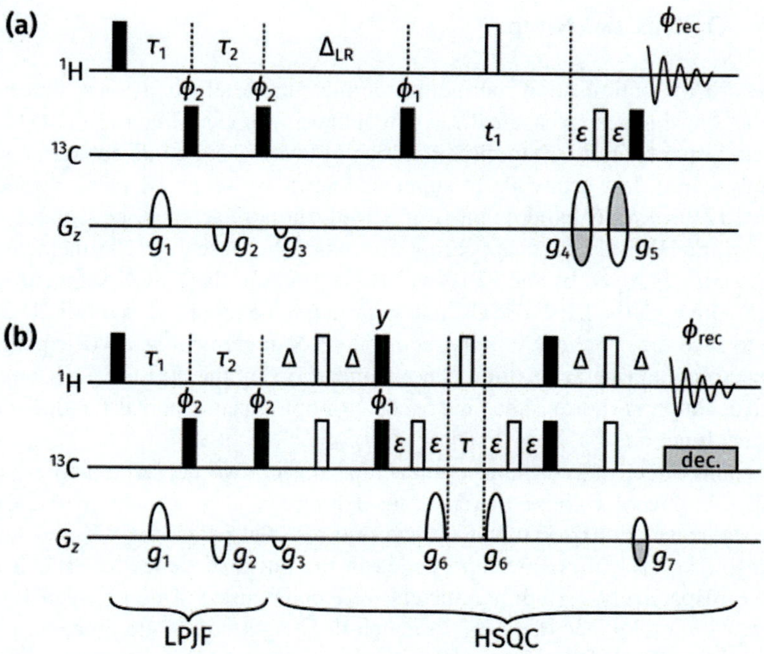

Fig. 3.18 **a** Standard 2D HMBC pulse sequence with a second-order LPJF **b** 1D LPJF–HSQC pulse sequence used in the POISE optimisation. In practice, only the first increment of this experiment was recorded. Delays are set as follows: $\Delta = 1/(4 \cdot {}^1J_{\mathrm{CH}})$; $\Delta_{\mathrm{LR}} = 1/(2 \cdot {}^nJ_{\mathrm{CH}})$; τ is as short as possible (several microseconds); and τ_1 and τ_2 are as described in the text. Phase cycling is performed with $\phi_1 = \phi_{\mathrm{rec}} = (x, -x)$, and $\phi_2 = (x, x, -x, -x)$. Gradient amplitudes are: $(g_1, g_2, g_3) = (15, -10, -5\%)$ for the LPJF, $(g_4, g_5) = (80\%, -47.85\%)$ (or vice versa) for EA selection in the HMBC experiment [70], and $(g_6, g_7) = (70, \pm 35.21\%)$ for EA selection in the LPJF–HSQC experiment

$$\tau_1 = \frac{1}{2}[J_{\mathrm{min}} + 0.146(J_{\mathrm{max}} - J_{\mathrm{min}})]^{-1}; \tag{3.31}$$

$$\tau_2 = \frac{1}{2}[J_{\mathrm{max}} - 0.146(J_{\mathrm{max}} - J_{\mathrm{min}})]^{-1}. \tag{3.32}$$

In practice, the TopSpin standard library HMBC sequence (`hmbcetgpl2nd`) uses delays of

$$\tau_1 = \frac{1}{2J_{\mathrm{min}}}; \quad \tau_2 = \frac{1}{2J_{\mathrm{max}}}. \tag{3.33}$$

In either case, though, these delays neglect the actual *distribution* of ${}^1J_{\mathrm{CH}}$: if many of the couplings are clustered around one particular value, then adjusting the delays to better suppress these values will yield better results. (Since ${}^1J_{\mathrm{CH}}$ roughly correlates with the hybridisation of the ^{13}C atom, [73] this can easily happen, for example, if the molecule is predominantly aliphatic or predominantly aromatic.)

3.4.7.1 Optimisation Setup

In order for the optimisation routine to minimise the artefacts, it must first have a cost function which can measure the extent of these artefacts. Given a full 2D HMBC spectrum, it is very difficult to discern which of the peaks are indeed artefacts (this is exactly why it is worthwhile to suppress them). However, by using a specially designed LPJF–HSQC experiment (Fig. 3.18b), the artefacts can be isolated without any of the HMBC signals appearing: this works because any $^{1}H^{C}$ magnetisation which is *not* destroyed by the LPJF will be sampled in the HSQC segment which follows. Although the LPJF–HSQC can in principle be recorded as a full 2D experiment, to save time, I chose to only record the first increment as a 1D experiment. ^{13}C decoupling during acquisition is not mandatory, but the boost in SNR obtained through decoupling shortens the time needed for optimisation and also reduces noise in the cost function.

We could directly perform an optimisation on the two delays τ_1 and τ_2. However, for simplicity, I chose to reuse the definitions of τ_1 and τ_2 in the Bruker library sequence (Eq. (3.33)) and instead optimise the parameters $J'_{\text{min}} = 1/(2\tau_1)$ and $J'_{\text{max}} = 1/(2\tau_2)$: these are represented in the pulse sequence as `CNST25` and `CNST26` respectively.[18] Both parameters were constrained to lie between 100 and 200 Hz, with a tolerance of 3 Hz. To match the magnitude-mode processing typically used for 2D HMBC experiments, the acquisition AU program was modified to also process the LPJF–HSQC spectrum in magnitude mode. The integral of the entire spectrum can therefore be used as the desired cost function for the POISE optimisation. More specifically, I chose to use a sum of squares:

$$f_{\text{sos}} = \sum_i |S_i|^2, \qquad (3.34)$$

which has the effect of penalising intense artefacts more strongly.

3.4.7.2 Optimisation Results

In previous sections, I have quoted ranges for the optima found in an effort to show that the optimisation results were reproducible. In multiple-parameter optimisations, it is no longer appropriate to report separate ranges for both parameters, since the parameters are in general *interdependent*, in that the spread of values obtained for the first parameter depends on the value obtained for the second, and vice versa. I therefore only provide ranges for the number of FEs, as well as the time taken. As for optima found, I will quote only the *best* of these, as measured by the reduction in the cost function. These optima should therefore not be viewed as a complete summary

[18] I use prime symbols here because J_{min} and J_{max} are *physical constants* which depend on the sample being studied, whereas J'_{min} and J'_{max} are *parameters* which can be chosen to have any value.

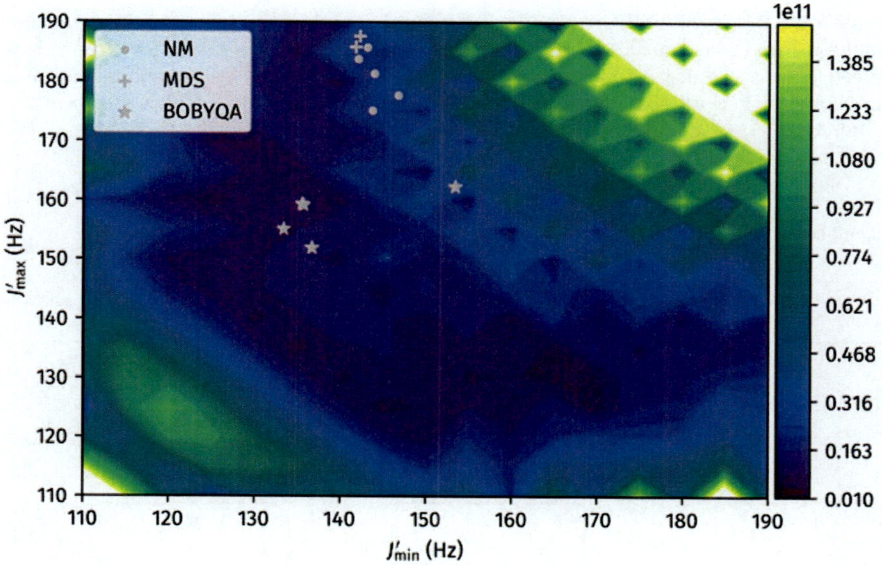

Fig. 3.19 Reference grid search for HMBC optimisation. Darker regions correspond to lower cost function values, i.e. better spectra. The results of the 15 optimisations (5 per algorithm) are plotted on the same axes. Note that the cost functions here are rather larger than in Table 3.13: this is solely because a different receiver gain was used. *Data code:* `7Z-210814`

of the five different runs, but rather as an *example* of the improvement which may be obtained through POISE.

Interestingly, BOBYQA finds an optimum which is different from—and substantially better than—the other two algorithms. As mentioned previously, it is not generally possible to predict which optimum an algorithm will converge to, and this is complicated even further by the presence of noise, which imparts some stochastic character to the optimisation. Nevertheless, a closer inspection of the other optimisation runs actually shows that the better performance of BOBYQA on this problem is reproducible: the optima found from each of the five rounds, as well as a reference grid search, are plotted in Fig. 3.19. It appears that the simplex-based methods get stuck in a suboptimal region, whereas BOBYQA is usually able to move leftwards and downwards into a more productive region. However, the exact reasons for this remain unclear. Also, BOBYQA does not always outperform NM and MDS in the other optimisations in this chapter. (This is possibly a manifestation of the 'no free lunch theorem' [74], which states that when averaged over all problems, no optimisation algorithm strictly outperforms another.)

In any case, when the best of these optimised values (i.e. entry 3 in Table 3.13) were used to run a standard 2D HMBC, a significant reduction in one-bond artefacts was observed (Fig. 3.20). It should be emphasised that these optimised values were obtained in a matter of several minutes, which is far shorter than a typical HMBC experiment requires. The performance of the optimised second-order LPJF

Table 3.13 Results of HMBC POISE optimisations. For each optimisation algorithm, the best optimum found is reported along with the value of the cost function f_{sos} (a lower value represents a better optimum). The number of FEs, as well as the time taken, is still reported as a range over the five optimisation runs performed per algorithm. The POISE routine used here is: `{"name": "lpjf", "pars": ["cnst25", "cnst26"], "lb": [100.0, 100.0], "ub": [200.0, 200.0], "init": [120.0, 180.0], "tol": [3.0, 3.0], "cf": "zerorealint_squared", "au": "poise_1d_mc"}`.
. *Data code:* `7Z-210814`

		Best optimum found			Aggregated results	
Entry	Algorithm	J'_{min} (Hz)	J'_{max} (Hz)	$f_{sos}/10^7$	FEs	Time taken (s)
1	NM	143.07	185.71	8.65	20–24	358–430
2	MDS	141.73	185.82	8.36	18	321–323
3	BOBYQA	133.38	155.03	6.42	14–17	251–304

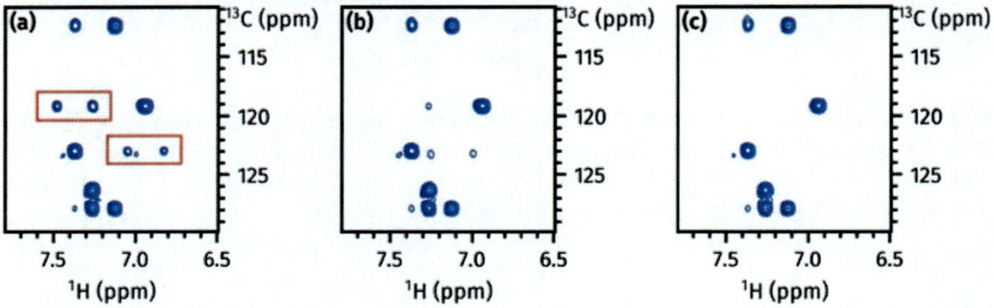

Fig. 3.20 **a** HMBC spectrum with 'compromise' values of 120 and 180 Hz for J'_{min} and J'_{max} respectively in the second-order LPJF. The one-bond artefacts are highlighted in red boxes. **b** HMBC spectrum with optimised second-order LPJF delays (J'_{min} = 133.38 Hz; J'_{max} = 155.03 Hz). **c** HMBC spectrum with unoptimised third-order LPJF. *Data code:* `7Z-210814`

(Fig. 3.20b) was in fact almost comparable to that of a third-order LPJF (Fig. 3.20c). However, using a third-order LPJF risks suppressing some of the desired HMBC signal as well, so if all else is equal, the optimised second-order LPJF should be preferred. In principle, the third-order LPJF itself can also be subjected to optimisation. This possibility was not explored in the present work but would be easy to set up using the methods outlined in this section (although, as there are three delays to optimise, it would require a longer time).

3.4.7.3 Application to NOAH HMBC

In NOAH supersequences, the HMBC experiment (or 'module') is typically modified by replacing the initial 90° excitation pulse with a zz-filter (Fig. 3.21) [75, 76]. This causes the $^1H^{!C}$ magnetisation used in the HMBC to be excited, but leaves $^1H^C$

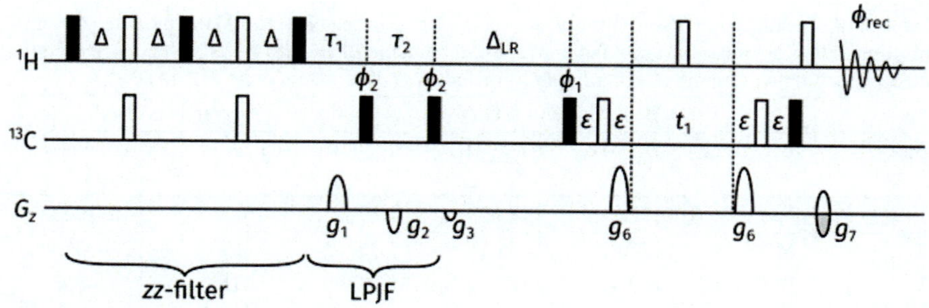

Fig. 3.21 NOAH HMBC module

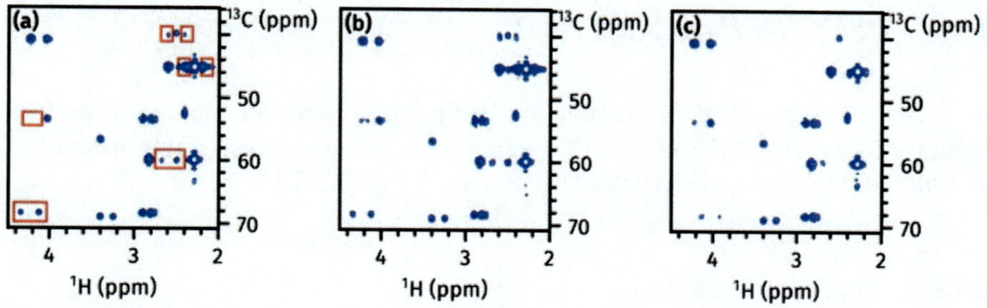

Fig. 3.22 **a** NOAH HMBC spectrum before POISE optimisation, using the default settings of 120 and 180 Hz. One-bond artefacts are highlighted in red boxes. **b** NOAH HMBC spectrum after POISE 'optimisation' (using J'_{min} = 125.2Hz, J'_{max} = 193.2 Hz, as per entry 2 of Table 3.14). **c** NOAH HMBC spectrum with no POISE optimisation, but instead adding a 90° ^{13}C pulse at the end of the zz-filter (see also Sect. 4.3.7). *Data code:* `7Z-210814`

magnetisation along the $+z$ axis so that it can be preserved for later modules in a supersequence. In an ideal situation, this means that the one-bond artefacts in the HMBC should in fact be suppressed even without an LPJF. However, significant one-bond artefacts are in fact still observed in NOAH HMBC spectra. I therefore sought to apply the same optimisation routine to the zz-HMBC module to see whether this could be improved.

Unfortunately, the results here were far less impactful: generally, the optimisation converged to values very close to the initial starting point, and in some cases the behaviour of the optimised spectrum was even worse than the unoptimised spectrum. (The deterioration is very marginal and likely arises only due to noise in the cost function during the optimisation.) This strongly suggests that the artefacts are not due to imperfect LPJF suppression, but instead stem from the zz-filter component of the sequence. This was indeed verified experimentally: the addition of a ^{13}C 90° pulse at the end of the zz-filter led to much better one-bond artefact suppression (Fig. 3.22c; this strategy is further explained in Sect. 4.3.7).

Table 3.14 Results of POISE optimisations on the LPJF in the NOAH HMBC module. All optimisation details are the same as in Table 3.13, except that the LPJF–HSQC pulse sequence used for the optimisation was modified accordingly to include a zz-filter. *Data code:* `7Z-210814`

		Best optimum found			Aggregated results	
Entry	Algorithm	J'_{min} (Hz)	J'_{max} (Hz)	$f_{\mathrm{sos}}/10^9$	FEs	Time taken (s)
1	NM	117.05	188.59	5.37	25–29	451–523
2	MDS	125.21	193.16	5.35	14–17	249–310
3	BOBYQA	119.49	182.74	5.78	9–16	164–293

3.4.8 *PSYCHE Pure Shift NMR*

In Sect. 2.4, I described the motivation behind, and early attempts towards, the optimisation of PSYCHE pure shift spectra. The content in this section is similar, except that it was performed within the framework of POISE.

3.4.8.1 Optimisation Setup

In this section, the standard double-saltire PSYCHE pure shift element was used (Fig. 3.23a) [77, 78]. As described in Sect. 2.4.1, the PSYCHE PSE can be described using six parameters; in this section, we investigate only three of these, namely the

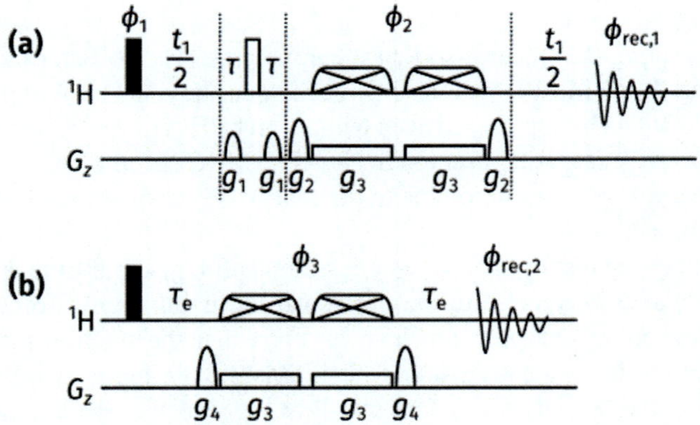

Fig. 3.23 **a** Pseudo-2D PSYCHE pure shift experiment. **b** J-resolved spin echo experiment (see also Sect. 2.4.1) using the PSYCHE PSE. Phase cycling is performed using: $\phi_1 = (x, -x)$, $\phi_2 = (x, x, y, y)$, $\phi_{\mathrm{rec},1} = (x, -x, -x, x)$ for the full PSYCHE experiment, and $\phi_3 = (x, y, -x, -y)$, $\phi_{\mathrm{rec},2} = (x, -x)$ for the JRSE experiment. CTP gradient amplitudes are $(g_1, g_2, g_4) = (35, 77, 50\%)$ (though the exact values are likely immaterial); the PSYCHE gradient g_3 was subjected to optimisation. The delay τ is set to $1/(4 \cdot T_{\mathrm{chunk}})$ for the PSYCHE experiment, and τ_e is 16 ms.

amplitude (i.e. flip angle β), bandwidth ΔF, and duration τ_p. In addition to this, the amplitude of the weak gradient during the PSE g_3 was also chosen as a fourth parameter to vary. As before, the quality of the PSE is evaluated using a JRSE experiment (Fig. 3.23b), which is then compared against a pulse–acquire spectrum: the cost function used is f_{diff}.

In the JRSE pulse sequence used for optimisation, the four parameters (flip angle, bandwidth, duration and gradient amplitude) are respectively `CNST20`, `CNST21`, `P40` and `GPZ10`. There is, however, a slight complication: whenever the bandwidth or duration is changed, the entire pulse must be re-created, because the x- and y-coefficients depend on these parameters. Therefore, a (TopSpin) Python script was written to generate the double-saltire pulse using the parameters above, and the POISE AU program was modified to call this Python script before acquiring the spectrum. Although the overall setup may perhaps be slightly confusing,[19] I contend that this is an excellent demonstration of the customisability that POISE provides.

One slightly odd behaviour noted with the PSYCHE optimisations was that at least one dummy scan was required for the optimisation to be robust: if no dummy scans were used, the *very first* FE in an optimisation run would yield a rather unreliable result, although subsequent FEs were unaffected. It is not obvious why this is the case, as FEs are already separated by a delay of several seconds, which is on the order of $5T_1$: thus, each FE should be starting from (almost) full equilibrium magnetisation. Nevertheless, to avoid this issue, all optimisations in this section were run using `DS=1` and `NS=2`.

The sample used was andrographolide: this is the same sample as was used in the dPSYCHE section (Sect. 2.6) and is fairly useful for pure shift studies as it contains quite diverse coupling patterns. The spectral region was limited to 1.15 ppm and above to exclude strong singlets from the methyl groups in andrographolide, which are irrelevant to pure shift NMR but disproportionately influence the value of the cost function (this issue was also discussed in Sect. 2.4.3).

3.4.8.2 Optimisation Results

I ran several different optimisations of increasing complexity:

- one-parameter: flip angle only
- two-parameter: flip angle and gradient amplitude
- three-parameter: flip angle, gradient amplitude and bandwidth
- four-parameter: flip angle, gradient amplitude, bandwidth and duration

Table 3.15 summarises the results obtained from all optimisations, whereas Tables 3.16, 3.17, 3.18, 3.19 show more detailed results for each individual set of optimisations, including a breakdown by algorithm. We can see from these results that—

[19] We have here a Python script (the POISE frontend) calling an AU program (for acquisition) which calls a Python script (to make the double saltire).

Table 3.15 Summary of all optimisations performed on the PSYCHE PSE. Unlike before, aggregated results are not grouped by optimisation algorithm; these ranges are therefore collected from a total of 15 runs (5 per algorithm). Parameters in parentheses were not optimised and were simply carried over from the initial point. A detailed breakdown of the results by algorithm, as well as the POISE routines used, are described in Tables 3.16, 3.17, 3.18, 3.19. *Data code:* `6A-200822`

	Aggregated results from all runs			Parameters from best optimum			
Description	f_{diff}	FEs	Time (s)	β (°)	g_3 (%)	ΔF (kHz)	τ_p (ms)
Initial point	0.353	–	–	(25.0)	(2.00)	(10.0)	(30.0)
1 parameter	0.340–0.343	5–10	84–168	17.5	(2.00)	(10.0)	(30.0)
2 parameters	0.325–0.338	12–33	205–565	16.5	1.00	(10.0)	(30.0)
3 parameters	0.320–0.328	18–77	315–1344	17.4	1.73	15.9	(30.0)
4 parameters	0.316–0.333	39–85	705–1504	13.9	1.45	15.9	36.0

Table 3.16 PSYCHE one-parameter (flip angle) optimisations. The POISE routine used was: `{"name": "psyche1", "pars": ["cnst20"], "lb": [10.0], "ub": [35.0], "init": [25.0], "tol": [2.0], "cf": "specdiff", "au": "poise_1d"}`. *Data code:* `6A-200822`

Entry	Algorithm	Optimum found (°)	FEs	Time (s)
1	NM	15.0–20.0	8–10	135–168
2	MDS	17.5–20.0	8	133–136
3	BOBYQA	18.4–19.9	5	84–85

Table 3.17 PSYCHE two-parameter (flip angle and gradient amplitude) optimisations. The POISE routine used was: `{"name": "psyche2", "pars": ["cnst20", "gpz10"], "lb": [10.0, 0.2], "ub": [35.0, 5.0], "init": [25.0, 2.0], "tol": [2.0, 0.2], "cf": "specdiff", "au": "poise_1d"}`. *Data code:* `6A-200822`

		Best optimum found			Aggregated results	
Entry	Algorithm	β (°)	g_3 (%)	f_{diff}	FEs	Time (s)
1	NM	16.48	1.00	0.3253	18–24	307–410
2	MDS	18.47	0.82	0.3261	25–33	424–565
3	BOBYQA	15.77	0.96	0.3279	12–16	205–271

Table 3.18 PSYCHE three-parameter (flip angle, gradient amplitude and bandwidth) optimisations. The POISE routine used was: `{"name": "psyche3", "pars": ["cnst20","gpz10", "cnst21"], "lb": [10.0, 0.2, 1000.0], "ub": [35.0, 5.0, 20000.0], "init": [25.0, 2.0, 10000.0], "tol": [2.0, 0.2, 500.0], "cf": "specdiff", "au": "poise_psyche"}`.

. *Data code:* `6A-200822`

		Best optimum found				Aggregated results	
Entry	Algorithm	β (°)	g_3 (%)	ΔF (kHz)	f_{diff}	FEs	Time (s)
1	NM	17.40	1.73	15.93	0.3196	33–43	576–770
2	MDS	16.75	1.71	16.33	0.3215	46–77	804–1344
3	BOBYQA	16.48	1.65	16.71	0.3200	18–31	315–540

Table 3.19 PSYCHE four-parameter (flip angle, gradient amplitude, bandwidth and pulse duration) optimisations. The POISE routine used was: `{"name":"psyche4", "pars": ["cnst20", "gpz10", "cnst21", "p40"], "lb": [10.0, 0.2,1000.0, 5000.0], "ub": [35.0, 5.0, 20000.0, 75000.0], "init": [25.0, 2.0,10000.0, 30000.0], "tol": [2.0, 0.2, 500.0, 2000.0], "cf": "specdiff", "au":"poise_psyche"}`.

. *Data code:* `6A-200822`

		Best optimum found					Aggregated results	
Entry	Algorithm	β (°)	g_3 (%)	ΔF (kHz)	τ_{p} (ms)	f_{diff}	FEs	Time (s)
1	NM	13.87	1.45	15.93	36.00	0.3163	40–47	733–845
2	MDS	19.15	1.06	11.19	36.52	0.3185	57–85	1006–1504
3	BOBYQA	17.59	1.20	12.76	43.06	0.3178	39–62	705–1130

perhaps unsurprisingly—optimising more parameters leads to greater improvements in the cost function, albeit at the cost of more FEs and more time.

An important question is whether these reductions in the cost function (measured on a JRSE experiment) do actually translate into improved pure shift spectra. The pure shift spectra, obtained with the optimised parameters in Table 3.15, are shown in Fig. 3.24. Of particular interest are the two strongly coupled protons at 2.5 ppm: in the pure shift spectrum, the two peaks on either side are genuine, but strong coupling artefacts appear in the middle of these peaks. On top of that, the decoupling performance for the peaks at 1.36 and 1.65 ppm is also a useful marker of pure shift performance.

Broadly speaking, all the optimised spectra provide better performance than the initial point in terms of decoupling quality. In particular, the four-parameter optimisation successfully suppresses some of the strong coupling artefacts: this may be explained by the fact that the pulse duration was increased from 30 to 36 ms, which

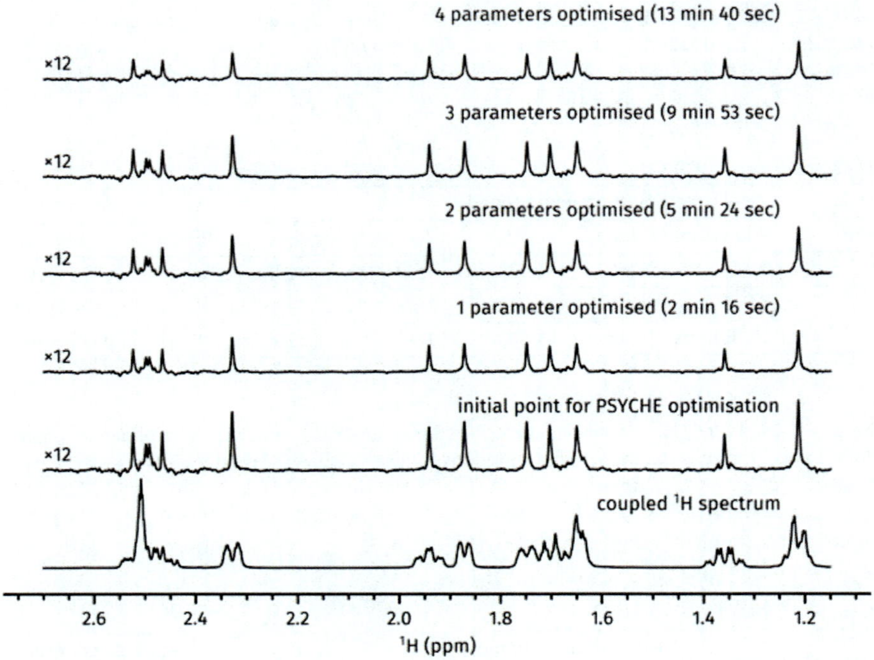

Fig. 3.24 Insets of PSYCHE spectra obtained using the best optimum from each set of optimisations in Table 3.15; the time required to obtain the optimised parameters is also indicated for each spectrum. The original coupled ^{1}H spectrum is shown as a reference. *Data code:* `6A-200823`

generally results in better spatiotemporal averaging, as was noted in Sect. 2.3.[20] Generally, this optimisation *does* come at a cost in sensitivity, which is particularly noticeable for the four-parameter optimised result. However, since the cost function f_{diff} also penalises sensitivity losses, it ensures that any drops in sensitivity are not excessive.

Given these improvements, one might wonder why it would ever be worth optimising fewer than four parameters. The first obvious drawback is the time required: in fact, it is probably more economical to use the TSE-PSYCHE experiment (which also has improved performance in the presence of strong coupling).[21] Furthermore, f_{diff} is a rather 'flat' cost function and optimisations using it are very susceptible to noise: this factor is specific to pure shift optimisations and was previously mentioned in Sect. 2.4.3. A closer inspection of the POISE logs for the three- and four-parameter

[20] This does raise the question of whether a two-parameter optimisation of just the flip angle and duration could yield similar results, but in a shorter time. I think it is possible, but I did not test this, as I simply did not have the appetite to try optimising *every* possible combination of parameters.

[21] Another thing I did not (thoroughly) investigate was to optimise the TSE-PSYCHE experiment. This could, in principle, be done using a TSE form of the JRSE experiment: 90°–chirp–τ–PSE–τ–chirp–detect. However, this has decreased sensitivity over the original JRSE experiment, which means even longer optimisations.

optimisations (which are not provided here, but can be accessed in the raw data for the POISE paper, available at https://doi.org/10.5281/zenodo.4698423) reveals that the optimisations are perhaps not as consistent as one would hope. Although *generally* the optimisation leads to similar results in that (for example) β is often decreased and τ_p increased, the extents of these changes are not quite uniform: for example, τ_p is optimised to anywhere between 36 ms and 55 ms, which suggests that there are *many* local minima and which of these the optimisation converges to depends on the noise in the cost function.[22]

Putting all of these factors together, it is (in my opinion) only really worthwhile to optimise two parameters at a time: either the flip angle plus gradient amplitude, or the flip angle plus the duration, appear to be sensible choices. That said, this *does* demonstrate that optimising multiple parameters at once—ordinarily a very challenging task for a human—does not actually require prohibitively long times when performed using a suitable algorithm.

3.4.9 Water Suppression

The suppression of solvent signals is one of the most important techniques in modern NMR, especially in cases where solvent deuteration is not a viable strategy (e.g. when exchangeable protons must be observed) [79–81]. Here, we specifically seek to optimise the 1D NOESY experiment with presaturation [82] (Fig. 3.25): although this sequence is popular, its performance can depend on at least four parameters, namely, the transmitter offset, presaturation power, the NOE mixing time and the presaturation duration or recovery delay. In this section, I use the symbols ν_{tx}, k_{ps}, τ_m and τ_r to respectively refer to these four parameters: the corresponding TopSpin parameters are `O1`, `CNST20`, `D8` and `D1`.

3.4.9.1 Optimisation Setup

In a similar style to the PSYCHE optimisations (Sect. 3.4.8), a series of optimisations with increasing numbers of parameters were run. However, in this case, the cost function is far easier to construct and far less noisy: we can simply integrate the water peak, which was defined to be the region of the (real) spectrum between 4.65 and 4.75 ppm, and use that (or its absolute value) as the cost function. In practice, because the phase of the water peak can vary rather unpredictably, I preferred to use a sum-of-squares cost function f'_{sos}:

[22] I still think, however, that it is misleading to quote these ranges in Tables 3.16, 3.17, 3.18, 3.19, because this gives the impression that there are no correlations between the optimised parameters. By providing access to the complete raw data, and explicitly pointing out this issue in the text, I hope that I have done enough to not be accused of cherry-picking data to present.

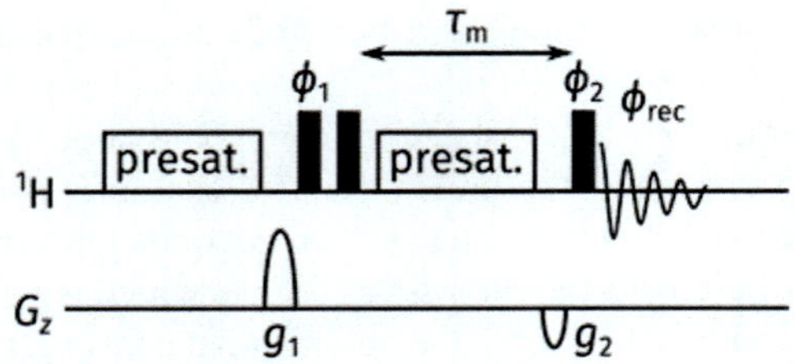

Fig. 3.25 1D NOESY pulse sequence used for water suppression. Presaturation of the water resonance is applied during the recovery delay as well as the NOE mixing time. Phase cycling is performed with $\phi_1 = (x, -x)$, $\phi_2 = (x, x, -x, -x, y, y, -y, -y)$, and $\phi_{\text{rec}} = (x, -x, -x, x, y, -y, -y, y)$. Gradient amplitudes are $(g_1, g_2) = (50, -10\%)$. Boxes labelled with 'presat'. indicates periods of weak presaturation

$$f'_{\text{sos}} = \sum_i S^2_{\text{re},i}. \tag{3.35}$$

This is very similar to f_{sos} (Eq. (3.34)), but uses only the real part of the spectrum instead of the magnitude-mode spectrum. Here, squaring each point of the spectrum not only accounts for the possible sign changes from out-of-phase peaks, but also more strongly penalises intense absorption-mode water peaks.[23]

One drawback of the water suppression optimisation is that the phase cycle of the 1D NOESY is integral to its performance: thus, several scans must be used for a reliable cost function to be obtained, making for relatively long optimisations. In this case, I used `DS=2` and `NS=4`.

For these optimisations, I used a sample of rodent urine in D2O, which was kindly provided by Abi Yates and Fay Probert (both University of Oxford).

3.4.9.2 Optimisation Results

As before, I first provide a 'summary' table comparing the optimisations with different numbers of parameters (Table 3.20). From this, we see a similar story to before, in that optimising more parameters takes more time but leads to larger decreases in the cost function. Individual details of each set of optimisations are given in Tables 3.21, 3.22, 3.23, 3.24.

The corresponding spectra post-optimisation are shown in Fig. 3.26. Clearly, the optimisations succeed in reducing the size of the water peak, especially the four-parameter optimisation. However, perhaps equally importantly, the doublet at 4.56 ppm is unaffected by the optimisation. Since the cost function does not actually check for the retention of peaks outside the target window, it is in principle possible that the

[23] Dispersion-mode peaks are possibly more problematic because of their long 'tails'. In hindsight, I could have considered trying the `minabsint` cost function instead (Eq. (3.19)). However, in practice this did not seem to matter very much: all of these cost functions ultimately share the same goal, i.e. to reduce the intensity of the water peak.

Table 3.20 Summary of optimisations on 1D NOESY/presaturation pulse sequence for water suppression. Parameters in parentheses were not optimised and were simply carried over from the initial point. A detailed breakdown of the results by algorithm, as well as the POISE routines used, are described in Tables 3.21, 3.22, 3.23, 3.24. *Data code:* `4P-210620`

Initial point	14.7	–	–	(1880.61)	(50.0)	(0.100)	(2.00)
1 parameter	1.85–2.49	6–12	259–520	1880.41	(50.0)	(0.100)	(2.00)
2 parameters	1.21–9.68	9–21	390–911	1880.20	51.94	(0.100)	(2.00)
3 parameters	1.24–4.20	19–26	825–1128	1879.90	47.79	0.118	(2.00)
4 parameters	0.165–1.65	25–53	1143–2314	1881.10	53.28	0.150	3.00

Table 3.21 Water suppression one-parameter (transmitter offset) optimisations. The POISE routine used was: `{"name": "solvsupp1", "pars": ["o1"], "lb": [1870.61], "ub": [1890.61], "init": [1880.61], "tol": [0.2], "cf": "zerorealint_squared", "au": "poise_1d_noapk"}`.

Data code: `4P-210620`

Entry	Algorithm	Optimum found (Hz)	FEs	Time taken (s)
1	NM	1880.24–1880.49	10–12	436–519
2	MDS	1880.24–1880.36	12	518–520
3	BOBYQA	1880.34–1880.47	6–7	259–303

Table 3.22 Water suppression two-parameter (transmitter offset and presaturation power) optimisations. The POISE routine used was: `{"name": "solvsupp2","pars": ["o1", "cnst20"], "lb": [1870.61, 10.0], "ub": [1890.61, 55.0],"init": [1880.61, 50.0], "tol": [0.2, 2.5], "cf": "zerorealint_squared","au": "poise_1d_noapk"}`.

Data code: `4P-210620`

		Best optimum found			Aggregated results	
Entry	Algorithm	ν_{tx} (Hz)	k_{ps} (Hz)	$f'_{sos}/10^{18}$	FEs	Time (s)
1	NM	1880.20	51.94	1.209	17–21	737–911
2	MDS	1880.85	50.81	2.600	15	649–652
3	BOBYQA	1881.12	50.03	2.887	9–14	390–608

Table 3.23 Water suppression three-parameter (transmitter offset, presaturation power and mixing time) optimisations. The POISE routine used was: `{"name": "solvsupp3", "pars": ["o1", "cnst20", "d8"], "lb": [1870.61, 10.0, 0.010], "ub": [1890.61, 55.0, 0.150], "init": [1880.61, 50.0, 0.100], "tol": [0.2, 2.5, 0.010], "cf": "zerorealint_squared", "au": "poise_1d_noapk"}`.

. *Data code:* `4P-210620`

		Best optimum found				Aggregated results	
Entry	Algorithm	ν_{tx} (Hz)	k_{ps} (Hz)	τ_m (ms)	$f'_{sos}/10^{18}$	FEs	Time (s)
1	NM	1879.90	47.79	117.7	1.238	23–26	1002–1128
2	MDS	1882.79	52.95	49.91	1.843	24–25	1034–1080
3	BOBYQA	1881.85	52.14	134.7	1.738	19–26	825–1133

Table 3.24 Water suppression four-parameter (transmitter offset, presaturation power, mixing time and presaturation duration) optimisations. The POISE routine used was: `{"name": "solvsupp4", "pars": ["o1", "cnst20", "d8", "d1"], "lb": [1870.61, 10.0, 0.010, 1.0], "ub": [1890.61, 55.0, 0.150, 3.0], "init": [1880.61, 50.0, 0.100, 2.0], "tol": [0.2, 2.5, 0.010, 0.1], "cf": "zerorealint_squared", "au": "poise_1d_noapk"}`.

. *Data code:* `4P-210620`

		Best optimum found					Aggregated results	
Entry	Algorithm	ν_{tx} (Hz)	k_{ps} (Hz)	τ_m (ms)	τ_r (s)	$f'_{sos}/10^{18}$	FEs	Time (s)
1	NM	1880.72	49.73	110.6	1.972	1.008	29–40	1250–1726
2	MDS	1882.65	53.41	69.46	2.137	1.199	34–53	1487–2314
3	BOBYQA	1881.10	53.28	150.0	3.000	0.1653	25–40	1143–1843

optimisation will converge to parameters which achieve excellent water suppression but also cause nearby peaks to be lost. This possibility is averted here by setting a conservative upper bound of 55 Hz on the presaturation power k_{ps}: this avoids inadvertently saturating nearby resonances.

In fact, it may be justifiable in this case to run a long, four-parameter optimisation (which takes around 30 min), especially if many samples are to be run using the same optimised water suppression parameters. However, for more routine usage, it is probably preferable to limit the number of parameters being optimised to two.

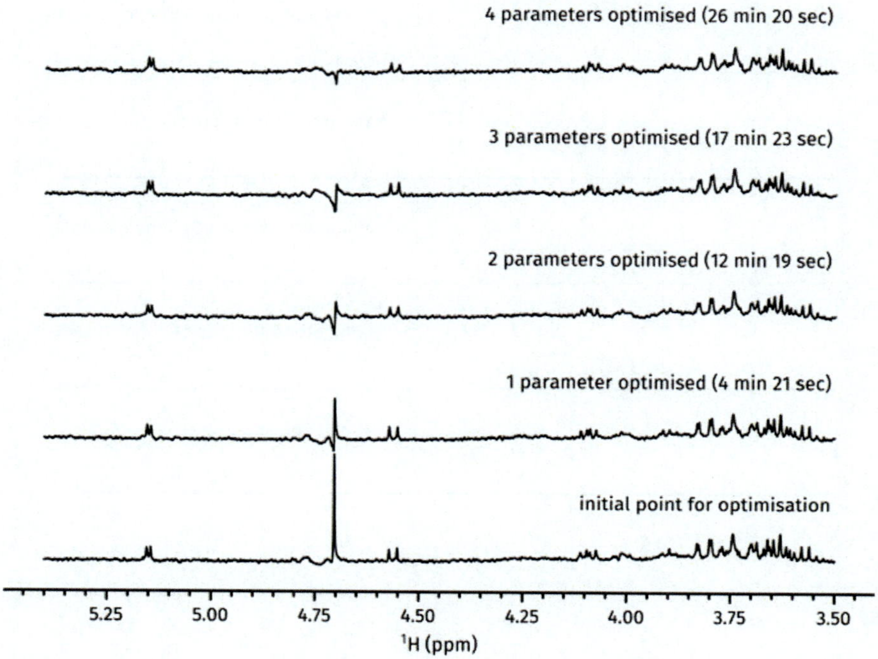

Fig. 3.26 Insets of 1D NOESY spectra acquired using optimised parameters from each set of optimisations in Table 3.15; the time required to obtain the optimised parameters is also indicated for each spectrum. The sample used was rodent urine in D2O. Note that, for the one-, three- and four-parameter optimisations, the best optimum from the 15 runs (i.e. the optima listed in Table 3.20) was used to acquire the spectra shown here. However, this was not the case for the two-parameter optimisation: I (inexplicably) used a different optimum. This was probably an oversight on my part. Nevertheless, it does not affect any of the conclusions drawn here. *Data code:* `4P-210620`

3.4.9.3 Optimisation on a Sucrose Sample

The water suppression optimisations shown above were in fact first developed and evaluated on a sample of sucrose in 90% H2O/10% D2O. This sample is less 'interesting' in that there are no resonances close to the water peak, and I also did not perform replicates of each optimisation run (this data never made it into the POISE paper). Nevertheless, I still believe it is still worth adding the results here, as they demonstrate that the optimisation procedure can be easily applied to different samples.

In this case, the one-parameter optimisation seemed to have no significant effect on the result. However, the two- through four-parameter optimisations all yielded significant improvements in the water suppression (Fig. 3.27). The parameters used for each of these spectra are listed in Table 3.25.

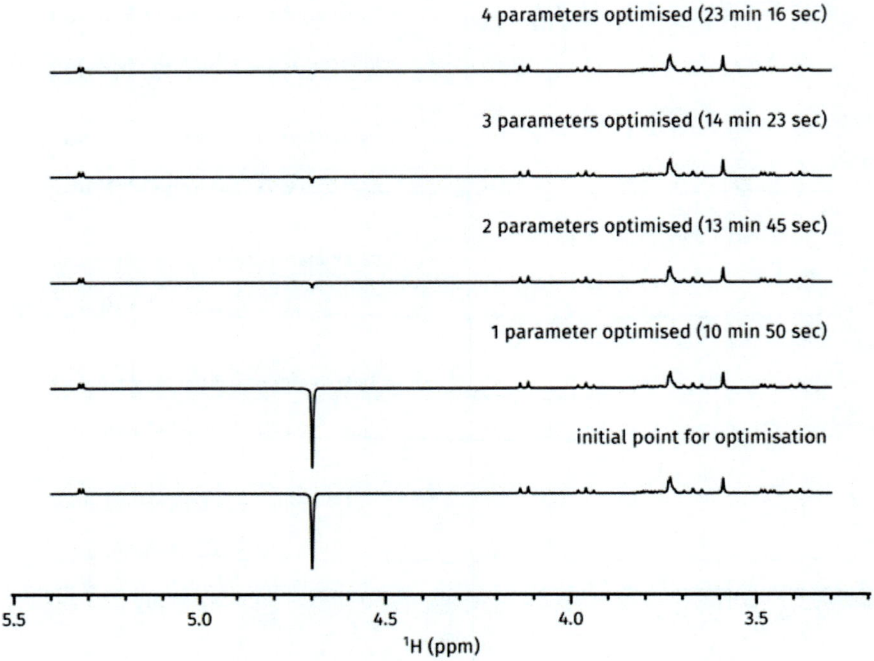

Fig. 3.27 Insets of 1D NOESY spectra acquired after optimisation of 1, 2, 3, or 4 parameters (the exact parameters being optimised are the same as in Fig. 3.26). The sample used was sucrose in 90% H2O/10% D2O. The time required to obtain the optimised parameters is also indicated for each spectrum. The parameters used for each of these spectra are in Table 3.25. *Data code:* `4S-210617`

Table 3.25 Optimised water suppression parameters for a sample of sucrose in 90% H2O/10% D2O. Parameters in parentheses were not optimised and were simply carried over from the initial point. These correspond to the spectra in Fig. 3.27. *Data code:* `4S-210617`

Description	$f'_{\text{sos}}/10^{10}$	FEs	Time (s)	ν_{tx} (Hz)	k_{ps} (Hz)	τ_{m} (s)	τ_{r} (s)
Initial point	7.74	–	–	(1880.61)	(50.0)	(0.100)	(2.00)
1 parameter	7.25	15	650	1882.11	(50.0)	(0.100)	(2.00)
2 parameters	3.65	19	825	1881.50	75.58	(0.100)	(2.00)
3 parameters	5.71	20	863	1881.40	78.48	0.045	(2.00)
4 parameters	4.62	29	1396	1881.09	58.54	0.107	2.91

3.4.10 Diffusion NMR

As the final example of a POISE optimisation, I demonstrate its application to diffusion-ordered spectroscopy (DOSY) experiments [83]. DOSY experiments measure molecular diffusion during a delay period, Δ; this delay is placed between two gradients, such as in the stimulated echo experiment (Fig. 3.28a), or the bipolar pulse pair version (Fig. 3.28b) which uses opposing gradients to refocus the lock signal during the sequence. In these sequences, the first 'encoding' gradient (or set thereof) imparts a spatially dependent phase, which—in the absence of diffusion—is perfectly refocused by the second 'decoding' gradient (or set thereof). However, if diffusion is present, this refocusing is not complete, leading to signal attenuation which is described by the Stejskal–Tanner equation:

$$I(G) = I(0)\exp\left[-(\gamma\delta G)^2 D\Delta'\right]. \quad (3.36)$$

Here, G is the amplitude of the encoding and decoding gradients, δ the duration of the gradients, and Δ' is a corrected diffusion delay.

In a pseudo-2D DOSY experiment, each increment is recorded with a different gradient amplitude G. The value of D can then be extracted through various means, most simply an exponential fitting of the measured intensities $I(G)$ for single species (or distinct species which do not overlap). In order to minimise the uncertainty in

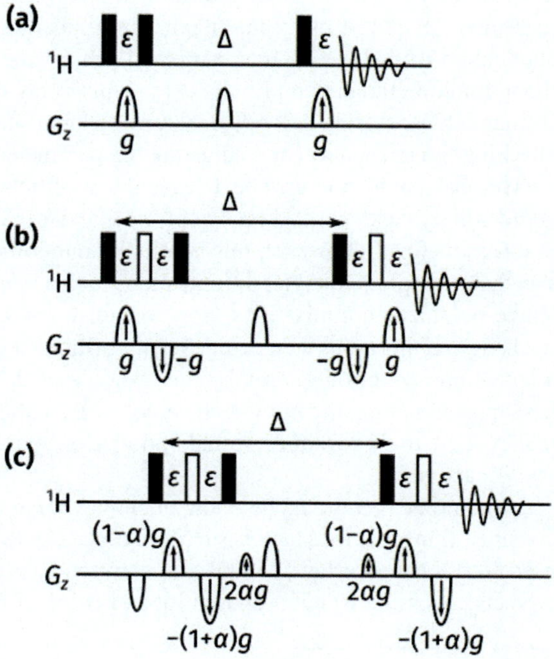

Fig. 3.28 **a** Stimulated echo DOSY pulse sequence. The diffusion-encoding and -decoding gradients, whose amplitudes are incremented, are marked with arrows. **b** Stimulated echo bipolar pulse pair DOSY pulse sequence. **c** Oneshot DOSY pulse sequence

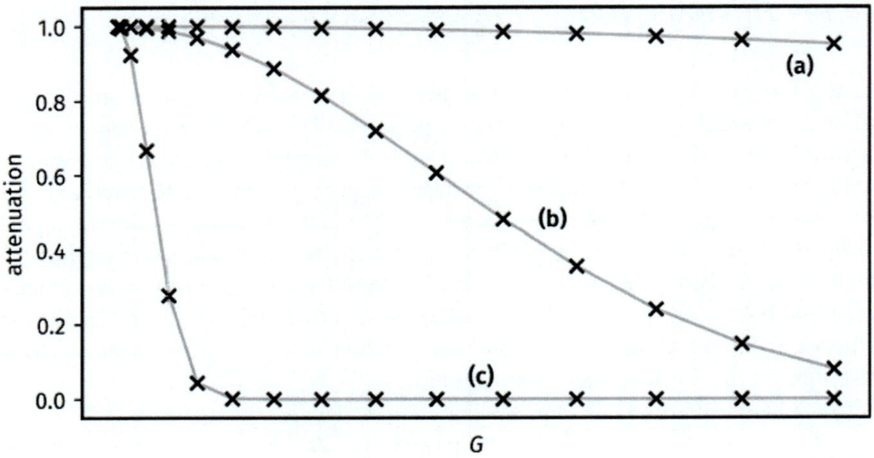

Fig. 3.29 Simulated diffusion profiles for species with different diffusion rates. **a** A slowly diffusing species: insufficient attenuation is observed with the given parameters, suggesting that Δ or δ need to be increased. **b** An 'ideal' diffusion profile where attenuation is neither too little nor too much. **c** A rapidly diffusing species, for which complete attenuation is already accomplished using low gradient amplitudes. This suggests that Δ or δ need to be decreased, or a new gradient ramp calculated

the calculated diffusion coefficients, the parameters Δ and δ, as well as the range of gradient amplitudes used (and possibly also their distribution), should be chosen in order to yield a diffusion profile of the form in Fig. 3.29b. Generally, the last increment (acquired using the maximum gradient strength, G_{max}) should have an attenuation of 5–35%[24] when compared against the first increment (acquired with the minimum gradient amplitude, G_{min}, which may be 0). Traditionally, this is done 'by hand' by acquiring individual increments of the pseudo-2D diffusion experiment, checking the attenuation and adjusting the parameters accordingly [83, 84].

This task could of course be delegated to a computer program such as POISE. We *could* simply tackle it head-on by optimising all of Δ, δ and G_{max} at once (assuming that G_{min} is fixed); however, this is actually rather inefficient. Firstly, we previously saw that multiple-parameter optimisations took a longer time than the corresponding single-parameter optimisations. However, on top of that, it is often not even necessary to change all three parameters: the Stejskal–Tanner equation shows that the diffusion attenuation is controlled only by the overall *product* $\delta^2 G_{\text{max}}^2 \Delta'$. Thus, at least to a first approximation, the individual values of Δ, δ and G_{max} do not actually matter. It makes sense to fix two of these and optimise one parameter, only changing the other two if necessary.

It should be mentioned here that obtaining *accurate* values of D requires careful consideration of various factors such as the exact DOSY sequence and gradient shapes used [85], peak overlap [86–90] and convection [91, 92]. This may influence the type of processing which is chosen for extraction of diffusion coefficients. However,

[24] Recommendations tend to vary.

these issues are not our concern here: the role of POISE is only to find a good set of acquisition parameters such that the resulting data are amenable towards further processing.

3.4.10.1 Optimisation Setup

The overall strategy for this 'optimisation' can be briefly summarised in a flowchart (Fig. 3.30). Note that this is *not* a classic optimisation which falls nicely into the POISE workflow (Fig. 3.1): thus, the entire procedure cannot simply be summarised into one POISE routine. However, it does contain two optimisation subproblems: one to adjust Δ if necessary, and one to find G_{max}. We can therefore create a wrapper script for the overall procedure (here called `dosy_opt.py`), which invokes POISE to solve each of the two individual subproblems. This approach is made possible by the command-line interface of POISE, which allows for the creation and execution of optimisation routines. We also exploit the fact that during the optimisation, POISE stores the value of the cost function as the `TI` parameter in TopSpin. This fact is often irrelevant as the value of the cost function is not often of great interest, but proves to be handy in this particular instance as it provides a mechanism by which POISE can pass information back to the wrapper script. Previously, we explored situations where the functionality of POISE could be extended from *within*, e.g. by invoking

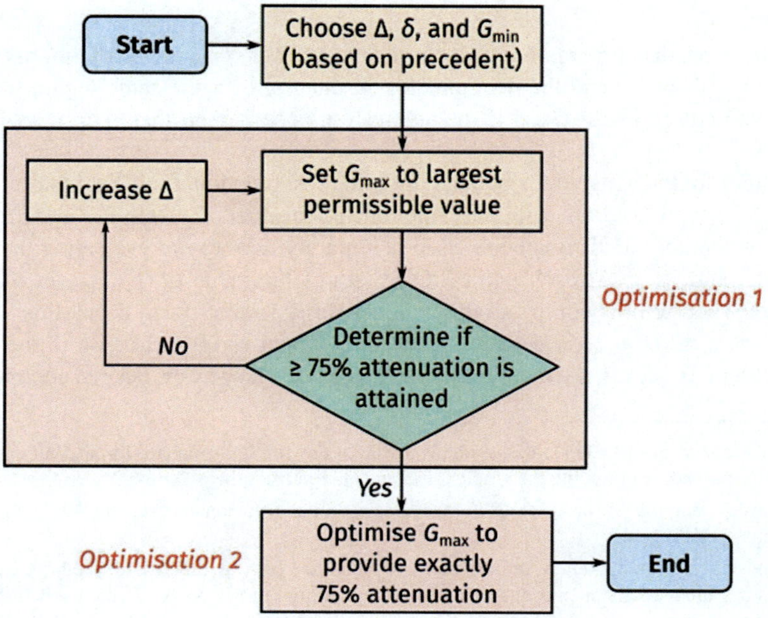

Fig. 3.30 Flowchart for DOSY parameter optimisation. This procedure consists of two subproblems which can be individually solved using POISE

different scripts in its acquisition AU program. In this case, we are going in the other direction, allowing POISE to be incorporated in other scripts.

The DOSY optimisation procedure uses a 1D pulse program throughout: this should essentially be a single increment of the desired pseudo-2D DOSY experiment.[25] In this work, I specifically used the Oneshot DOSY sequence [93] (Fig. 3.28c): this sequence has the advantage of not requiring onerously long phase cycles, which would otherwise lead to very long FEs and optimisations.[26] In principle, though, the optimisation can be run with any DOSY variant, including convection-compensated sequences [94–96]: the `dosy_opt.py` wrapper script is sufficiently general that only minor modifications are needed to use it with other DOSY sequences.

The first optimisation subproblem is the determination of an appropriate diffusion delay Δ. Since larger diffusion delays lead to a greater loss of signal through T_1 relaxation, as well as more convection [91], Δ should always be kept as short as possible.[27] Since the procedure in Fig. 3.30 only *increases* Δ, the wrapper script should always be started with the smallest value of Δ which is sensible for the sample under study: in the present case, I used an initial value of 50 ms. The next step is to determine whether using the full range of available gradient amplitudes leads to sufficient attenuation (which I defined to be 75%), relative to the first increment of the DOSY acquired using G_{min}. This involves setting G_{max} to the largest permissible value (for the Oneshot sequence, this is 80%), then calculating a cost function $f_{\text{dosy,aux}}$:

$$f_{\text{dosy,aux}} = \frac{\sum_i S_{\text{re},i}(G_{\text{max}})}{\sum_i S_{\text{re},i}(G_{\text{min}})} - 0.25. \tag{3.37}$$

In the fraction, the numerator represents an integral over some region of the spectrum acquired with G_{max}, and the denominator an integral over the same region for the spectrum acquired with G_{min}. If the attenuation is sufficient, then $f_{\text{dosy,aux}}$ will be negative.

In order for the wrapper script to monitor this cost function, POISE is instructed to perform an 'optimisation' with only one function evaluation, using the `--maxfev 1` flag: essentially, POISE is being used here purely as a way of evaluating the cost function $f_{\text{dosy,aux}}$.[28] At each point, POISE stores the value of $f_{\text{dosy,aux}}$ in the `TI` parameter and returns control to the wrapper script so that it can check this value. If the value of $f_{\text{dosy,aux}}$ is negative, then the wrapper script moves on to the next subproblem. However, if it is still positive, then there is no way the desired attenuation

[25] 1D versions of most DOSY variants are available in the TopSpin standard library, but can also be easily obtained by removing the gradient incrementation from the pseudo-2D versions.

[26] However, using *too few* scans leads to inaccuracies in the diffusion coefficients [93]. I chose to use 4 scans per FE here.

[27] In hindsight, it would likely have been better to increase δ instead: this would also have a larger effect as the attenuation depends quadratically on δ but only linearly on Δ. Care must be taken to not increase δ beyond what the spectrometer can tolerate, though.

[28] Of course, this cost function evaluation could just be done in the wrapper script itself. In fact, it would likely be simpler to do so. But there is one clear benefit of doing it in this rather circuitous route: POISE allows the user to write cost functions in Python 3.

can be accomplished using the gradient strengths available. Thus, Δ is increased by a fixed amount (50 ms in this case), and the sub-optimisation step repeated.[29]

Once a suitable value of Δ has been found, the script then searches for the correct value of G_{max} in order to provide *exactly* the 'correct' amount of attenuation (in this case, 75%). This is done through a conventional POISE optimisation, where the cost function is

$$f_{\text{dosy}} = |f_{\text{dosy,aux}}| = \left| \frac{\sum_i S_{\text{re},i}(G_{\text{max}})}{\sum_i S_{\text{re},i}(G_{\text{min}})} - 0.25 \right|. \tag{3.38}$$

The absolute value here makes sure that the optimisation converges to the value of G_{max} which gives *precisely* a 75% attenuation, no more and no less. At this point, the wrapper script reports the appropriate values of Δ and G_{max}, and then exits; these values can then be used to set up a DOSY experiment (e.g. using the `dosy` TopSpin AU program).

Before presenting some results, it is worth contemplating what would happen if the sample under study contained a mixture of different molecules, as-is often the case in DOSY experiments. Assuming that the integrals $\sum_i S_{\text{re},i}$ are taken over the entire spectrum, the parameters thus obtained would then be 'compromise values' which are weighted more heavily towards the major component(s) in the sample. This spectral region being evaluated can be changed by the user (via the `dpl` command), but that relies on some knowledge of the sample being studied. Thus, in this case, our best hope is really that the 'compromise' parameters obtained are *good enough* for all the components in the mixture. Obviously, this can hardly be guaranteed, and the larger the difference in the diffusion coefficients of the constituent species, the more prone this approach is to failing.

In my defence, any script which seeks to optimise DOSY parameters must account for the presence of mixtures in some way, for example by identifying rapidly and slowly diffusing species. This is therefore not truly an issue with POISE *per se*, but rather with the approach taken in the wrapper script. (For example, if the wrapper script can identify distinct resonances which have different attenuation profiles, then POISE could be used to perform individual optimisations on each of these.) Nevertheless, for this reason, I refrain from claiming that the work in this section constitutes a *full* method for automating DOSY.[30] I am more comfortable labelling this as a demonstration of how POISE can be integrated into larger procedures—for example, in the setup of single-component DOSY experiments.

[29] If sample spinning is used, then this incrementation of Δ needs to be chosen carefully, as the diffusion delay should always correspond to an integral number of rotations.

[30] Doing such an endeavor proper justice would anyway have required rather more time in my DPhil and more space in this thesis.

3.4.10.2 Optimisation Results

The DOSY optimisation procedure described above was tested on the sample of andrographolide (in DMSO-d_6). Despite this being a single-component sample, it does at least have some mildly interesting behaviour, in that the exchangeable OH protons have a larger apparent diffusion coefficient compared to the CH protons.

As described previously, the procedure was initialised with a value of $\Delta = 50ms$. The gradient duration δ, which is not modified, was set to 2 ms. The first sub-optimisation always increased the value of the diffusion delay by one step to 100 ms. Since this part only uses one function evaluation in POISE, the optimisation algorithm used does not actually matter. However, the algorithm does matter for the second sub-optimisation (which searches for the best value of G_{max}). This part was therefore performed five times per algorithm, in line with previous optimisations in this chapter (Table 3.26).

In this case, all three algorithms converged to the same result of 75%. The full Oneshot DOSY spectrum was then acquired using these parameters of $\Delta = 100$ms, $\delta = 2$ms, and $G_{\text{max}} = 75\%$ (nominally 49.3 G cm^{-1} on the spectrometer used here; see table 0.1). The overall time taken was approximately five minutes: two for the first sub-optimisation to find Δ, and three for the second sub-optimisation (as shown in Table 3.26). This figure may of course be larger depending on the number of scans used (here I used `DS=2` and `NS=4`), as well as the number of iterations required in the first sub-optimisation.

The resulting attenuation profiles for one CH and one OH peak are shown in Fig. 3.31. The apparently more slowly diffusing CH peak is attenuated to 28% of its original intensity, whereas the exchanging OH peak is attenuated to 20%. This is certainly a case where the 'different components' have diffusion coefficients which are not too dissimilar, allowing reasonable results to be obtained. Gaussian fitting using the modified Stejskal–Tanner equation

$$I(G) = I(0)\exp\left\{-(\gamma\delta G)^2 D\left[\Delta - \frac{\delta(5-3\alpha^2)}{16} - \frac{\tau(1-\alpha^2)}{2}\right]\right\} \tag{3.39}$$

Table 3.26 Results of maximum gradient amplitude sub-optimisations for DOSY parameter setup. The POISE routine used here is: `{"name": "dosy", "pars": ["gpz1"],"lb": [20.0], "ub": [80.0], "init": [50.0],"tol": [2.0], "cf": "dosy", "au": "poise_1d"}`.

Data code: `6A-200823`

Entry	Algorithm	Optimum found (%)	FEs	Time taken (s)
1	NM	75.00	9	195–197
2	MDS	75.00	9	196–197
3	BOBYQA	75.00–75.60	8–9	175–197

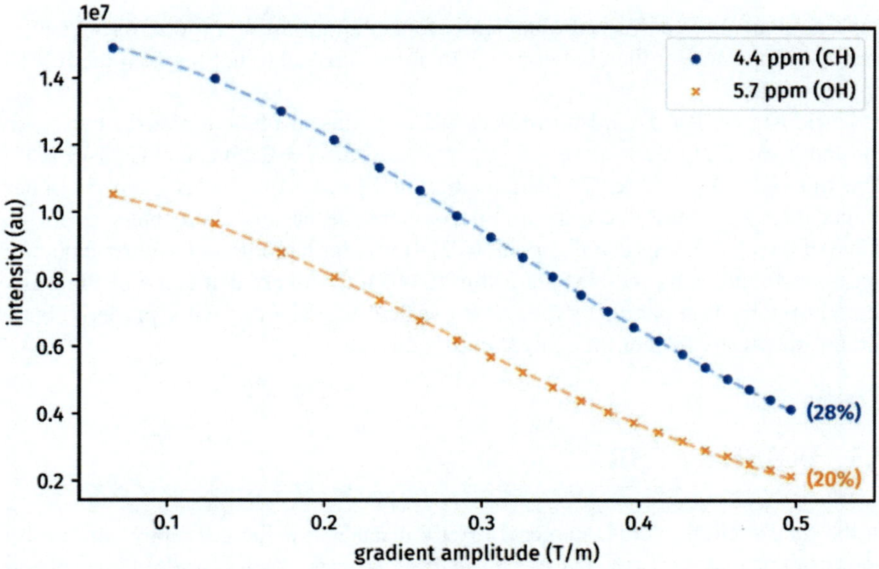

Fig. 3.31 Diffusion profiles obtained using a Oneshot experiment for a CH peak (blue, circles) and an OH peak (orange, crosses) in andrographolide. The dashed lines represent the Gaussian curves obtained through non-linear least-squares fitting (`scipy.optimise.curve_fit`). *Data code:* `6A-200823`

yielded *apparent* diffusion coefficients of $(7.411 \pm 0.013) \cdot 10^{-}9\mathrm{m}^2\mathrm{s}^{-}1$ for the CH peak and $(9.369 \pm 0.049) \cdot 10^{-}9\mathrm{m}^2\mathrm{s}^{-}1$ for the OH peak. (The equation was obtained from the pulse program provided on the University of Manchester website; here τ represents the time between the midpoints of each gradient within the bipolar pulse pair, and α the gradient imbalance factor as shown in Fig. 3.28c.)

3.4.10.3 Simultaneous Optimisation of Δ and G_{max}

It is possible—albeit inefficient, as previously mentioned—to dispense with the wrapper script entirely and perform a single optimisation of both Δ and G_{max}. In order to ensure that Δ is not increased too much, we must add a penalty term to the cost function:

$$f_{\mathrm{dosy,2p}} = f_{\mathrm{dosy}} + (\Delta/s) = \left| \frac{\sum_i S_{\mathrm{re},i}(G_{\mathrm{max}})}{\sum_i S_{\mathrm{re},i}(G_{\mathrm{min}})} - 0.25 \right| + (\Delta/s), \tag{3.40}$$

where ‘2p’ stands for two-parameter. One significant drawback of this is that the reference spectrum $\mathbf{S}(G_{\mathrm{min}})$ must be reacquired every time Δ is changed: so, each FE requires twice as long as before. For example, using the NM algorithm, the optimisation converged to $\Delta = 91.6\mathrm{ms}$ and $\mathrm{G}_{\mathrm{max}} = 79.3\%$. Although this result is quite similar to what we had obtained before, the optimisation took over 26 min, over five times longer than previously. I therefore elected to not perform the full 15 optimisations.

Curiously, BOBYQA did not work as well with this cost function: in the two times I tested it, BOBYQA converged to an 'optimum' of $\Delta = 225\text{ms}$ and $G_{max} = 50\%$. This did yield the expected 75% attenuation in the spectrum, but Δ is clearly rather longer than we would like it to be. Furthermore, because of the penalty term, the value of the cost function at this point (0.229) was far larger than the corresponding optimum found using the NM algorithm (0.095). t is likely that some of the trust-region optimisation parameters must be tweaked to make this work properly, but I did not spend any further time investigating this.

3.5 POISE for ESR

In the final section of this chapter, I briefly discuss how the concept of on-the-fly optimisation may be extended to pulsed ESR spectroscopy as well. Much of the underlying code from NMR-POISE was recycled for this, including the implementation of the three optimisation algorithms (NM, MDS, and BOBYQA). I added only one small difference in ESR-POISE, namely, a way to control the initial size of the simplex (for NM and MDS) or the trust-region radius (for BOBYQA): in NMR-POISE, this is fixed at ten times the desired optimisation tolerance. The overall optimisation framework (Fig. 3.1), as well as the concept of the optimisation routine (Sect. 3.2.1), are retained: this preserves the generality of POISE, which is probably its greatest strength.

Naturally, TopSpin-specific sections of the code had to be rewritten to instead be compatible with Bruker's Xepr software. This in fact proved to be easier than anticipated: most of the TopSpin code could be completely deleted, because Xepr provides an API which can be accessed from external programs such as Python 3. Thus, ESR-POISE is in fact completely written in Python 3, and there is no need for an artificial separation between 'frontend' and 'backend' (which also neatly circumvents most of the issues discussed in Sect. 3.2.5). The Xepr-facing code was written by Jean-Baptiste Verstraete (University of Oxford) and David Goodwin (University of Southampton, formerly Oxford).

A number of applications in ESR-POISE were explored. All ESR experimental work was done by Jean-Baptiste Verstraete and William Myers (University of Oxford); thus, I only provide very brief descriptions here. The examples include:

- optimisation of the signal phase in a simple spin echo experiment, by maximising the intensity of the detected echo;
- similarly, calibration of 90° and 180° pulse widths and powers in a spin echo;
- optimisation of an inversion pulse used on the pump channel in a DEER experiment [97], in order to increase the modulation depth of the resulting DEER traces (which reveal dipolar couplings between electrons);

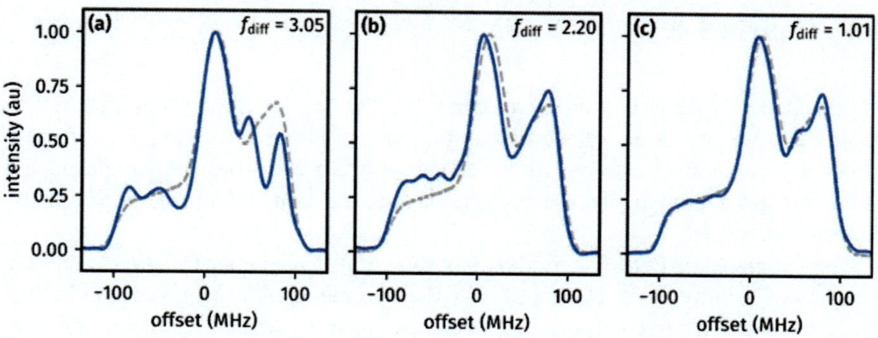

Fig. 3.32 Comparison between CHORUS spectrum (blue, solid line) and field sweep profile (grey, dashed line): **a** without any optimisation of CHORUS, **b** after optimisation of the CHORUS amplitudes, and **c** using a compensated CHORUS pulse obtained through determination of the resonator transfer function. The value of f_{diff} is indicated on each plot. The sample used was a bisnitroxide; further experimental details are provided in the paper [2]. The data used for this figure were acquired by Jean-Baptiste Verstraete.

- calibration of shaped pulse amplitudes for CHORUS broadband excitation [98, 99];
- compensation for resonator distortions when applying shaped pulses, again demonstrated using the CHORUS sequence.

An example of the results obtained via the latter two optimisations is shown in Fig. 3.32. In Fig. 3.32a, the CHORUS broadband excitation experiment was run without any optimisations: the resulting spectrum (blue, solid) is compared against that obtained via a standard field sweep (grey, dashed). Clearly, there are substantial mismatches. Figure 3.32b shows the improved performance of CHORUS after a simple optimisation of the shaped pulse amplitudes, using the spectral intensity as the cost function (or the negative thereof, since we seek to maximise it). It can be seen that the POISE optimisation allows for a much more accurate spectrum to be obtained.

This can be further improved, however, by compensating for distortions in the pulse shape caused by the resonator. Typically, this is a time-consuming process, requiring the determination of a transfer function: in this case, however, it can be performed in a fully automated fashion using POISE. The coefficients in the transfer function are used as optimisation parameters and are used within each function evaluation to back-calculate a compensated pulse shape: the cost function used is the difference between the compensated CHORUS spectrum and the field sweep (f_{diff}, as defined in Eq. (2.33)). At the point where the cost function is minimised, the transfer function parameters will have been accurately determined. The final compensated CHORUS spectrum is shown in Fig. 3.32c.

3.6 Conclusion

In this chapter, I described the development of POISE, a Python program for optimising any set of (numeric) acquisition parameters in Bruker's TopSpin software. This allows experiments to be tailored to the sample and instrument being used and can be run in a completely automated fashion once the requisite information has been defined in a routine.

The examples of POISE optimisations shown in this chapter (Sect. 3.4) include a variety of solution-state 1D and 2D NMR experiments. However, because POISE was intentionally designed to be *general*, it is easy to adapt it to essentially any NMR experiment, as long as a suitable cost function can be constructed. Thus, it is not difficult to envision POISE being used in other fields such as biomolecular and solid-state NMR, and extensions in this direction are worth pursuing: the idea of applying POISE to the cross-polarisation technique [100] was briefly floated, but never actually explored. More broadly speaking, real-time experimental optimisation can be extended to other forms of spectroscopy: the successful development of ESR-POISE (Sect. 3.5) bears witness to this, although this is rather more involved as it entails writing instrument- and manufacturer-specific code.

Returning to the confines of solution-state NMR, though, it is worth considering the types of situations in which POISE is likely to find most use. In this chapter, I have asserted that POISE is *fast*, being able to run optimisations in a matter of seconds to minutes. However, perhaps with the exception of the pulse width calibration in Sect. 3.4.1, I did not perform any comparisons (either in terms of time, or the quality of the optimum found) against a 'skilled user' such as an experienced spectroscopist. Without such data, I cannot justly claim that POISE should be used to replace an expert. Nonetheless, there are many situations in which experts are not available: most notably, when samples are run in a high-throughput 'automation' manner. In my view, it is these situations in which an optimisation program can best be used: the POISE routine need only be set up once and can thereafter be incorporated into data acquisition procedures. A simple example would be to determine T_1 values for a compound (which, as shown in Sect. 3.4.3, takes 2–3 min), so that recovery delays in 2D experiments can be appropriately adjusted. Other caveats regarding the usage of POISE have already been covered in Sect. 3.3. Even bearing these in mind, though, the results in this chapter form a strong argument for the inclusion of POISE in a spectroscopist's toolbox.

References

1. Yong JRJ, Foroozandeh M (2021) On-the-fly, sample-tailored optimization of NMR experiments. Anal Chem 93(10735–10739):767. https://doi.org/10.1021/acs.analchem.1c01
2. Verstraete J-B, Yong JRJ, Goodwin DL, Myers WK, Foroozandeh M (2022) Towards fully optimised and automated ESR spectroscopy. Chem Commun 58:10715–10718. https://doi.org/10.1039/d2cc02742a

3. Bardeen CJ, Yakovlev VV, Wilson KR, Carpenter SD, Weber PM, Warren WS (1997) Feedback quantum control of molecular electronic population transfer. Chem Phys Lett 280:151–158. https://doi.org/10.1016/S0009-2614(97)01081-6
4. Schiano JL, Routhier T, Blauch AJ, Ginsberg MD (1999) Feedback optimization of pulse width in the SORC sequence. J Magn Reson 140:84–90. https://doi.org/10.1006/jmre.1999.1824
5. Schiano JL, Blauch AJ, Ginsberg MD (2000) Optimization of NQR pulse parameters using feedback control. Z Naturforsch A: Phys Sci 55:67–73. https://doi.org/10.1515/zna-2000-1-213
6. Monea C (2020) Optimization of NQR excitation sequences using black-box techniques. J Magn Reson 321:106858. https://doi.org/10.1016/j.jmr.2020.106858
7. Goodwin DL, Myers WK, Timmel CR, Kuprov I (2018) Feedback control optimisation of ESR experiments. J Magn Reson 297:9–16. https://doi.org/10.1016/j.jmr.2018.09.009
8. De Paëpe G, Hodgkinson P, Emsley L (2003) Improved heteronuclear decoupling schemes for solid-state magic angle spinning NMR by direct spectral optimization. Chem Phys Lett 376:259–267. https://doi.org/10.1016/S0009-2614(03)00966-7
9. Elena B, de Paëpe G, Emsley L (2004) Direct spectral optimisation of proton-proton homonuclear dipolar decoupling in solid-state NMR. Chem Phys Lett 398:532–538. https://doi.org/10.1016/j.cplett.2004.09.122
10. Salager E, Dumez J-N, Stein RS, Steuernagel S, Lesage A, Elena-Herrmann B, Emsley L (2010) Homonuclear dipolar decoupling with very large scaling factors for highresolution ultrafast magic angle spinning 1H solid-state NMR spectroscopy. Chem Phys Lett 498:214–220. https://doi.org/10.1016/j.cplett.2010.08.038
11. Skinner TE, Reiss TO, Luy B, Khaneja N, Glaser SJ (2003) Application of optimal control theory to the design of broadband excitation pulses for high-resolution NMR. J Magn Reson 163:8–15. https://doi.org/10.1016/s1090-7807(03)00153-8
12. Kobzar K, Skinner TE, Khaneja N, Glaser SJ, Luy B (2004) Exploring the limits of broadband excitation and inversion pulses. J Magn Reson 170:236–243. https://doi.org/10.1016/j.jmr.2004.06.017
13. Khaneja N, Reiss T, Kehlet C, Schulte-Herbrüggen T, Glaser SJ (2005) Optimal control of coupled spin dynamics: design of NMR pulse sequences by gradient ascent algorithms. J Magn Reson 172:296–305. https://doi.org/10.1016/j.jmr.2004.11.004
14. Kobzar K, Skinner TE, Khaneja N, Glaser SJ, Luy B (2008) Exploring the limits of broadband excitation and inversion: II. Rf-power optimized pulses. J Magn Reson 194:58–66. https://doi.org/10.1016/j.jmr.2008.05.023
15. Kobzar K, Ehni S, Skinner TE, Glaser SJ, Luy B (2012) Exploring the limits of broadband 90° and 180° universal rotation pulses. J Magn Reson 225:142–160, https://doi.org/10.1016/j.jmr.2012.09.013
16. Schilling F, Warner LR, Gershenzon NI, Skinner TE, Sattler M, Glaser SJ (2014) Next-generation heteronuclear decoupling for high-field biomolecular NMR spectroscopy. Angew Chem, Int Ed 53:4475–4479. https://doi.org/10.1002/anie.201400178
17. Glaser SJ, Boscain U, Calarco T, Koch CP, Köckenberger W, Kosloff R, Kuprov I, Luy B, Schirmer S, Schulte-Herbrüggen T, Sugny D, Wilhelm FK (2015) Training Schrödinger's cat: quantum optimal control. Eur Phys J D 69:279. https://doi.org/10.1140/epjd/e2015-60464-1
18. Geen H, Wimperis S, Freeman R (1989) Band-selective pulses without phase distortion: a simulated annealing approach. J Magn Reson 85:620–627. https://doi.org/10.1016/0022-2364(89)90254-0
19. Emsley L, Bodenhausen G (1990) Gaussian pulse cascades: new analytical functions for rectangular selective inversion and in-phase excitation in NMR. Chem Phys Lett 165:469–476. https://doi.org/10.1016/0009-2614(90)87025-m
20. Geen H, Freeman R (1991) Band-selective radiofrequency pulses. J Magn Reson 93:93–141. https://doi.org/10.1016/0022-2364(91)90034-q
21. Nuzillard JM, Freeman R (1994) Band-selective pulses designed to accommodate relaxation. J Magn Reson, Ser A 107:113–118. https://doi.org/10.1006/jmra.1994.1056

22. Kupce E, Freeman R (1995) Band-selective correlation spectroscopy. J Magn Reson, Ser A 112:134–137. https://doi.org/10.1006/jmra.1995.1023
23. Kupce E, Boyd J, Campbell ID (1995) Short selective pulses for biochemical applications. J Magn Reson, Ser B 106:300–303. https://doi.org/10.1006/jmrb.1995.1049
24. Shaka AJ, Barker PB, Freeman R (1985) Computer-optimized decoupling scheme for wide-band applications and low-level operation. J Magn Reson 64:547–552. https://doi.org/10.1016/0022-2364(85)90122-2
25. Freeman R, Xili W (1987) Design of magnetic resonance experiments by genetic evolution. J Magn Reson 75:184–189. https://doi.org/10.1016/0022-2364(87)90331-3
26. Bechmann M, Clark J, Sebald A (2013) Genetic algorithms and solid state NMR pulse sequences. J Magn Reson 228:66–75. https://doi.org/10.1016/j.jmr.2012.12.015
27. Ehni S, Luy B (2014) Robust INEPT and refocused INEPT transfer with compensation of a wide range of couplings, offsets, and B 1 -field inhomogeneities (COB3). J Magn Reson 247:111–117. https://doi.org/10.1016/j.jmr.2014.07.010
28. Lapin J, Nevzorov AA (2020) De novo NMR pulse sequence design using Monte-Carlo optimization techniques. J Magn Reson 310:106641. https://doi.org/10.1016/j.jmr.2019.106641
29. Sakellariou D, Lesage A, Hodgkinson P, Emsley L (2000) Homonuclear dipolar decoupling in solid-state NMR using continuous phase modulation. Chem Phys Lett 319:253–260. https://doi.org/10.1016/s0009-2614(00)00127-5
30. Nocedal J, Wright SJ (2006) Numerical optimization, 2nd edn. Springer, New York
31. Nelder JA, Mead R (1965) A simplex method for function minimization. Comput J 7:308–313. https://doi.org/10.1093/comjnl/7.4.308
32. Torczon VJ (1989) Multidirectional search: A direct search algorithm for parallel machines, Ph.D. Rice University
33. Dennis JE Jr, Torczon V (1991) Direct search methods on parallel machines. SIAM J Optim 1:448–474. https://doi.org/10.1137/0801027
34. Powell MJD (2009) The BOBYQA algorithm for bound constrained optimization without derivatives, Technical report DAMTP 2009/NA06. University of Cambridge
35. Cartis C, Fiala J, Marteau B, Roberts L (2019) Improving the flexibility and robustness of model-based derivative-free optimization solvers. ACM Trans Math Softw 45:1–41. https://doi.org/10.1145/3338517
36. Spendley W, Hext GR, Himsworth FR (1962) Sequential application of simplex designs in optimisation and evolutionary operation. Technometrics 4:441–461. https://doi.org/10.1080/00401706.1962.10490033
37. Torczon V (1991) On the convergence of the multidirectional search algorithm. SIAM J Optim 1:123–145. https://doi.org/10.1137/0801010
38. Powell MJD (2003) On trust region methods for unconstrained minimization without derivatives. Math Program 97(605–623):30–6. https://doi.org/10.1007/s10107-003-04
39. Cartis C, Roberts L, Sheridan-Methven O (2022) Escaping local minima with local derivativefree methods: a numerical investigation. Optimization 71:2343–2373. https://doi.org/10.1080/02331934.2021.1883015
40. Song Y-Q, Tang Y, Hürlimann MD, Cory DG (2018) Real-time optimization of nuclear magnetic resonance experiments. J Magn Reson 289:72–78. https://doi.org/10.1016/j.jmr.2018.02.009
41. Tang Y, Song Y-Q (2019) Realtime optimization of multidimensional NMR spectroscopy on embedded sensing devices. Sci Rep 9. https://doi.org/10.1038/s41598-019-53929-1
42. Eghbalnia HR, Bahrami A, Tonelli M, Hallenga K, Markley JL (2005) High-resolution iterative frequency identification for NMR as a general strategy for multidimensional data collection. J Am Chem Soc 127:12528–12536. https://doi.org/10.1021/ja052120i
43. Hansen AL, Brüschweiler R (2016) Absolute minimal sampling in high-dimensional NMR spectroscopy. Angew Chem Int Ed 55:14169–14172. https://doi.org/10.1002/anie.201608048
44. Corporation B (2022) Advanced acquisition software application | NMR software https://www.bruker.com/en/products-and-solutions/mr/nmr-software/smartdrivenmr.html. Accessed 22 Aug 2022

45. Keifer PA (1999) 90° pulse width calibrations: how to read a pulse width array. Concepts Magn Reson 11:165–180, https://doi.org/10.1002/(SICI)1099-0534(1999)11:3<165::AIDCMR4>3.0.CO;2-D
46. Wu PSC, Otting G (2005) Rapid pulse length determination in high-resolution NMR. J Magn Reson 176:115–119. https://doi.org/10.1016/j.jmr.2005.05.018
47. Bloch F (1946) Nuclear induction. Phys Rev 70:460–474. https://doi.org/10.1103/physrev.70.460
48. Ernst RR, Anderson WA (1966) Application of Fourier transform spectroscopy to magnetic resonance. Rev Sci Instrum 37:93–102. https://doi.org/10.1063/1.1719961
49. Waugh JS (1970) Sensitivity in Fourier transform NMR spectroscopy of slowly relaxing systems. J Mol Spectrosc 35:298–305. https://doi.org/10.1016/0022-2852(70)90205-5
50. Traficante DD (1992) Optimum tip angle and relaxation delay for quantitative analysis. Concepts Magn Reson 4:153–160. https://doi.org/10.1002/cmr.1820040204
51. Reynolds WF, Enríquez RG (2002) Choosing the best pulse sequences, acquisition parameters, postacquisition processing strategies, and probes for natural product structure elucidation by NMR spectroscopy. J Nat Prod 65:221—244. https://doi.org/10.1021/np010444o
52. Burns DC, Reynolds WF (2021) Minimizing the risk of deducing wrong natural product structures from NMR data. Magn Reson Chem 59:500–533. https://doi.org/10.1002/mrc.4933
53. Pauli GF, Jaki BU, Lankin DC (2005) Quantitative 1H NMR: development and potential of a method for natural products analysis. J Nat Prod 68:133–149. https://doi.org/10.1021/np0497301
54. Giraudeau P (2014) Quantitative 2D liquid-state NMR. Magn Reson Chem 52:259–272. https://doi.org/10.1002/mrc.4068
55. Christensen KA, Grant DM, Schulman EM, Walling C (1977) Optimal determination of relaxation times of Fourier transform nuclear magnetic resonance: determination of spin-lattice relaxation times in chemically polarized species. J Phys Chem 78:1971–1977. https://doi.org/10.1021/j100612a022
56. Homer J, Beevers MS (1985) Driven-equilibrium single-pulse observation of ..1 relaxation: a reevaluation of a rapid "new" method for determining NMR spin-lattice relaxation times. J Magn Reson 63:287–297. https://doi.org/10.1016/0022-2364(85)90318-x
57. Loening NM, Thrippleton MJ, Keeler J, Griffin RG (2003) Single-scan longitudinal relaxation measurements in high-resolution NMR spectroscopy. J Magn Reson 164:321–328. https://doi.org/10.1016/s1090-7807(03)00186-1
58. Smith PES, Donovan KJ, Szekely O, Baias M, Frydman L (2013) Ultrafast NMR ..1 relaxation measurements: probing molecular properties in real time. Chem Phys Chem 14:3138–3145. https://doi.org/10.1002/cphc.201300436
59. Wei R, Dickson CL, Uhrin D, Lloyd-Jones GC (2021) Rapid estimation of ..1 for quantitative NMR. J Org Chem 86:9023–9029. https://doi.org/10.1021/acs.joc.1c01007
60. Schulze-Sunninghausen D, Becker J, Luy B (2014) Rapid heteronuclear single quantum correlation NMR spectra at natural abundance. J Am Chem Soc 136:1242–1245. https://doi.org/10.1021/ja411588d
61. Schulze-Sunninghausen D, Becker J, Koos MRM, Luy B (2017) Improvements, extensions, and practical aspects of rapid ASAP-HSQC and ALSOFAST-HSQC pulse sequences for studying small molecules at natural abundance. J Magn Reson 281:151–161. https://doi.org/10.1016/j.jmr.2017.05.012
62. Koos MRM, Luy B (2019) Polarization recovery during ASAP and SOFAST/ALSOFASTtype experiments. J Magn Reson 300:61–75. https://doi.org/10.1016/j.jmr.2018.12.014
63. Frydman L, Scherf T, Lupulescu A (2002) The acquisition of multidimensional NMR spectra within a single scan. Proc Natl Acad Sci USA 99:15858–15862. https://doi.org/10.1073/pnas.252644399
64. Pelupessy P (2003) Adiabatic single scan two-dimensional NMR spectrocopy. J Am Chem Soc 125:12345–12350. https://doi.org/10.1021/ja034958g

65. Frydman L, Lupulescu A, Scherf T (2003) Principles and features of single-scan two-dimensional NMR spectroscopy. J Am Chem Soc 125:9204–9217. https://doi.org/10.1021/ja030055b
66. Tal A, Frydman L (2010) Single-scan multidimensional magnetic resonance. Prog Nucl Magn Reson Spectrosc 57:241–292. https://doi.org/10.1016/j.pnmrs.2010.04.001
67. Giraudeau P, Frydman L (2014) Ultrafast 2D NMR: an emerging tool in analytical spectroscopy. Annu Rev Anal Chem 7(129–161):213–020208. https://doi.org/10.1146/annurev-anchem-071
68. Gouilleux B, Rouger L, Giraudeau P (2018) Ultrafast 2D NMR: methods and applications. Annu Rep NMR Spectrosc 75–144. https://doi.org/10.1016/bs.arnmr.2017.08.003
69. Kupče E, Frydman L, Webb AG, Yong JRJ, Claridge TDW (2021) Parallel nuclear magnetic resonance spectroscopy. Nat Rev Methods Prim 1:27. https://doi.org/10.1038/s43586-021-00024-3
70. Cicero DO, Barbato G, Bazzo R (2001) Sensitivity enhancement of a two-dimensional experiment for the measurement of heteronuclear long-range coupling constants, by a new scheme of coherence selection by gradients. J Magn Reson 148:209–213. https://doi.org/10.1006/jmre.2000.2234
71. Nielsen NC, Bildsoe H, Jakobsen HJ, Sorensen OW (1986) Pulse techniques for calibration of the decoupler radiofrequency field strength. J Magn Reson 66:456–469. https://doi.org/10.1016/0022-2364(86)90189-7
72. Meissner A, Sorensen OW (2000) Economizing spectrometer time and broadband excitation in small-molecule heteronuclear NMR correlation spectroscopy: broadband HMBC. Magn Reson Chem 38:981–984. https://doi.org/10.1002/1097-458x(200011)38:11<98 1::aid-mrc778>3.0.co;2-l
73. Krivdin LB (2018) Theoretical calculations of carbon-hydrogen spin-spin coupling constants. Prog Nucl Magn Reson Spectrosc 108(17–73):002. https://doi.org/10.1016/j.pnmrs.2018.10
74. Wolpert DH, Macready WG (1997) No free lunch theorems for optimization. IEEE Trans Evol Computat 1:67–82. https://doi.org/10.1109/4235.585893
75. Kupče E, Claridge TDW (2018) Molecular structure from a single NMR supersequence. Chem Commun 54:7139–7142. https://doi.org/10.1039/c8cc03296c
76. Kupče E, Claridge TDW (2019) New NOAH modules for structure elucidation at natural isotopic abundance. J Magn Reson 307(106568):6568. https://doi.org/10.1016/j.jmr.2019.10
77. Foroozandeh M, Adams RW, Meharry NJ, Jeannerat D, Nilsson M, Morris GA (2014) Ultrahigh-resolution NMR spectroscopy. Angew Chem, Int Ed 53:6990–6992. https://doi.org/10.1002/anie.201404111
78. Foroozandeh M, Morris GA, Nilsson M (2018) PSYCHE pure shift NMR spectroscopy. Chem Eur J 24:13988–14000. https://doi.org/10.1002/chem.201800524
79. Hore PJ (1983) Solvent suppression in Fourier transform nuclear magnetic resonance. J Magn Reson 55:283–300. https://doi.org/10.1016/0022-2364(83)90240-8
80. Zheng G, Price WS (2010) Solvent signal suppression in NMR. Prog Nucl Magn Reson Spectrosc 56:267–288. https://doi.org/10.1016/j.pnmrs.2010.01.001
81. Giraudeau P, Silvestre V, Akoka S (2015) Optimizing water suppression for quantitative NMR-based metabolomics: a tutorial review. Metabolomics 11:1041–1055. https://doi.org/10.1007/s11306-015-0794-7
82. Mckay RT (2011) How the 1D-NOESY suppresses solvent signal in metabonomics NMR spectroscopy: an examination of the pulse sequence components and evolution. Concepts Magn Reson 38A:197–220. https://doi.org/10.1002/cmr.a.20223
83. Johnson Jr CS (1999) Diffusion ordered nuclear magnetic resonance spectroscopy: principles and applications. Prog Nucl Magn Reson Spectrosc 34:203–256. https://doi.org/10.1016/s0079-6565(99)00003-5
84. Claridge TDW (2016) High-resolution NMR techniques in organic chemistry, 3rd edn. Elsevier, Amsterdam
85. Sinnaeve D (2012) The Stejskal-Tanner equation generalized for any gradient shape–an overview of most pulse sequences measuring free diffusion. Concepts Magn Reson 40A:39–65. https://doi.org/10.1002/cmr.a.21223

86. Antalek B, Windig W (1996) Generalized rank annihilation method applied to a single multicomponent pulsed gradient spin echo NMR data set. J Am Chem Soc 118:10331–10332. https://doi.org/10.1021/ja962172v
87. Windig W, Antalek B (1997) Direct exponential curve resolution algorithm (DECRA): a novel application of the generalized rank annihilation method for a single spectral mixture data set with exponentially decaying contribution profiles. Chemom Intell Lab Syst 37:241–254. https://doi.org/10.1016/s0169-7439(97)00028-2
88. Nilsson M, Connell MA, Davis AL, Morris GA (2006) Biexponential fitting of diffusion-ordered nmr data: practicalities and limitations. Anal Chem 78:3040–3045. https://doi.org/10.1021/ac060034a
89. Nilsson M, Morris GA (2008) Speedy component resolution: an improved tool for processing diffusion-ordered spectroscopy data. Anal Chem 80:3777–3782. https://doi.org/10.1021/ac7025833
90. Colbourne AA, Morris GA, Nilsson M (2011) Local covariance order diffusion-ordered spectroscopy: a powerful tool for mixture analysis. J Am Chem Soc 133:7640–7643. https://doi.org/10.1021/ja2004895
91. Swan I, Reid M, Howe PWA, Connell MA, Nilsson M, Moore MA, Morris GA (2015) Sample convection in liquid-state NMR: why it is always with us, and what we can do about it. J Magn Reson 252:120–129. https://doi.org/10.1016/j.jmr.2014.12.006
92. Barbosa TM, Rittner R, Tormena CF, Morris GA, Nilsson M (2016) Convection in liquid-state NMR: expect the unexpected. RSC Adv 6:95173–95176. https://doi.org/10.1039/c6ra23427e
93. Pelta MD, Morris GA, Stchedroff MJ, Hammond SJ (2002) A one-shot sequence for high-resolution diffusion-ordered spectroscopy. Magn Reson Chem 40:S147–S152. https://doi.org/10.1002/mrc.1107
94. Jerschow A, Müller N (1997) Suppression of convection artifacts in stimulated-echo diffusion experiments: double-stimulated-echo experiments. J Magn Reson 125:372–375. https://doi.org/10.1006/jmre.1997.1123
95. Sørland GH, Seland JG, Krane J, Anthonsen HW (2000) Improved convection compensating pulsed field gradient spin-echo and stimulated-echo methods. J Magn Reson 142:323–325. https://doi.org/10.1006/jmre.1999.1941
96. Nilsson M, Morris GA (2005) Improving pulse sequences for 3D DOSY: convection compensation. J Magn Reson 177:203–211. https://doi.org/10.1016/j.jmr.2005.07.019
97. Pannier M, Veit S, Godt A, Jeschke G, Spiess HW (2000) Dead-time free measurement of dipole-dipole interactions between electron spins. J Magn Reson 142:331–340. https://doi.org/10.1006/jmre.1999.1944
98. Foroozandeh M, Nilsson M, Morris GA (2019) Improved ultra-broadband chirp excitation. J Magn Reson 302:28–33. https://doi.org/10.1016/j.jmr.2019.03.007
99. Verstraete J-B, Myers WK, Foroozandeh M (2021) Chirped ordered pulses for ultra-broadband ESR spectroscopy. J Chem Phys 154:094201. https://doi.org/10.1063/5.0038511
100. Pines A, Gibby MG, Waugh JS (1972) Proton-enhanced nuclear induction spectroscopy: a method for high resolution NMR of dilute spins in solids. J Chem Phys 56:1776–1777. https://doi.org/10.1063/1.1677439

Chapter 4
NOAH

'The power of doing any thing with quickness is always much prized by the possessor, and often without any attention to the imperfection of the performance.'

— JANE AUSTEN, *Pride and Prejudice*

This final chapter describes my work on *NOAH* (NMR by Ordered Acquisition using ^{1}H detection) *supersequences*, pulse sequences which record multiple 2D datasets (*'modules'*) in the time required for one. This is an attractive NMR technique for several reasons: the time savings are clearly a key factor, but the flexibility of being able to combine almost any set of modules also makes NOAH supersequences applicable to a variety of contexts.

I begin by introducing the concepts underlying NOAH supersequences, as well as a general discussion of the time savings (and sensitivity per unit time benefits) thus realised. I then describe the GENESIS (GENEration of Supersequences In Silico) website, written by me, which allows users to generate Bruker pulse programs for almost every imaginable NOAH supersequence. After this, my work on various aspects of the actual sequences themselves is described, with a special focus on newly developed and/or improved modules. Finally, the design of 'parallel' supersequences, which comprise multiple interleaved supersequences, is discussed.

This work was done in close collaboration with Ēriks Kupče (Bruker UK). However, all results and analysis shown in this chapter are mine. The work in this chapter forms the subject of several publications:

- Yong, J. R. J.; Hansen, A. L.; Kupče, Ē.; Claridge, T. D. W. Increasing sensitivity and versatility in NMR supersequences with new HSQC-based modules. *J. Magn. Reson.* **2021**, *329*, 107027, https://doi.org/10.1016/j.jmr.2021.107027.
- Kupče, Ē.; Yong, J. R. J.; Widmalm, G.; Claridge, T. D. W. Parallel NMR Supersequences: Ten Spectra in a Single Measurement. *JACS Au* **2021**, *1*, 1892–1897, https://doi.org/10.1021/jacsau.1c00423.

J. Yong, *Optimising NMR Spectroscopy Through Method and Software Development*, Springer Theses, https://doi.org/10.1007/978-3-031-46684-7_4

- Yong, J. R. J.; Kupče, Ē.; Claridge, T. D. W. Modular Pulse Program Generation for NMR Supersequences. *Anal. Chem.* **2022**, *94*, 2271–2278, https://doi.org/10.1021/acs.analchem.1c04964.
- Yong, J. R. J.; Kupče, Ē.; Claridge, T. D. W. A general scheme for generating NMR supersequences combining high- and low-sensitivity experiments. *Chem. Commun.* **2023**, *59*, 7827–7830, https://doi.org/10.1039/d3cc01472j.

The material in the introductory sections also closely follows two reviews which I have contributed to:

- Kupče, Ē.; Frydman, L.; Webb, A. G.; Yong, J. R. J.; Claridge, T. D. W. Parallel nuclear magnetic resonance spectroscopy. *Nat. Rev. Methods Primers* **2021**, *1*, No. 27, https://doi.org/10.1038/s43586-021-00024-3.
- Yong, J. R. J.; Kupče, Ē.; Claridge, T. D. W. In *Fast 2D solution-state NMR: concepts and applications*, Giraudeau, P., Dumez, J.-N., Eds.; Royal Society of Chemistry: London, UK, 2023, https://doi.org/10.1039/BK9781839168062-00084.

4.1 Introduction

The characterisation of small molecules and biomolecules by NMR spectroscopy relies on a suite of standard 2D NMR experiments, which seek to detect heteronuclear scalar couplings (e.g. HSQC and HMBC), homonuclear scalar couplings (e.g. COSY and TOCSY) or through-space interactions (e.g. NOESY and ROESY). Although such 2D experiments provide far superior resolution and information content compared to 1D spectra, they also require substantially longer experimental durations, as the indirect dimension must be constructed through the acquisition of many t_1 increments. This problem is further exacerbated by the fact that structural elucidation or verification often requires a combination of information from multiple different 2D experiments.

The acceleration of 2D NMR has thus proven to be a popular area of research. We may broadly categorise the existing techniques into two classes.[1] First come the methods which seek to directly speed up the acquisition of *individual* 2D spectra: these include non-uniform sampling (NUS) [7–10], fast pulsing (i.e. shortening of recovery delays) [11–14], ultrafast NMR [5, 15–19], Hadamard encoding [20, 21] and spectral aliasing [22–25].

On the other hand, *multiple-FID experiments* aim to collect two or more 2D spectra in the time required for one. The corresponding FIDs may either be detected at the same time (as in time-shared NMR [26–28] and multiple-receiver NMR [29–31]), or *sequentially*: this is the category which NOAH supersequences [5, 32, 33] fall under. NOAH supersequences are furthermore set apart from other sequential-acquisition

[1] These are by no means mutually exclusive: many of the techniques here can be combined to provide even greater efficiency.

experiments [34–39] by their 'modular' nature: they can be constructed from building blocks, or 'modules', which (usually) contain one FID each. Typically, each of these building blocks corresponds to a specific 2D experiment, as will be shown below. It should be noted that there are several other multiple-FID experiments which, while not explicitly advertised as 'NOAH experiments', are built from 'modules' and are conceptually identical [40–43]; I do not discuss these here.

The time savings provided by NOAH experiments stems from the elision of *recovery delays*—the time required for spins to relax to their equilibrium state (ρ_0 in Eq. (1.39)), such that the next transient or t_1 increment can be recorded. These recovery delays are responsible for most of the experimental duration of a typical 2D NMR experiment. In a NOAH supersequence, their removal is made possible by designing each module to draw on a different *magnetisation pool*: this refers to a source of magnetisation belonging to a specific isotopologue. For example, ^{13}C-bound protons and ^{12}C-bound protons are two different magnetisation pools. This allows modules to be directly concatenated, without the addition of extra recovery delays between them; only one overall recovery delay is required for the entire supersequence, where *all* magnetisation pools simultaneously recover.

In this section, I first provide a formal overview of how the time savings for NOAH experiments may be quantified and analysed. This includes an analysis of the conditions under which multiple-FID techniques such as NOAH are most valuable. Following this, I turn to the design of NOAH pulse sequences, using specific case studies to illustrate the general principles underlying these experiments.

4.1.1 Time Savings and Sensitivity Analyses

4.1.1.1 Time Savings in the Sampling-Limited Regime

The duration of an NMR experiment, τ_{exp}, can be expressed as a sum of its parts:

$$\tau_{\mathrm{exp}} = \tau_{\mathrm{ps}} + \tau_{\mathrm{acq}} + d_1, \tag{4.1}$$

where τ_{ps} is the time required for the pulse sequence itself (typically several milliseconds), τ_{acq} is the acquisition time (several hundred milliseconds), and d_1 is the recovery delay (one or more seconds). These timings are illustrated in Fig. 4.1 for a NOAH supersequence formed from two modules (HSQC and COSY). The *time-saving factor* for a NOAH supersequence, as compared to a series of conventional standalone experiments, is denoted ρ_t and may be expressed as

$$\rho_t = \frac{\sum_i \tau_{\mathrm{conv}}^{(i)}}{\tau_{\mathrm{NOAH}}} = \frac{\sum_i (\tau_{\mathrm{ps}}^{(i)} + \tau_{\mathrm{acq}}^{(i)} + d_1^{(i)})}{d_1 + \sum_i (\tau_{\mathrm{ps}}^{(i)} + \tau_{\mathrm{acq}}^{(i)})}, \tag{4.2}$$

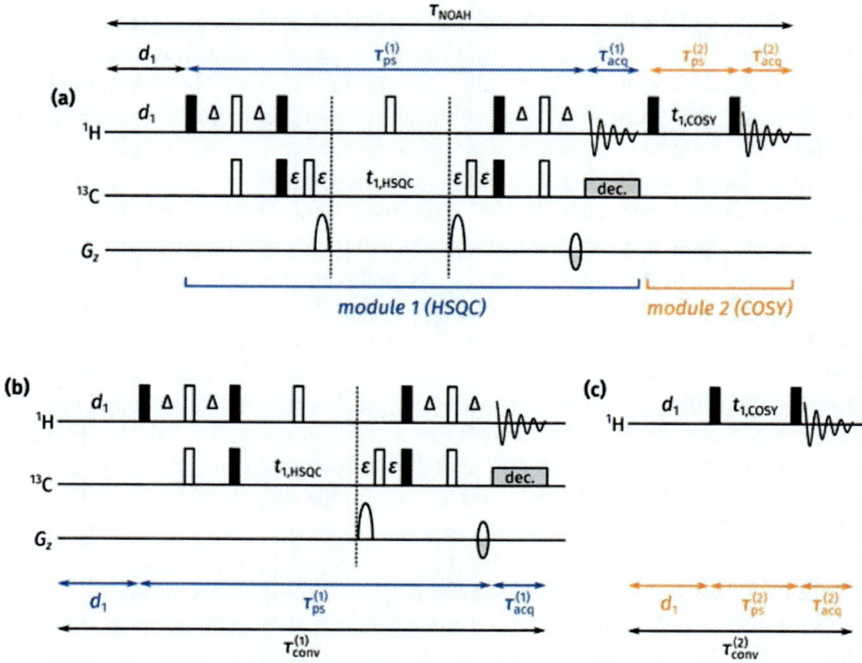

Fig. 4.1 **a** NOAH-2 SC supersequence, comprising HSQC and COSY modules (see Table 4.1 for an explanation of the single-letter module codes used). **b** 'Conventional' echo–antiecho HSQC (the same as in Fig. 1.7). **c** 'Conventional' COSY. The timings referred to in the text are highlighted for all three experiments; d_1 for each experiment is assumed to be the same. Note that the lengths are not to scale: d_1 is typically far longer than τ_{ps} and τ_{acq}

where τ_{NOAH} is the duration of the NOAH supersequence, τ_{conv} is the duration of a conventional experiment, and the superscript (i) represents the i-th module or conventional experiment being acquired.[2] The sum runs from $i = 1$ to N, where N is the number of modules. If we assume that $d_1^{(i)} = d_1$ is the same for all N conventional experiments and the supersequence, then in the limit where

$$d_1 \gg \sum_i \tau_{\text{ps}}^{(i)} + \tau_{\text{acq}}^{(i)}, \tag{4.3}$$

we have that $\rho_t \to N d_1/d_1 = N$, i.e. an N-fold time saving.

This statement makes plenty of assumptions and is not entirely valid in practice. For example, *each* τ_{acq} is often around 5–10% of d_1, so is not entirely negligible, especially when N is larger. Furthermore, for some modules, τ_{ps} itself can be substantial:

[2] This notation implies that $\tau_{\text{ps}}^{(i)}$ is the same in both the conventional experiment and the corresponding NOAH module. This is not strictly true: NOAH modules are often modified from the original, as will be explained shortly. However, the differences in timings are so minute that they can be completely ignored, especially when compared against d_1.

most notable is the NOESY module, which contains a mixing time of several hundred milliseconds. (HMBC, TOCSY and ROESY modules are also lesser offenders.)

These factors serve to reduce ρ_t from its idealised value of N; generally, this deviation is larger as N increases, because Eq. (4.3) becomes less and less valid. Despite this, it remains true that the time savings are approximately proportional to N, as will be shown in Table 4.2.

For relatively concentrated samples, where sensitivity is not an issue, these time savings are the sole relevant point. In this *sampling-limited regime*, the minimum 2D experiment duration is dictated purely by the number of t_1 increments needed to obtain sufficient resolution in the indirect dimension, as well as the minimum phase cycle required for artefact suppression.[3] NOAH supersequences are identical to conventional experiments in both these aspects, but come with the added benefit of time savings. The development of modern NMR instrumentation (such as high-field magnets and cryogenic probes) plays an important role in extending the sampling-limited regime to ever lower concentrations.

4.1.1.2 Sensitivity Comparisons

However, the opposite *sensitivity-limited regime* is still very commonly encountered. This can happen, for example, with naturally insensitive experiments (e.g. ADEQUATE), low-field benchtop NMR, or most simply, dilute samples.[4] In such cases, the benefits of NOAH supersequences must be much more carefully considered: in particular, it becomes mandatory to compare the SNRs of the NOAH modules and conventional experiments. To do so, we define for each module an *SNR factor* A, which is the SNR of the NOAH module divided by the SNR of a conventional experiment, acquired with the same parameters.[5] The SNR factor of the i-th module in a supersequence may be denoted as $A^{(i)}$.

In general, we have that $A \leq 1$, because (as will be explained in Sect. 4.1.2) NOAH modules frequently contain small modifications with respect to conventional experiments. For sensitivity-limited samples, we must balance these losses against the time savings obtained, by defining a *gain in sensitivity per unit time*, ε as

$$\varepsilon = A\sqrt{\rho_t}. \tag{4.4}$$

[3] With modern gradient-enhanced experiments, the minimum phase cycle may not even be a 'cycle', as one scan per increment may suffice; see also Fig. 1.8.

[4] If the SNR factor $A^{(i)}$ as discussed below is *very small*, then it is possible that even concentrated samples may be shifted into the sensitivity-limited regime. This is never really the case in practice, though, as the case studies in Sect. 4.1.3 will illustrate.

[5] The relative SNR will likely vary from peak to peak in the spectrum, and A should in theory be quoted either as an average over all peaks, or as a range. This is what I have done in this thesis. Furthermore, the exact values calculated for A will depend on the sample used for the comparison. Any values given here should therefore be assumed to be valid only for similar samples. Of course, my hope is that the samples chosen are reasonably representative of 'typical' organic molecules.

Here, the square root accounts for the fact that SNR scales only as the square root of the number of scans, or equivalently, the number of times the experiment can be repeated in a given time. If $\varepsilon^{(i)} > 1$, as is frequently the case, this means that the NOAH supersequence provides greater sensitivity per unit time in the i-th module compared to a standalone experiment. Equivalently, performing a NOAH experiment allows data of sufficient sensitivity to be obtained in less time.[6]

4.1.1.3 Time Savings in the Sensitivity-Limited Regime

An important—yet often ignored—caveat with sensitivity-limited samples is that the time savings obtained are not *truly* on the order of N as advertised above. This is because each module in a NOAH supersequence is run with the same number of scans. In contrast, with conventional experiments, it is common to run more insensitive experiments with a larger number of scans but to use fewer scans for sensitive experiments. In this case, the *effective* time savings provided by NOAH experiments are smaller:

$$\rho_{t,\text{eff}} = \frac{\sum_i \tau_{\text{conv}}^{(i)}}{\tau_{\text{NOAH}}} = \frac{\sum_i S^{(i)}(\tau_{\text{ps}}^{(i)} + \tau_{\text{acq}}^{(i)} + d_1^{(i)})}{Sd_1 + S\sum_i(\tau_{\text{ps}}^{(i)} + \tau_{\text{acq}}^{(i)})}, \tag{4.5}$$

where each standalone experiment is acquired with $S^{(i)}$ scans, and the NOAH experiment with S scans. We could also analogously define $\varepsilon_{\text{eff}} = A\sqrt{\rho_{t,\text{eff}}}$.

Typically, S is simply the largest of the $S^{(i)}$, as this ensures sufficient data quality for the most insensitive module in the supersequence. This means that generally, $\rho_{t,\text{eff}} < \rho_t$. In such a situation, it is probably more appropriate to describe a NOAH supersequence as 'measuring the most insensitive module and getting the others for free'. Indeed, in the extreme case where $S = S^{(i)} \gg S^{(j \neq i)}$, then 'the other' modules require almost no time to measure (relative to the least sensitive, i-th, module), and $\rho_{t,\text{eff}}$ tends towards 1, meaning that even the time-saving utility of NOAH vanishes. A corollary of this is that NOAH supersequences are generally most useful when constructed from modules which have similar intrinsic sensitivities and hence similar $S^{(i)}$'s: this is not necessary in the sampling-limited regime where $S^{(i)} = S$ for all i (and Eq. (4.5) simply reduces to Eq. (4.2)), but is crucial in the sensitivity-limited regime.

As the reader will no doubt appreciate by now, the comparison of NOAH and conventional spectra is fraught with subtleties. In fact, it is very much possible to construct yet more edge cases which challenge the framework used in this analysis. For example, one may not want to acquire all the individual spectra 'conventionally': for example, NUS may be used for an HSQC experiment but not for others; or d_1

[6] The use of $\varepsilon = A\sqrt{\rho_t}$ ignores the fact that experiments can only be repeated an integer number of times. For example, if $\rho_t = 2.5$, this does not necessarily imply that the NOAH experiment can be repeated 2.5 times more to get a $\sqrt{2.5} \approx 1.58\times$ boost in sensitivity. However, it is impossible to account for this in a general manner, so I will ignore it.

may be varied for different experiments. These will have an impact on both the durations of the experiments, as well as their sensitivities. To make any meaningful comparisons or generalisations, it is therefore necessary to restrict the discussion to values of ρ_t, A, and ε, which can be objectively calculated. These should, however, be read with the understanding that other factors may, depending on the context, lead to *some*—but never a *complete*—decrease in the utility of NOAH experiments.

4.1.2 Magnetisation Pools

Having dealt with this relatively dry material, I now turn to exactly how NOAH supersequences are constructed. Ordinarily, if the recovery delay is removed from an NMR experiment, its sensitivity will be greatly reduced because insufficient magnetisation will have recovered between repetitions; or in other words, $A^{(i)}$ will be very small. Such experiments would only be useful if sensitivity was greatly abundant.

As briefly mentioned earlier, the key to avoiding this in NOAH supersequences is to make sure that *each module samples a different source of magnetisation*. These 'sources' are referred to as *magnetisation pools*, and arise because of the different isotopologues present in natural-abundance samples (Fig. 4.2a). For example, an HSQC module can be designed to only sample magnetisation of protons directly bonded to the 1.1%-natural abundance ^{13}C, and leave all other proton magnetisation untouched. Immediately following this, the remainder of the proton magnetisation can then be used to record (say) a COSY module, without needing a separate recovery delay. Figure 4.2b shows a more elaborate example, which includes both ^{15}N and ^{13}C HSQC experiments.

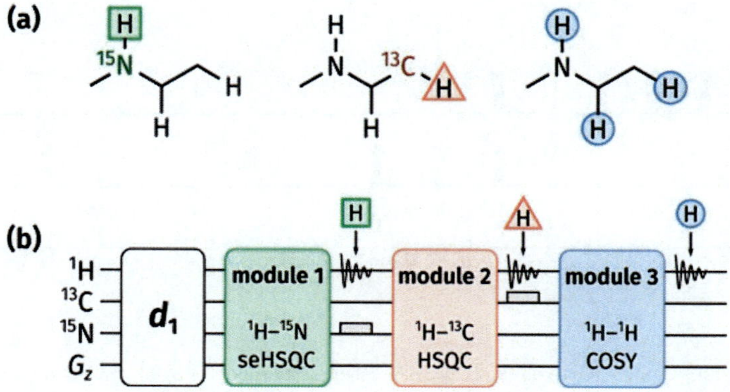

Fig. 4.2 **a** An illustration of the different magnetisation pools used in a typical NOAH supersequence: the three pools here (^{15}N-bound protons, ^{13}C-bound protons, and all others) would be referred to as $^{1}H^{N}$, $^{1}H^{C}$ and $^{1}H^{!X}$, respectively. **b** A NOAH supersequence consisting of three modules, which each consume one of the above magnetisation pools

Using the notation of Orts and Gossert [44], the magnetisation of ^{13}C-bound protons is denoted as ^{1}H^C, and that of ^{15}N-bound protons is ^{1}H^N. The magnetisation of protons *not* bonded to ^{13}C is denoted as ^{1}H$^{!C}$. Protons not directly bonded to *any* NMR-active heteronucleus are labelled ^{1}H$^{!X}$ and are often referred to as 'bulk' magnetisation, since the majority of protons in natural-abundance samples fall into this category.

Since standard 2D experiments typically seek to *destroy*—instead of *preserve*—unused magnetisation, NOAH modules often require some modifications compared to standard experiments. For example, compared to the echo–antiecho HSQC (discussed in Sect. 1.4.5), the NOAH HSQC module [32] adds an extra CTP gradient. This ensures that the bulk magnetisation is rephased after t_1, and ultimately returned to the $+z$ equilibrium state (Fig. 4.3a). (This is largely identical to the 'symmetrised' ASAP-HSQC experiment [45].) Sometimes, the modifications required are more extensive, such as in the HMBC module. In order for this module to preserve ^{1}H^C

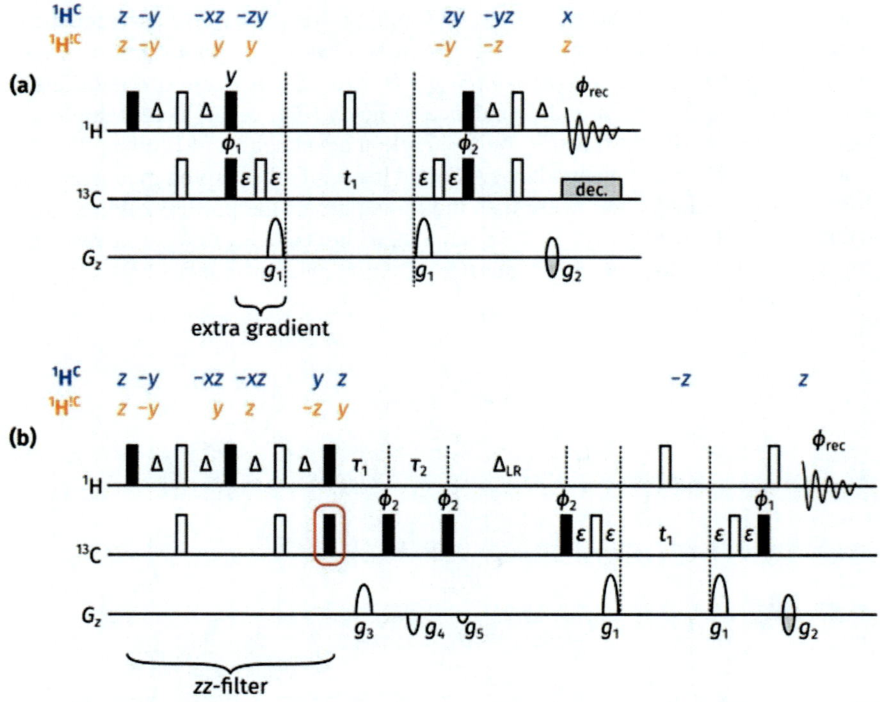

Fig. 4.3 **a** NOAH HSQC module. **b** NOAH HMBC module. The 90° pulse highlighted in red is described in Sect. 4.3.7. Delays are set as: $\Delta = 1/(4 \cdot {}^1J_{\mathrm{CH}})$; $\Delta_{\mathrm{LR}} = 1/(2 \cdot {}^nJ_{\mathrm{CH}})$; $\tau_1 = 1/(2 \cdot {}^1J_{\mathrm{CH,max}})$; $\tau_2 = 1/(2 \cdot {}^1J_{\mathrm{CH,min}})$ (see also Sect. 3.4.7 for the LPJF). Phase cycling is performed with $\phi_1 = (x, -x)$, $\phi_2 = (x, x, -x, -x)$ and $\phi_{\mathrm{rec}} = (x, -x, -x, x)$. Gradient amplitudes are $(g_1, g_2, g_3, g_4, g_5) = (80\%, \pm 40.2\%, 15\%, -10\%, -5\%)$. Product operator analysis is provided above both modules for both the ^{1}H^C and ^{1}H$^{!C}$ magnetisation pools; the notation for this is explained in the *Preface*

magnetisation (e.g. for a later HSQC module), the initial 90° excitation pulse must be replaced with a *zz*-filter (Fig. 4.3b). This performs an *isotope-selective rotation* in that $^1H^C$ magnetisation is stored along the *z*-axis, but $^1H^{!C}$ magnetisation is excited and subsequently detected.

In general, sequences which are thus modified are placed at the front of supersequences, so that they can preserve magnetisation for later modules. These modules tend to have lower sensitivities (i.e. $A < 1$) than the 'original' sequences from which they were derived. This is partly because of imperfect manipulation of magnetisation by the extra pulse sequence elements, and also increased losses due to relaxation during these extended sequences. In contrast, modules placed towards the *end* of a supersequence do not need to be modified, as they do not need to preserve any magnetisation. This includes virtually all homonuclear modules, which are allowed to simply consume any remaining magnetisation. Although this makes their implementation very straightforward, in general these modules will *also* suffer some losses in sensitivity, because the preceding modules do not perfectly retain all magnetisation.

It is therefore fairly rare for a NOAH module to have $A = 1$: for this to happen, the module must be placed first in the supersequence *and* not have undergone any modifications relative to the standalone experiment.[7] Such cases are very rare, and it is thus necessary to accept some decreases in A, which are often fairly small (on the order of 10–20%). For sensitive (typically homonuclear) modules, such sensitivity losses are often perfectly tolerable, as even with this sensitivity penalty they are still more intense than the other (heteronuclear) modules in the supersequence. However, for insensitive modules, it is imperative to make sure that A is not decreased by too much.

4.1.3 Case Studies

Using all of the concepts described in the previous sections, we now look at a few 'typical' supersequences to understand their construction. Before beginning, a quick note about the nomenclature of NOAH supersequences is warranted. Supersequences are labelled by the number of modules N, plus a series of single-letter codes, each corresponding to a specific module (Table 4.1). Occasionally, superscripts or subscripts are used to qualify the modules involved.[8] Thus, a NOAH supersequence containing three modules—say ^{15}N HMQC, ^{13}C HSQC and CLIP-COSY—would

[7] Of course, this also depends on exactly *what* standalone experiment the NOAH supersequence is being compared against. Sometimes, in the literature, the NOAH experiment has been compared against its constituent modules acquired in a standalone fashion; in this case, the first module will always have $A = 1$, since by definition it is not modified. This tells us how much we gain through the act of concatenating modules, but is less meaningful in the 'real world' where one is interested in how useful NOAH is relative to 'typical' optimised 2D experiments.

[8] With the increasing number of modules, and the variety of modern NMR experiments which could be incorporated into NOAH supersequences, keeping these abbreviations short yet meaningful has been a challenge.

Table 4.1 A (non-exhaustive) list of single-letter module codes for NOAH modules. In the literature, the ^{15}N HMQC module has been referred to simply by 'M', since the HSQC module is preferred for ^{1}H–^{13}C correlations. In this thesis, I include the subscript N throughout to avoid any ambiguity

^{1}H–^{15}N modules		^{1}H–^{13}C modules		^{1}H–^{1}H modules	
Module	Code	Module	Code	Module	Code
HMQC	M_N	HSQC	S	COSY	C
HSQC	S_N	seHSQC	S^+	CLIP-COSY	C^c
seHSQC	S_N^+	HSQC-TOCSY	S^T	DQF-COSY	D
HMBC	B_N	HSQC-COSY	S^C	TOCSY	T
		2BOB	O	NOESY	N
		HMBC	B	ROESY	R
		ADEQUATE	A	PSYCHE	P
				TSE-PSYCHE	P^T
				PSYCHE 2DJ	J

be referred to as a NOAH-3 $M_N SC^c$. Table 4.2 provides values of ρ_t and A for the example supersequences used in this section; these values which will be rationalised in the text which follows.

4.1.3.1 NOAH-2 SC: HSQC + COSY

We begin with perhaps the simplest example of a NOAH supersequence, one containing the HSQC and COSY modules: this is labelled as a NOAH-2 SC experiment (entry 1, Table 4.2). Using the implementation shown in Fig. 4.3a, the HSQC module only samples $^1H^C$ magnetisation, and leaves $^1H^{!C}$ magnetisation along the $+z$-axis. This modification to the HSQC module is very small, so its sensitivity is practically unaffected as compared to a 'standard' HSQC ($A = 0.97$). Furthermore, the COSY module retains *most* of its sensitivity ($A = 0.90$). The small loss here is because the HSQC module does not *perfectly* preserve the $^1H^{!C}$ magnetisation: for example, evolution of homonuclear couplings as well as relaxation occur during the HSQC pulse sequence, both of which were ignored in the product operator analysis of Fig. 4.3a.

Furthermore, the value of the time-saving factor, $\rho_t = 1.87$, is very close to the theoretical limit of $N = 2$. This reflects the fact that the pulse sequence itself, τ_{ps}, is fairly short for both the HSQC and COSY modules; the deviation therefore chiefly arises from the acquisition time, τ_{acq}. In all respects, this is therefore an example of an 'ideal' NOAH supersequence, where the combination of two modules provides an almost-twofold time saving without compromising on sensitivity.

Table 4.2 Sensitivity and time-saving analyses of several typical NOAH supersequences. All experiments were acquired with 2 scans per increment, 256 t_1 increments, an acquisition time of 67 ms and a recovery delay of 1.5 s. The HMBC module used here includes the extra ^{13}C 90° pulse described later in Sect. 4.3.7: this has no significant impact on the SNR, and is only mentioned as a technicality. The ^{15}N seHSQC module is that described in Sect. 4.3.3. The CT module here was run with States indirect-dimension quadrature detection, and the individual C module (in entry 1) with echo–antiecho. The following Bruker library sequences were used as the 'conventional' experiments: `hmbcetgpl2nd`, `hsqcetf3gpsi2`, `hsqcetgpsp.2`, `cosygpqf` and `dipsi2gpphzs`. *Data code:* `7Z-220224`

Entry	Sequence	τ_{NOAH}	ρ_t	A				
				HMBC	seHSQC	HSQC	COSY	TOCSY
1	SC	15 min 0 s	1.87			0.97	0.90	
2	SCT	16 min 25 s	2.60			1.01	0.99	0.79
3	BS	15 min 40 s	1.82	0.93		0.87		
4	SB	15 min 35 s	1.83	0.99		0.96		
5	BSCT	17 min 48 s	3.22	0.95		0.90	0.36	0.28
6	BS_N^+SCT	18 min 57 s	3.74	0.95	0.71	0.66	0.38	0.30
7	S_N^+BSCT	18 min 56 s	3.75	0.76	0.79	0.74	0.33	0.26

It is worth pointing out that the order of the modules cannot be reversed: the COSY module cannot be (easily) modified to preserve $^1\text{H}^\text{C}$ magnetisation. In a hypothetical NOAH-2 CS supersequence, the later HSQC module would only be able to use magnetisation recovered during the COSY FID, leading to a substantial decrease in sensitivity.

A final point to consider would be whether the NOAH data has comparable spectral quality in terms of (for example) artefacts. In this case, the answer is yes: the NOAH HSQC spectrum is virtually identical to the standalone (Fig. 4.4a and c; both spectra have low-level artefacts of different kinds, which do not seriously impede the interpretation and are not shown). On the other hand, the NOAH COSY spectrum seems to actually *improve* on the standalone COSY, in that it better suppresses off-diagonal artefacts (Fig. 4.4b and d). These artefacts likely arise in the standalone COSY because of accidental refocusing of magnetisation which has not completely relaxed between successive t_1 increments [46]. In contrast, the NOAH COSY module has an extra set of HSQC gradients between every repetition of the COSY, so accidental refocusing is less likely. (Similar artefacts have previously been noted in the DQF-COSY experiment [47, 48], and have also shown to be attenuated in the corresponding NOAH module [49].) That said, such improvements are not always guaranteed. It is possible for specific artefacts to arise exclusively in NOAH exper-

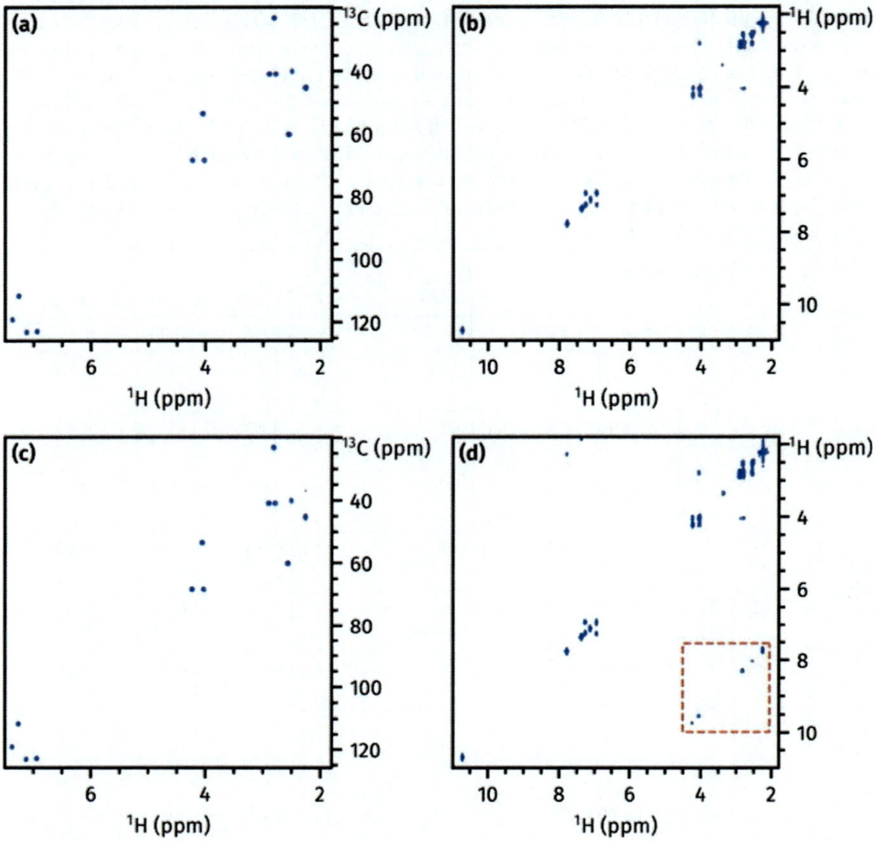

Fig. 4.4 **a** HSQC from NOAH-2 SC supersequence. **b** COSY from NOAH-2 SC supersequence. **c** Standalone HSQC. **d** Standalone COSY; off-diagonal artefacts are highlighted in the red box. *Data code:* `7Z-220224`

iments; some of these, and methods of suppressing them, will be discussed in the following sections.

4.1.3.2 NOAH-2 SCT: HSQC + COSY + TOCSY

Evidently, the fact that the HSQC preserves almost all $^{1}H^{!C}$ magnetisation means that *any* homonuclear module can be placed after it. This means that supersequences such as SC^{c}, SD and ST can be constructed in a very similar way. It would be very useful if we could go beyond this and place *multiple* homonuclear modules in a single supersequence.

However, since every homonuclear module tends to consume all remaining bulk magnetisation, creating combinations of homonuclear modules which do not ‘com-

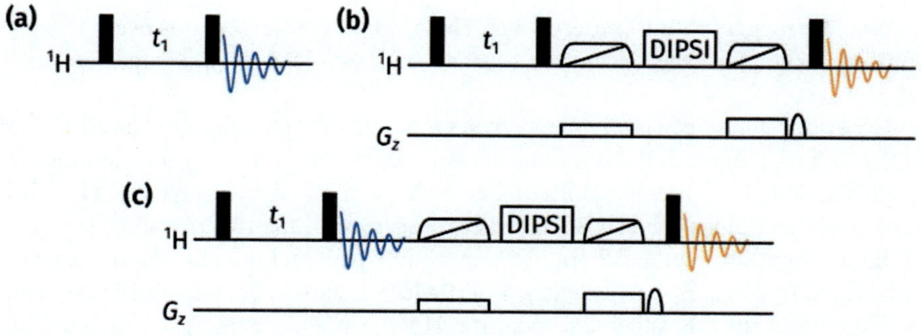

Fig. 4.5 **a** COSY module. **b** TOCSY module; zero-quantum suppression is employed before and after the isotropic mixing period. **c** Combined COSY/TOCSY module, where the COSY FID is acquired immediately before the TOCSY mixing

pete' for this bulk magnetisation is a difficult task. The only notable exceptions to this are 'COSY/X' combinations, where X can be NOESY, ROESY or TOCSY: instead of concatenating the COSY and X modules, the COSY pulse sequence can instead be nested *within* the X module, as was first demonstrated with X = NOESY [34, 35].

Here, we use the COSY/TOCSY combination [37] as an example (Fig. 4.5). The 90°–t_1–90° COSY pulse sequence leads to a mixture of both transverse and longitudinal magnetisation which is frequency-labelled in t_1. The transverse component is sampled during the COSY acquisition period, whereas the longitudinal component (usually discarded in a COSY experiment) is subjected to isotropic mixing and subsequently sampled as part of the TOCSY. Unlike the HSQC and COSY in the previous example, the COSY and TOCSY in Fig. 4.5c cannot be clearly identified as being distinct sections of the pulse sequence: they in fact share a common t_1 period. Nevertheless, it is common to still refer to these as separate modules in a supersequence.

As shown in entry 2 of Table 4.2, this nesting of the COSY module does not materially affect the TOCSY sensitivity. A small loss of approximately 20% is observed, which is partly due to the imperfect magnetisation preservation by the HSQC, and perhaps also due to relaxation during the COSY acquisition period. As for the time-saving factor, a slightly larger deviation ($\rho_t = 2.60$) is observed from the ideal value of 3. This reflects the addition of an extra acquisition period and the TOCSY mixing period, which make the assumption in Eq. (4.3) less valid.

4.1.3.3 NOAH-2 BS: HMBC + HSQC

As mentioned previously, the HMBC module shown in Fig. 4.3b is designed to retain $^1H^C$ magnetisation through the addition of the *zz*-filter. This can be used in a subsequent HSQC module in a NOAH-2 BS supersequence. Entry 3 of Table 4.2 shows that the addition of the *zz*-filter to the HMBC causes a relatively small 7% decrease in sensitivity; on the other hand, the HSQC loses 13% of its sensitivity because of incomplete magnetisation preservation. These sensitivity losses, however, are reasonably acceptable considering the time savings provided by this supersequence ($\rho_t = 1.82$).

Generally, it has been recommended that less sensitive modules are placed earlier in the supersequence so that they can access a larger proportion of the equilibrium magnetisation. This principle does not, however, actually apply to the NOAH-2 BS supersequence. Since the HMBC is the less sensitive of the two modules, this rule of thumb suggests that the BS arrangement should be better than the alternative SB. In fact, the opposite is true, as entry 4 of Table 4.2 shows. This can be understood as follows: in the SB supersequence, the HSQC module experiences a boost in sensitivity because it is placed first in the supersequence, and no longer needs to rely on the $^{1}H^{C}$ magnetisation preserved by the HMBC; and the HMBC benefits because the *zz*-filter modification is no longer needed.[9] This may just be the exception that proves the rule, but in my view, this provides some justification for considering the ordering of modules on a case-by-case basis.

4.1.3.4 NOAH-4 BSCT: HMBC + HSQC + COSY + TOCSY

We now move on to a longer supersequence containing four modules, with a correspondingly larger ρ_t value of 3.22. The sensitivity of the HSQC module is practically the same as in the NOAH-2 BS supersequence just described: however, the COSY and TOCSY modules expose one weakness of the HMBC module which has so far been overlooked. In principle, the HMBC module should only excite magnetisation of protons which have long-range couplings to ^{13}C (which we could, for example, denote as $^{1}H^{C(LR)}$). This magnetisation pool is distinct from both the directly coupled protons ($^{1}H^{C}$), as well as protons which are not coupled to any ^{13}C at all ($^{1}H^{!C}$). Unfortunately, this is not the case: it is not actually possible to separate the $^{1}H^{C(LR)}$ and $^{1}H^{!C}$ magnetisation pools. The HMBC excites both of these magnetisation sources, dephases the latter using CTP gradients and detects the signal arising from the former.

This means that the COSY and TOCSY modules, which rely on $^{1}H^{!C}$ magnetisation, will have substantially lower sensitivities. The signal detected in these two modules derives only from polarisation that has recovered during the previous two acquisition periods, as shown in entry 5 of Table 4.2 (the values of A for the COSY and TOCSY are 0.36 and 0.28, respectively). That said, this is in fact not likely to be an issue *even* for sensitivity-limited samples. Because the intrinsic sensitivity of the HMBC is orders of magnitude lower than that of the COSY and TOCSY, the COSY and TOCSY spectra still have greater intensities than the HMBC, even after these large losses in sensitivity. Thus, as long as the entire supersequence is acquired with enough scans such that the HMBC SNR is sufficient, the COSY and TOCSY will *also* have acceptable SNR. This is proven by the spectra shown in Fig. 4.6.

[9] In fact, the final 180° pulse in the HMBC module could also be removed: this is likely to give a further boost in SNR, as discussed in Sect. 4.3.7. However, this was not done here.

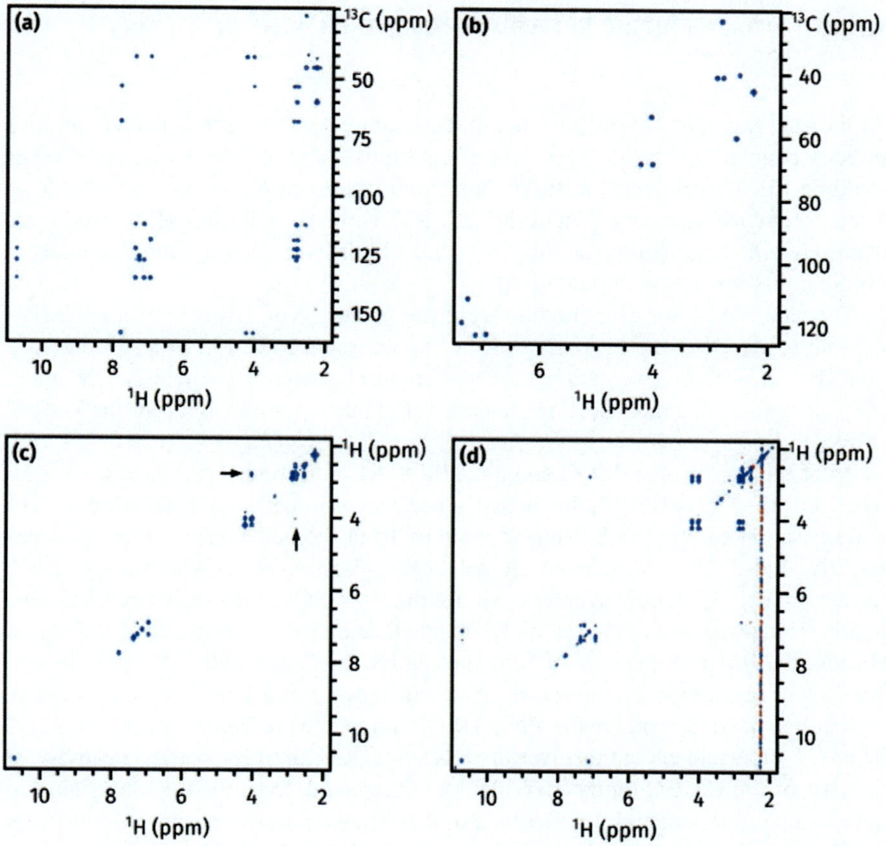

Fig. 4.6 Spectra obtained from a NOAH-4 BSCT supersequence. **a** HMBC. **b** HSQC. **c** COSY; a pair of crosspeaks with asymmetric intensities is highlighted with black arrows. **d** TOCSY (60 ms DIPSI-2 mixing). Despite the COSY and TOCSY having only around 30% of their sensitivity compared to standalone experiments, the intensity of the spectra obtained is still perfectly acceptable (the contour levels chosen are 1–2 orders of magnitude larger than for the HMBC) *Data code:* `7Z-220224`

A rather more insidious problem is that different signals relax at different rates: thus, the COSY and TOCSY spectra (or indeed, any homonuclear module) will have uneven intensities and are frequently asymmetric about the main diagonal. This can be seen in the COSY spectrum, where a pair of asymmetric crosspeaks are highlighted. Adding a period of isotropic mixing before the COSY module [50] can ameliorate this to some extent (although the spectra in Fig. 4.6 were not acquired with this).

4.1.3.5 NOAH-5 BS_N^+SCT: HMBC + ^{15}N seHSQC + HSQC + COSY + TOCSY

As the final example, we add a further magnetisation pool to the mix, namely protons directly coupled to ^{15}N (i.e. $^1H^N$). As of the time of writing, the implementation of multiple-FID experiments on Bruker spectrometers limits N to a maximum of 5, so a supersequence such as this NOAH-5 BS_N^+SCT is the current limit. (However, there is no *scientific* argument forbidding $N > 5$, and it is likely that in future versions of TopSpin this restriction will be lifted.)

The values of A for each module are given in entry 6 of Table 4.2. If the HMBC module is placed at the beginning of the supersequence, then in order to preserve *both* $^1H^N$ and $^1H^C$ magnetisation, the zz-filter must be extended to include ^{15}N pulses [51]. As before, this modification leads to a slight drop in sensitivity for the HMBC (5%).

After this, the ^{15}N seHSQC and ^{13}C HSQC modules both suffer drops in sensitivity. For the ^{15}N seHSQC, this is partly because of imperfect preservation of $^1H^N$ magnetisation by the HMBC but also stems from the addition of the zz isotope-selective pulse (ZIP) element to the seHSQC pulse sequence, which is described further in Sect. 4.3.3. On the other hand, for the ^{13}C HSQC, the sensitivity loss stems purely from imperfect retention of $^1H^C$ magnetisation by previous modules. Finally, because the HMBC dephases $^1H^{!X}$ magnetisation, the COSY and TOCSY at the end have lower sensitivities: however, as discussed above, this is not an issue in practice.

It is also possible to move the ^{15}N seHSQC module to the front to form a NOAH-5 S_N^+BSCT supersequence: this gives the ^{15}N seHSQC a slightly greater sensitivity, at the cost of the HMBC (entry 7, Table 4.2). In general, these two modules tend to have comparable sensitivity, and which of these two arrangements is better depends on which module the sensitivity needs to be optimised for.

Finally, the value of ρ_t given here of 3.74 represents an effective upper limit on the time-saving factor. Although ρ_t increases with N, the extent to which it deviates from the ideal value of N also increases: it is very difficult to obtain $\rho_t > 4$, even with five modules in the supersequence. Of course, it is possible to increase ρ_t further by lengthening the recovery delay d_1 used for the experiments: for example, if d_1 is increased to 2 s from its present value of 1.5 s, ρ_t increases to 3.94. Obviously, this can only be pushed so far before it becomes meaningless.

4.2 GENESIS: Automated Pulse Program Creation

In this section, I discuss the development of the GENESIS (GENEration of Supersequences In Silico) website (Fig. 4.7).[10] As its name suggests, GENESIS is a tool which programmatically generates (Bruker) pulse programs for arbitrary NOAH

[10] Although placing this section first may at first glance appear to be nonchronological, in that the GENESIS paper [3] was published later than much of the other work in this chapter, an early version of the GENESIS tool was in fact created much earlier (by July 2020).

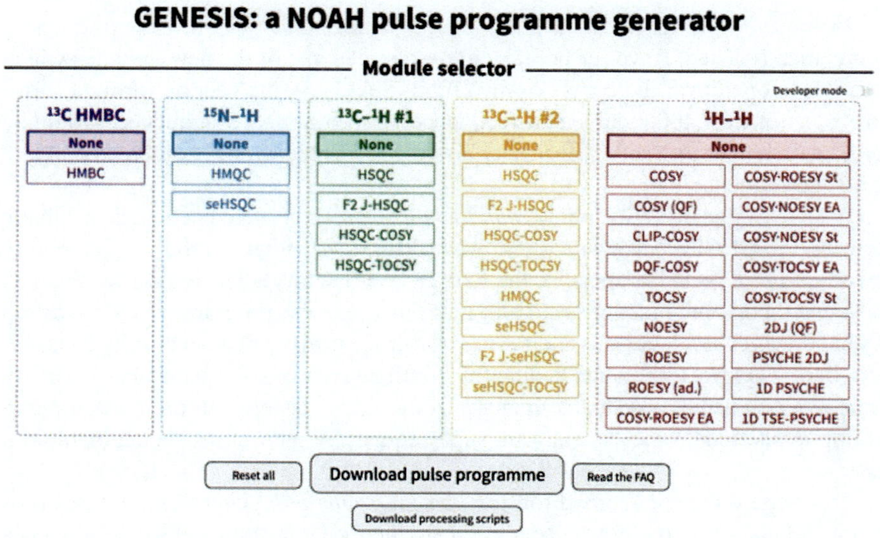

Fig. 4.7 The front page of the GENESIS website (https://nmr-genesis.co.uk), as of 10 September 2022

supersequences. The website also provides extensive instructions on acquiring and processing NOAH data. The present version of GENESIS is available at https://nmr-genesis.co.uk; the source code can also be obtained from https://github.com/yongrenjie/genesis.

4.2.1 Motivation

The preceding discussion in Sect. 4.1 makes clear how supersequences may be constructed from a large variety of modules. Furthermore, modules which consume (and preserve) the same magnetisation pools may be directly swapped out for one another. Thus, for example, the NOAH-2 SC supersequence can be generalised to any ${}^{1}\mathrm{H}^{\mathrm{C}}$ module plus any ${}^{1}\mathrm{H}^{!\mathrm{C}}$ module. Very broadly speaking, we may define a generic supersequence as having any or all of the following:

- an HMBC module, which actually uses ${}^{1}\mathrm{H}^{!\mathrm{X}}$ magnetisation but can be placed at the front as discussed in the NOAH-3 BSC example above;
- a ${}^{1}\mathrm{H}^{\mathrm{N}}$ module;
- one or more ${}^{1}\mathrm{H}^{\mathrm{C}}$ modules (it is possible to partition the ${}^{1}\mathrm{H}^{\mathrm{C}}$ magnetisation pool between two modules, as will be discussed in Sect. 4.3.4);
- finally, a ${}^{1}\mathrm{H}^{!\mathrm{X}}$ module (or a COSY/X combination) which consumes all remaining bulk magnetisation.

In the first NOAH paper in 2017 [32], a total of 285 'viable' supersequences were already listed. If we further take into account some of the new modules which were developed over the course of my DPhil (Sect. 4.3), the generic formula above provides for over 4000 viable supersequences.[11] ('Non-viable' sequences would be those which have unwanted sensitivity losses, such as the wrongly ordered NOAH-2 CS supersequence.)

In spite of this diversity, *only around 45 pulse programs* had been made available prior to the GENESIS website (these were distributed either via the supplementary information of NOAH papers, or the Bruker User Library). The reason for this gap between theory and practice is simple: traditionally, pulse programs must be written by hand, which is a laborious and fairly error-prone process made worse by the sheer length of NOAH experiments. Doing this for thousands of supersequences is clearly impractical. Furthermore, each time a new module is developed, or an old module is improved, updating every relevant supersequence would in itself be a mammoth task.

To bridge this gap, I turned towards the *programmatic* generation of pulse program.[12] This not only allows users to create their own customised supersequences but also provides an easy way for pulse sequence updates to be rapidly disseminated to the NMR community, as new pulse programs can be accessed immediately upon the deployment of updated source code. Furthermore, the website can serve as a 'one-stop' shop where—after downloading pulse programs—users may download associated NOAH processing scripts and also access instructions on how to run NOAH experiments. This information did already exist, but was scattered across several different websites and/or supplementary information documents, and would have been needlessly confusing to a new user (not even taking into account the different versions of scripts available in different publications).

4.2.2 Implementation Details

I will now describe a few features of GENESIS pulse programs, as well as how their generation is implemented in code. GENESIS is written in TypeScript; during website deployment, this is compiled to JavaScript, which can then be directly executed in the client's web browser. No server-side code is employed, meaning that the GENESIS web page is actually a static site; it is currently hosted using GitHub Pages.

[11] This is also ignoring the 'parallel' supersequences, which are discussed in Sect. 4.5. The support for parallel supersequences in GENESIS is not complete: integrating these fully would require substantial changes to the user interface, which I have not had time to do.

[12] This is actually a bit of a lie: GENESIS was initially created *for my own convenience*. Throughout this chapter, I have had to perform many comparisons of different supersequences, and this tool spared me from having to write everything by hand (and—more often than not—subsequently discover mistakes which invalidated the results). Of course, it soon became apparent that it could find much wider use.

Listing 4.1: Abridged GENESIS pulse program for the NOAH-2 SC supersequence shown in Fig. 4.1a

```
/* PREAMBLE */
; gns_noah2-SCqf
; 13C HSQC
; 1H magnitude-mode COSY
"d4          = 0.25s/cnst2"   ; 13C INEPT
"in0         = inf1/2"        ; 13C HSQC increment
"in11        = 2*dw"          ; COSY increment
; ...
define delay DC_HSQCa
"DC_HSQCa    = d4-p14/2"
"l0          = td1/2"         ; TD1/NBL

/* MAIN SECTION */
1 ze
4 d1 st0
  ; MODULE 1 - HSQC
  (p1 ph0):f1
  ; ...
  goscnp ph30 cpd2:f2
  50u do:f2
  2m st
  ; MODULE 2 - COSY
  (p1 ph12):f1
  ; ...
  go=2 ph26
  1m iu1        ; TD1/NBL counter
  1m igrad EA   ; echo-antiecho gradients
  1m id11       ; COSY t1
  30m wr #0 if #0 zd
if "l1 % 2 == 0" {
  1m id0        ; HSQC t1
}
  lo to 4 times l0
exit

/* EPILOGUE */
ph0=0
;gpnam4: SMSQ10.100
;gpz4: 70% (13C CTP)
;cpd2:wvm:wudec: cawurst_d-20(220 ppm, 1.4 ms; L2H)
;d4: 1/4J(CH)
;auprog: noah_hsqc:noah_cosy QF
;module identifiers: C_HSQC H_COSY_QF
;pulse programme created by genesis-v2.2.2, https://nmr-genesis.co.uk
;Sun Sep 11 2022 16:04:54 GMT+0800 (Malaysia Time)
```

4.2.2.1 Overall Structure

The structure of a pulse program can loosely be separated into three parts, which are shown in Listing 4.1:

1. the *preamble*, which consists of everything up until the beginning of the actual pulse sequence (signified by the `ze` command). This includes header comments as well as definitions of parameters, such as delays and pulse widths;
2. the *main section*, which contains the actual pulse sequence;
3. the *epilogue*, which contains phase cycle information as well as footer comments describing each parameter. Instructions for generating shaped pulses using Bruker's WaveMaker software are also included here, as well as instructions for automatic processing (Sect. 4.2.3), and comments indicating how the pulse program was generated (for reproducibility purposes).

The algorithm used for pulse program construction can be broken up into similar sections. The construction of the preamble and main section is largely accomplished through the collation of *module-specific information*, the most important of which are:

- descriptive information about the modules themselves, which go into the header comments;
- parameter definitions, which are collated to form the preamble. Duplicates must be removed here to avoid errors; and
- the pulse programs themselves, which are directly concatenated to form the main section.

These, as well as other smaller bits of information (e.g. relevant citations, appropriate processing scripts), are stored within `NOAHModule` objects (an example is provided in Listing 4.2). Each distinct module corresponds to one such object. Therefore, if one wants to add a new module to GENESIS, most of the work can be completed by simply defining a new `NOAHModule` object: no changes to the algorithm itself are needed.

Listing 4.2: An excerpt of the NOAHModule object for the HSQC module shown in Fig. 4.3a (internal code `C_HSQC`)

```
let shortDescription = `; 13C HSQC`

let preamble = `
"d4       = 0.25s/cnst2"                ; 13C INEPT
; ...
"D[ID]a  = d4-p14/2"
`    // [ID] is later replaced with the module code

let pulprog = `
  ; 13C-1H HSQC

  ; INEPT
  (p1 ph0):f1
  ; ...
  goscnp ph30 cpd2:f2
  50u do:f2
`

const mod = new NOAHModule(
      "C_HSQC",              // internal module code
      "c13",                 // module category
      "S",                   // single-letter code, Table 4.1
      [],                    // relevant citations (if any)
      "noah_hsqc",           // AU programme for processing
      shortDescription,      // short description
      [AF_EDIT],             // available acquisition flags
      preamble,              // preamble text
      pulprog,               // pulse programme text
      1,                     // number of FIDs
      false                  // flag for 'parallel' modules
);
export default mod;
```

To put the epilogue together, the pulse program constructed so far is scanned for pulse phases, shaped pulses and all other parameters. Using predefined lookup tables, GENESIS then creates pulse phase definitions, WaveMaker directives (where appropriate) and comments containing textual descriptions of each parameter. These comments are not merely cosmetic; they are also displayed in the `ased` screen when setting up an experiment. Finally, instructions for automatic processing of the NOAH data (to be explained in Sect. 4.2.3) are added to the bottom, together with a timestamp and the version number of GENESIS for reproducibility purposes.

4.2.2.2 Phase/Delay Incrementation and Looping

Since NOAH experiments are 2D experiments, there is one additional complication: the pulse program must contain appropriate looping statements, together with pulse phase and delay incrementation, in order to correctly generate the indirect dimension.

In many existing NOAH pulse programs, looping in 2D experiments was written using the equivalent of nested `for` loops (Listing 4.3, *left*). Although this structure suffices for the vast majority of supersequences, whenever any parameter must be incremented in a different manner (e.g. in parallel supersequences where multiple supersequences are interleaved, or when using the PSYCHE module which has a different number of t_1 increments), the nested loop structure must be modified, which is not easy to reason about. I therefore opted to change the structure to use only one loop, and to control the phase and delay incrementation using modular arithmetic (Listing 4.3, *right*). The outcome is entirely equivalent, but this strategy allows for other cases to be implemented simply by adding another check on the loop counter `L1`.

Listing 4.3: Implementation of phase/delay incrementation and looping in previous NOAH sequences (left, using nested loops) and in GENESIS (right, using modular arithmetic)

```
"l0 = td1/4"


1 ze
3 1m
4 d1 st0

  ; ... (pulse sequence goes here)

  ; in inner loop
  1m igrad EA    ; HSQC gradients
  1m id11        ; COSY t1
  30m wr #0 if #0 zd
  lo to 4 times 2

  ; in outer loop
  1m ip5*2       ; HSQC 13C 90
  1m ip30*2      ; HSQC receiver
  1m id0         ; HSQC t1
  lo to 3 times l0


end
```

```
"l0 = td1/2"
"l1 = 0"

1 ze
4 d1 st0


  ; ... (pulse sequence goes here)

  ; on every pass
  1m iu1         ; loop counter
  1m igrad EA    ; HSQC gradients
  1m id11        ; COSY t1
  30m wr #0 if #0 zd

  ; on every second pass
  if "l1 % 2 == 0" {
    1m ip5*2       ; HSQC 13C 90
    1m ip30*2      ; HSQC receiver
    1m id0         ; HSQC t1
  }
  lo to 4 times l0

end
```

4.2.2.3 Parameter Standardisation

Each NOAH module contains a number of parameters, including pulse widths, delays, gradient amplitudes and shaped pulse waveforms. Since different modules are stored separately (as different `NOAHModule` objects), directly concatenating their pulse programs may lead to conflicting parameter definitions if appropriate care is not taken. GENESIS avoids this by maintaining a global table of parameter definitions which are applicable to all modules: when new modules are added, they must be checked against this to ensure that there are no inconsistencies.

In general, where possible, these parameters are chosen to be consistent with pulse programs in the Bruker standard library: thus, for example, `P1` is the ^{1}H 90° pulse width, and `CNST2` is the $^{1}J_{CH}$ value used for calculating INEPT (and other) delays. This makes it easy to read in parameters either from the *prosol* (probe and solvent) relation tables in TopSpin, or from other existing parameter sets. Some delays are module-specific and do not need to be reused, and in standard library sequences, are often labelled as `DELTA1`, `DELTA2` and so on. To avoid conflicting definitions and also to improve readability, I renamed these such that they include the name of the module: thus, in a ^{13}C HSQC these may be labelled `DC_HSQC_1`. Here, `C_HSQC` is the name associated with the `NOAHModule` object.

If combined with some caution when adding new modules, these measures ensure that there will be no parameter clashes between the modules *within a given supersequence*: we may view this as a *local uniqueness* of parameters. However, the impact of this design choice is even more far-reaching: since parameters are stored globally, they will always have the same value in *all possible supersequences* (or in other words, the parameters are *globally unique*). Thus, `CNST2` in a NOAH-2 BS supersequence has the same meaning as `CNST2` in a NOAH-3 BSC supersequence (and so on). This makes it exceptionally easy to set up multiple different supersequences in TopSpin using GENESIS pulse programs, as virtually all of the parameter values may simply be copied from a previous NOAH dataset.

One potential issue with this strategy is that TopSpin provides only a finite number of named pulse widths (for example). Thus, there are only so many different parameters which can be stored in a global table before running into inevitable conflicts. A workaround would be to sacrifice the global uniqueness of each parameter, and only have it be unique within a given supersequence. Fortunately, this situation has not (yet) surfaced.[13]

4.2.2.4 Parameter Descriptions

At the epilogue of the pulse program, extra comments are added for every parameter present in the pulse program. Most of these are purely textual in nature, and appear in the TopSpin `ased` parameter setup screen; this naturally helps to make the pulse programs as easy to use as possible. However, some of these comments have special meanings: gradient amplitudes and shapes, for example, are specified in a way which

[13] The number of *pulse phases* in GENESIS is in fact dangerously close to the maximum number of 32. However, the global uniqueness criterion is not really important for pulse phases, because—unlike, say, delays—pulse phases are hardcoded in the pulse program, and cannot be copied from one dataset to another. So, if necessary, I could dispense with the global uniqueness for pulse phases only, at the cost of some increased code complexity. I did briefly contemplate this possibility, but since I am at the end of my DPhil and am unlikely to add any new modules soon, this will likely remain a hypothetical.

allows them to be automatically populated using the `gppp` Python script packaged with TopSpin. Furthermore, WaveMaker directives are also specified for some shaped pulses, allowing them to be created in an on-the-fly manner using the `wvm` command. This means that the user (generally) need not separately download and install a set of shaped pulses.

4.2.2.5 Module Choice

One final issue which must be overcome is the fact that some NOAH modules may be implemented differently depending on the supersequence which it is being used in. The HMBC module described in Sect. 4.1.3 is one such example: the form of the *zz*-filter depends on whether the HMBC module is required to retain $^{1}H^{C}$ and/or $^{1}H^{N}$ magnetisation for subsequent modules. Thus, in the NOAH-2 BS supersequence the *zz*-HMBC must be used, but in the NOAH-2 SB supersequence the 'original' HMBC without a *zz*-filter is preferable. Since these have different pulse programs, each of these 'versions' of the HMBC are described by *separate* `NOAHModule` objects (`C_HMBC_CF` and `C_HMBC_NOF`, respectively).

However, it is unlikely that the majority of users would want to manually configure the supersequence in such detail by selecting `NOAHModule` objects. Thus, the GENESIS web interface actually hides these different versions from the user, only showing one button labelled 'HMBC'. Under the hood, the logic above is used to deduce the correct version of the HMBC based on what other modules the user has chosen (Fig. 4.8).

Should the user indeed want to control the exact module used, the website also offers a 'developer mode' switch: turning this on allows the user to directly choose the desired module. (Enabling developer mode also reveals a handful of extra modules which were used exclusively for my DPhil and are not generally of interest.)

4.2.2.6 Tests

One risk inherent in any NMR experiment is the possibility of causing spectrometer damage when executing malformed pulse programs. To minimise the chances of this, each version of GENESIS is checked against a series of automatic tests before being released. The most important tests ensure that every module containing decoupling statements also turns off decoupling immediately after acquisition. On top of this, there are also a series of regression tests where (some representative) supersequences are checked against the same ones produced by previous versions of GENESIS: any differences in these are automatically flagged for review. The live website is only updated if all tests are passed.

However, none of this can actually *stop* GENESIS from creating wrong pulse programs. For example, if a module with an incorrect pulse program is created, then any supersequence containing that module will not be fully functional. There is no adequate way to check for this, as GENESIS cannot verify the correctness of an

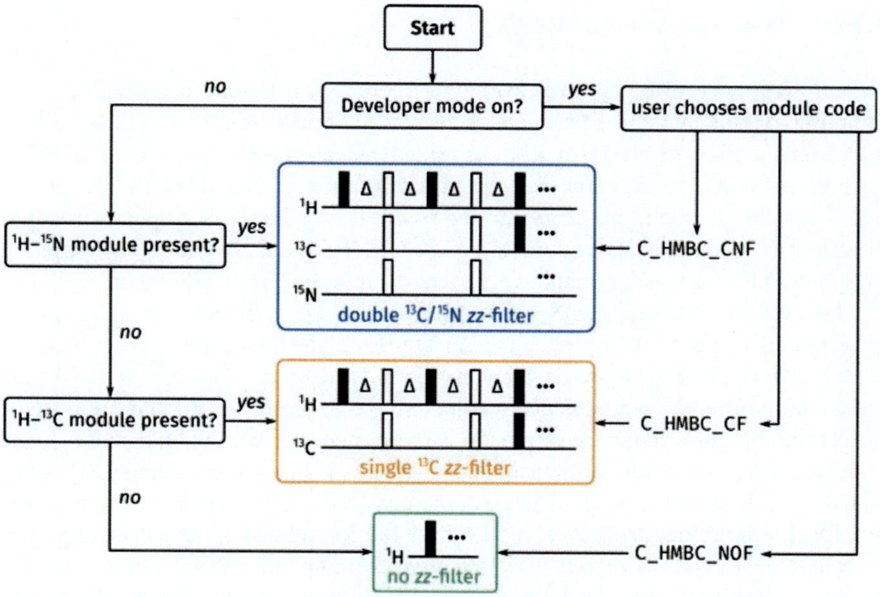

Fig. 4.8 Flowchart showing how the correct 'version' of the HMBC module is determined when constructing a supersequence using GENESIS. When developer mode is on, the user directly chooses module codes corresponding to each version of the HMBC. When it is off, the appropriate version is chosen based on which other modules are present in the supersequence

NMR experiment. However, the inclusion of the tests described above minimises the chances of creating pulse programs which are outright dangerous; any mistakes will at worst lead to the wrong (or no) data being obtained. It should be mentioned that this is not a *new* problem which GENESIS introduces: errors in pulse programs can arise equally easily (if not *even more easily*) when they are handwritten.

4.2.2.7 Reproducibility

The final section of a GENESIS pulse program consists of comments intended solely for reproducibility purposes. These comments indicate exactly which `NOAHModule` objects were used to create the pulse program, as well as a GENESIS *version number*. Since GENESIS has been updated with some regularity during my DPhil, each release is assigned a version number, which broadly follows the principles of *semantic versioning*. Old versions of GENESIS may be accessed by navigating to the specific URL https://nmr-genesis.co.uk/X/Y/Z for version `X.Y.Z`. Then, after enabling developer mode, the specific list of `NOAHModule` objects may be input in order to create the corresponding pulse program. Thus, these two pieces of information together allow all GENESIS pulse programs to be exactly regenerated whenever necessary: this is important in ensuring that the data thus acquired are reproducible.

4.2.2.8 How Smart is GENESIS?

Having written several pages of text about the *features* present in GENESIS, it is tempting to think that it is 'intelligent' in its design of pulse programs. I have, at various points in time, received suggestions for extensions to the concept: in GENESIS, the building blocks used for pulse program construction are NOAH modules, but one could envision breaking up these into the smallest possible units of pulses, delays, gradients and so on, and using something like GENESIS to construct *arbitrary* pulse programs. This was even speculatively mentioned in the GENESIS paper [3].

However, in truth, GENESIS is a long way from being able to do such things. In particular, it does not have any actual understanding of the Bruker pulse programming syntax: *almost* all of the pulse program instructions are hardcoded as strings.[14] Thus, the creation of pulse programs is not *truly* being done from the ground up: instead of 'combining pulse sequence elements', a more accurate description is that the *text* corresponding to different pulse sequence elements is being concatenated. There is actually a substantial amount of brute force involved, and this is also why it is not possible for the automated tests to really check the correctness of the pulse programs.

Of course, it is hardly a trivial task to write something more sophisticated: I would need to construct abstract representations of each pulse sequence element (pulses, delays, etc.), and functions which could translate these into actual pulse program commands. I feel that I would need rather more knowledge in computer science, especially a course on compilers. However, this was something I would have very much liked to do had I had more time! I will also leave open the possibility of—although I do not *commit to*—expanding the GENESIS code in this direction as a future personal project.

4.2.3 Processing Improvements

Finally, I touch on a few small improvements in the processing of NOAH data. These are not strictly part of the GENESIS website, but having a unified source for pulse programs and processing scripts does make it substantially easier to propagate these developments to NMR users.

4.2.3.1 AU Processing Scripts

In the processing of NOAH experiments, the first step is to split up the FIDs of different modules into different files: this is performed by the `splitx_au` AU program. However, after that is done, `splitx_au` *also* calls on auxiliary AU programs to process each of the resulting datasets. Each module is associated with a specific

[14] There are several exceptions where some parameters (for example, loop counters in DIPSI mixing) are dynamically generated, mainly to avoid clashes between different parts of the supersequence.

auxiliary AU program: for example, HSQC data are processed with `noah_hsqc`, COSY data with `noah_cosy` and so on.

Previously, these auxiliary AU programs had to be specified manually using the `USERP1` through `USERP5` processing parameters, which, respectively, apply to the first through fifth modules in a supersequence. Although this was a perfectly serviceable setup, I realised that it was a source of annoyance when running multiple different supersequences, as modules would often be processed incorrectly if these processing parameters were inadvertently copied from an old dataset. Even when just running a single supersequence, this setup step would nonetheless represent an extra cognitive load, especially for users who are not familiar with NOAH experiments.

Since the pulse program specifies which modules are run, and the 'correct' processing parameters also depend on the modules, it stands to reason that the processing parameters can be encoded in the pulse program itself. Specifically, the auxiliary AU programs are part of the `NOAHModule` objects (see Listing 4.2), which makes this information accessible to the GENESIS algorithm. Using this, GENESIS inserts a line near the bottom of every pulse program, specifying which AU programs are to be used for each of the modules present. I furthermore modified the `splitx_au` AU program to parse the pulse program for this extra information. The overall effect is that the `USERP1` through `USERP5` parameters no longer need to be specified, as long as a GENESIS pulse program and a recent version of `splitx_au` are used. However, if the parameters *are* specified, they will override the 'default' processing instructions found in the pulse program. This maintains backwards compatibility with, for example, parameter sets which have been previously saved; it also allows advanced users to customise the processing, if so desired.

4.2.3.2 Non-uniform Sampling

Another small improvement relates to the implementation of non-uniform sampling (NUS) in NOAH supersequences. Because the indirect dimension in NOAH pulse programs is generated through explicit looping, instead of the `mc` macro used in most 2D experiments, the standard NUS implementation in TopSpin cannot be used. Instead, one must explicitly define a list of t_1 increments to sample, which is stored as the `VCLIST` parameter. This is true regardless of which looping implementation in Listing 4.3 is used.

On top of this, the original NUS implementation [49] used a Python script (`noah_nus.py`) to modify the NOAH pulse program such that it only sampled t_1 values from this list. This, however, necessitated storing two copies of the same pulse program, one with NUS and one without.

There is no way to circumvent the requirement for the `VCLIST` parameter, but in GENESIS pulse programs, I opted to implement NUS as an acquisition flag which could be toggled. This means that there is no need to store two versions of the same pulse program, and also makes it easier to seamlessly switch between non-uniform and uniform sampling as desired.

Unfortunately, this change is slightly more problematic than the `splitx_au` change: this new NUS implementation is not backwards-compatible, so requires a new NUS setup script, which I have called `noah_nus2.py`. This script cannot be used with old pulse programs, and the old NUS script cannot be used with GENESIS pulse programs. Nevertheless, it does represent a real improvement over the previous implementation: the GENESIS website also helps to make the changeover as smooth as possible.[15]

4.3 Discussion of Individual Modules

The GENESIS website makes it easy to implement new modules and update old ones: simply changing the underlying `NOAHModule` objects is sufficient to propagate changes to all supersequences generated using those modules. In this section, I discuss a number of new pulse sequence developments made in the course of my DPhil; all of these have been successfully implemented in GENESIS and are available to download.

With regards to sensitivity analyses, since this chapter discusses the design of *individual modules*, I have chosen to focus almost entirely on comparing the SNRs of the modules. This is related to the SNR factor A (although my basis for comparison is not always the standalone experiments, so the definition is not exactly the same). I ignore the quantities ρ_t and ε_t, as these depend on the *supersequence* within which the module is used, and can be trivially improved by adding more modules to the supersequence under consideration. These numbers reflect the utility of the NOAH *technique*, but are not relevant to the individual modules from which they are constructed.

4.3.1 ^{13}C Sensitivity-Enhanced HSQC

The first new module is the sensitivity-enhanced HSQC (seHSQC) experiment, which provides up to 2× increased SNR over a standard (echo–antiecho) HSQC.[16] In the original version of the seHSQC (Fig. 4.9a), developed by Cavanagh, Rance and Kay ('CRK') [53, 54], this is accomplished through the so-called *preservation of equivalent pathways* (PEP) technique [55], which I now provide a product operator analysis of.

---

[15] That said, there were still some bugs in the new NUS implementation right up until August 2022, so this change was not as smooth-sailing as I had hoped for.

[16] Since a States HSQC has $\sqrt{2}$ times the SNR of an EA HSQC (as shown in Fig. 1.8), this also means that the seHSQC has a $\sqrt{2}$ SNR improvement over a States HSQC. The literature can be somewhat confusing on this point: sometimes the gain in signal is even conflated with the gain in SNR. The clearest exposition I have found is that provided by Kontaxis et al. [52].

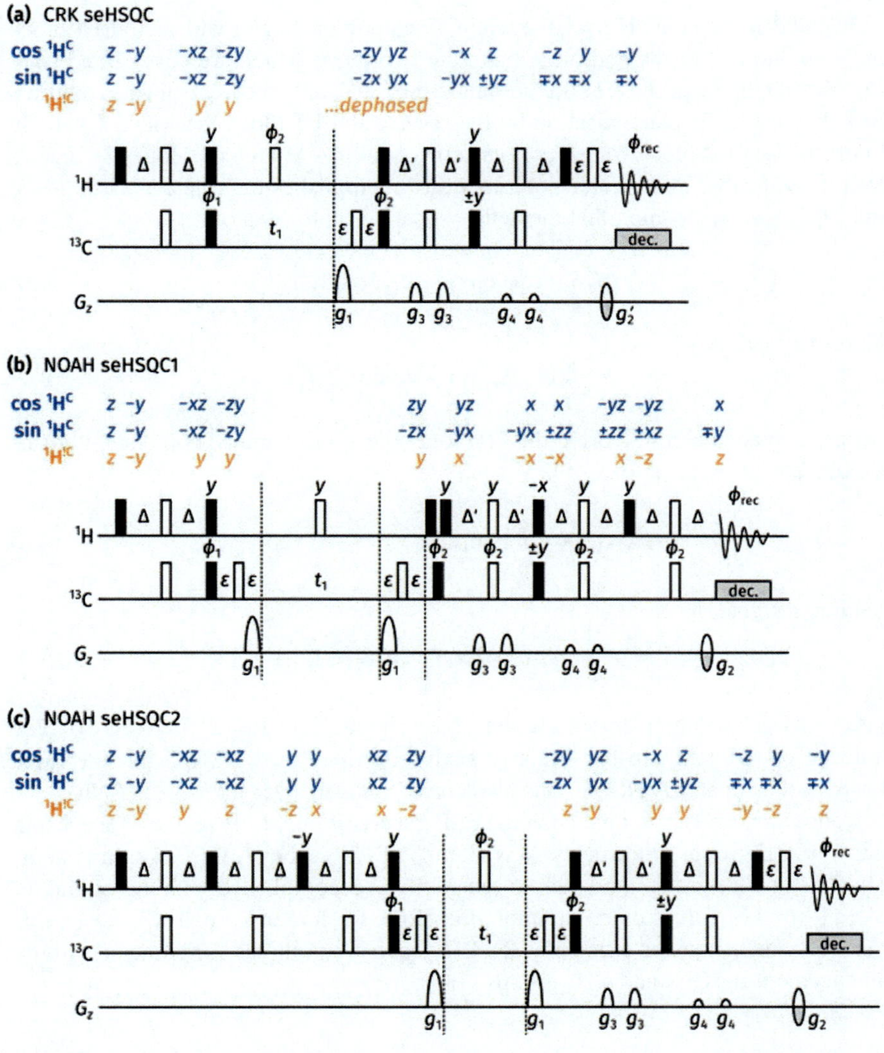

Fig. 4.9 Sensitivity-enhanced HSQC sequences discussed in this section, along with product operator analysis. This analysis is provided only for the first step of the phase cycle and assumes a CH spin pair with $\Delta' = 1/(4 \cdot {}^1J_{\mathrm{CH}})$. 'cos $^1\mathrm{H}^\mathrm{C}$' refers to the component of the $^1\mathrm{H}^\mathrm{C}$ magnetisation which is cosine-modulated during t_1. **a** Cavanagh–Rance–Kay seHSQC. **b** NOAH seHSQC1 module. **c** NOAH seHSQC2 module. Phase cycling is performed with $\phi_1 = (x, -x)$, $\phi_2 = (x, x, -x, -x)$ and $\phi_{\mathrm{rec}} = (x, -x, -x, x)$. The delay Δ is set to $1/(4 \cdot {}^1J_{\mathrm{CH}})$; see the text for a discussion of Δ'. A value of 145 Hz was used for $^1J_{\mathrm{CH}}$. The pulses marked with a phase of $\pm y$ are applied with a phase of y in the echo experiment and $-y$ in the antiecho. Gradient amplitudes are $(g_1, g_2, g_2', g_3, g_4) = (70\%, \pm 35.2\%, \pm 17.6\%, 11\%, -5\%)$

Just like in a standard HSQC, the seHSQC experiment begins with an INEPT block and t_1 evolution. At the end of t_1, there are two terms which are cosine- and sine-modulated with respect to Ω_S. In the standard HSQC, only the $2I_zS_y$ term is returned into observable ^{1}H magnetisation by the reverse INEPT block (see Sect. 1.4.4). In contrast, the PEP block transfers *both* terms back to ^{1}H and subsequently detects both. Specifically, in the echo experiment it accomplishes the transfer $2I_zS_y \rightarrow I_y$ and $2I_zS_x \rightarrow I_x$, meaning that the density operator at the end of t_1

$$-2I_zS_y\cos(\Omega_S t_1) - 2I_zS_x\sin(\Omega_S t_1) \tag{4.6}$$

is transformed into

$$-I_y\cos(\Omega_S t_1) - I_x\sin(\Omega_S t_1) \tag{4.7}$$

just prior to acquisition. During the FID, only the -1-coherence component of this is detected:

$$\frac{1}{2\mathrm{i}}I_-\cos(\Omega_S t_1) - \frac{1}{2}I_-\sin(\Omega_S t_1) = \frac{1}{2\mathrm{i}}I_-\exp(-\mathrm{i}\Omega_S t_1), \tag{4.8}$$

yielding a signal of

$$\frac{1}{2\mathrm{i}}\exp(-\mathrm{i}\Omega_S t_1)\exp(\mathrm{i}\Omega_I t_2), \tag{4.9}$$

which has a $2\times$ larger amplitude than the original EA HSQC (Eq. (1.96)). The antiecho experiment can be similarly analysed. Since the seHSQC has the same noise level as in an EA HSQC, this also corresponds to a $2\times$ increase in SNR.

It should, however, be noted that the PEP transfer is only fully realised for IS spin pairs (i.e. CH groups) if the delay Δ' is set to $1/(4 \cdot {}^1J_{\mathrm{CH}})$. For I_2S or I_3S spin systems (i.e. CH_2 or CH_3 groups), no gain in sensitivity is accomplished with this setting of $\Delta' = 1/(4 \cdot {}^1J_{\mathrm{CH}})$. It is more common, therefore, to shorten Δ' to $1/(8 \cdot {}^1J_{\mathrm{CH}})$: this sacrifices some transfer efficiency for CH groups, but allows for some sensitivity enhancement for CH_2 and CH_3 groups (Table 4.3).

4.3.1.1 NOAH seHSQC Versions

Just like the original EA HSQC (Fig. 1.7), the CRK seHSQC must be modified in order to be compatible with NOAH supersequences. In particular, the CTP gradients in the CRK seHSQC dephase the bulk $^1\mathrm{H}^{!\mathrm{C}}$ magnetisation pool: we would very much like this to be instead returned to $+z$, so that it can be sampled in later homonuclear modules (or, indeed, an HMBC module). This experiment is more tricky to adapt than the HSQC, because there are *three* different magnetisation components to juggle: the cosine-modulated $^1\mathrm{H}^{\mathrm{C}}$, sine-modulated $^1\mathrm{H}^{\mathrm{C}}$ and $^1\mathrm{H}^{!\mathrm{C}}$. Nevertheless, there are at least two ways of doing so; these two modified modules are labelled seHSQC1 and seHSQC2, respectively.

Table 4.3 Theoretical sensitivity enhancements for IS, I_2S and I_3S spin systems in the seHSQC, as a function of the delay Δ'. The values are taken from Schleucher et al. [56]

Spin system	Theoretical sensitivity enhancement		
	$\Delta' = 1/(4 \cdot J_{IS})$	$\Delta' = 1/(8 \cdot J_{IS})$	$\Delta' = 1/(12 \cdot J_{IS})$
IS	2	1.71	1.5
I_2S	1	1.41	1.37
I_3S	1	1.21	1.25

The seHSQC1 module (Fig. 4.9b) was developed by me.[17] It retains the same general structure of the CRK seHSQC up until t_1, but immediately after t_1 a composite ^{1}H pulse is used to effect the transformations

$$I_z \rightarrow I_y; \qquad I_y \rightarrow I_x, \tag{4.10}$$

where the former is required for the $^1H^C$ pool and the latter for $^1H^{!C}$. This has the effect of storing the sine-modulated term as $2I_zS_z$ magnetisation for one spin echo (as opposed to the CRK seHSQC, which stores the cosine-modulated term as I_z). At the beginning of the final spin echo, this is transformed into antiphase magnetisation of the form $2I_xS_z$; thus, this spin echo must be lengthened to a total duration of $2\Delta = 1/(2 \cdot {}^1J_{CH})$ in order to fully refocus $^1J_{CH}$. The final modification involves the addition of an extra gradient immediately before t_1: this ensures that the bulk $^1H^{!C}$ magnetisation is not dephased, just like in the NOAH HSQC module (Fig. 4.3a).

The seHSQC2 module (Fig. 4.9c), on the other hand, was first developed by Hansen et al. [57] In this pulse sequence, the initial $^1H\ 90^\circ_x$ excitation pulse is replaced with a zz isotope-selective pulse (ZIP) element, which is very similar to the zz-filter but has different pulse phases. The effect of this on the $^1H^C$ magnetisation pool amounts to a 90°_{-x} pulse, and thus the same signals are ultimately detected from this magnetisation pool (save for a trivial 180° phase shift). However, on $^1H^{!C}$ magnetisation, the ZIP element acts as a 90°_y pulse; it turns out that this modification alone is sufficient to return the bulk $^1H^{!C}$ magnetisation to $+z$ at the end of the sequence. The ZIP element therefore represents an *isotope-specific rotation* on protons, where the rotation axis depends on whether the proton is directly coupled to ^{13}C or not.

In my work, I made one change to the experiment, namely the addition of an extra gradient echo prior to t_1 (thus leading to a symmetric gradient scheme similar to that in seHSQC1). Instead of this, the original paper had inserted a purge gradient between the ^{1}H and ^{13}C 90° pulses just after the INEPT spin echo (Fig. 4.10). The scheme I used leads to similar results, but has one advantage in that it allows the amplitude of the final CTP gradient (g_2 in Fig. 4.9) to be twice as large: this is particularly relevant to ^{15}N experiments, as will be discussed in Sect. 4.3.2.

Naturally, the primary consideration when evaluating the different seHSQC versions in Fig. 4.9 is sensitivity. Figure 4.11 compares the sensitivities of the various

[17] Through a great deal of trial and error, and certainly *not* intelligent design.

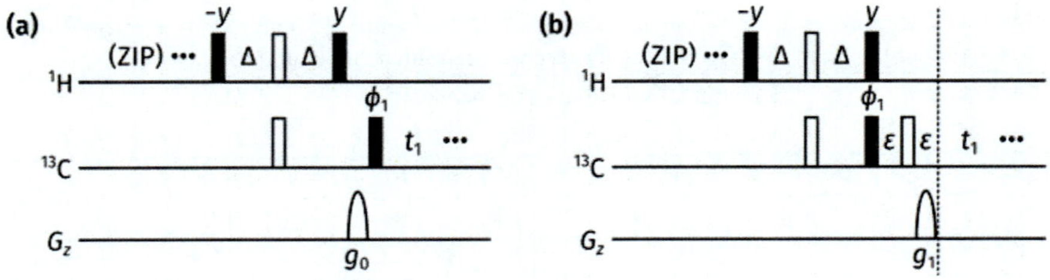

Fig. 4.10 Comparison of CTP gradient schemes in seHSQC2 module. **a** As reported in Hansen et al. [57] g_0 is a purge gradient with arbitrary amplitude; the amplitude of the final CTP gradient must be halved. **b** The version used in this work (corresponding to Fig. 4.9c)

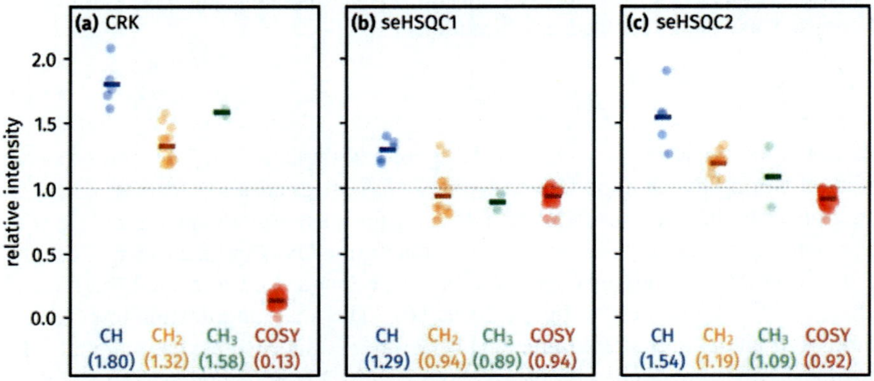

Fig. 4.11 Sensitivities of seHSQC and CLIP-COSY modules in the NOAH-2 S^+C^c supersequences (run using $\Delta' = 1/(8 \cdot {}^1J_{CH})$). Intensities are reported relative to the corresponding peaks in a NOAH-2 SC^c supersequence; HSQC peaks are further broken down by their multiplicity. Each dot represents one peak; the solid lines as well as the numbers at the bottom of each plot represent averages over all peaks. **a** Using the CRK seHSQC. **b** Using the seHSQC1 module. **c** Using the seHSQC2 module. *Data code:* `7A-201115`

possible NOAH-2 S^+C^c supersequence, against a NOAH-2 SC^c experiment. As can be seen, all three seHSQC implementations yield improvements in the sensitivities of CH peaks: the original CRK seHSQC has the largest effect here, followed by seHSQC2 and seHSQC1. For CH_2 and CH_3 peaks, the CRK seHSQC and seHSQC2 perform better than the standard HSQC, but the relative sensitivities in seHSQC1 have dipped slightly below 1. This is not entirely surprising in light of the discussion in Sect. 4.1.1: the modified seHSQC1 and seHSQC2 sequences are expected to have lower sensitivities than the original experiment from which they are derived. In this case, it appears that the modifications made to the seHSQC1 sequence cause sensitivity losses which outweigh the benefits of using a sensitivity-enhanced experiment.

In the context of a NOAH supersequence, it is also important to consider how well each module preserves ${}^1H^{!C}$ magnetisation for subsequent homonuclear module(s). In Fig. 4.11, the CLIP-COSY module is used as the 'reporter' module, but the conclusions drawn here are applicable to *any* homonuclear module (or the HMBC). The

CRK seHSQC does poorly in this respect, as it dephases the bulk magnetisation: this leads to almost an order-of-magnitude sensitivity loss in the CLIP-COSY. Both the other NOAH seHSQC modules, however, perform almost identically to the original HSQC module, as expected from the product operator analysis in Fig. 4.9.

From Fig. 4.11, we can draw some very simple conclusions: if the seHSQC is to be used at (or near) the *beginning* of a NOAH supersequence, where it needs to preserve $^{1}H^{!C}$ magnetisation, then the seHSQC2 module is the best choice. On the other hand, if the seHSQC module is placed at the *end* (say in a NOAH-2 BS^{+} supersequence), then the CRK seHSQC is optimal. This logic is used by GENESIS to choose the best seHSQC implementation based on other modules in the requested supersequence.

4.3.1.2 COSY-Type Artefacts

One (minor) way in which the seHSQC1 module outperforms seHSQC2 is in the suppression of COSY-type artefacts in the seHSQC spectrum [58]. In the CRK seHSQC, these arise due to evolution of $^{n}J_{HH}$ during the second-last spin echo. Specifically, if we assume the presence of a separate ^{1}H spin K which is coupled to spin I, the sine-modulated $2I_yS_z$ term can evolve under both $^{1}J_{IS}$ and $^{n}J_{IK}$ into $2I_yK_z$ and the last ^{1}H 90° pulse transfers this coherence to the spin K. This leads to a peak correlating spin S with spin K. Since the seHSQC2 module shares the same product operator analysis as the CRK seHSQC, these artefacts are also visible in seHSQC2 spectra.

However, for the seHSQC1 module, the corresponding term giving rise to these artefacts would be the cosine-modulated I_x term (at the beginning of the second-last spin echo). During this spin echo, this can again evolve under both J-couplings into $4I_yK_zS_z$, and the final 90° pulse would transform this to $-4I_zK_yS_z$. However, crucially, this term is antiphase with respect to the heteronucleus S: therefore, when decoupling is applied during acquisition, this term should not be observed. This analysis can be verified experimentally by inspection of the seHSQC spectra thus obtained. The COSY-type artefacts, labelled with red boxes in the seHSQC2 spectrum (Fig. 4.12b), are largely attenuated in the seHSQC1 spectrum (Fig. 4.12a). Some artefacts still remain, which perhaps arise due to pulse imperfections (or perhaps more complicated spin systems than the three-spin system considered here; I did not analyse this in any further detail).

4.3.1.3 Wing Artefacts

One point worth considering is whether the extra gradient which I placed in the seHSQC2 module (Fig. 4.10b) is truly needed. In the HSQC and seHSQC1 modules, the bulk magnetisation is placed along the transverse axis during t_1; therefore, if a pair of symmetric gradients are not used, this magnetisation will be dephased (as in the CRK seHSQC). However, in the seHSQC2 module, the bulk magnetisation is longitudinal, meaning that this gradient is not actually required for $^{1}H^{!C}$ preservation.

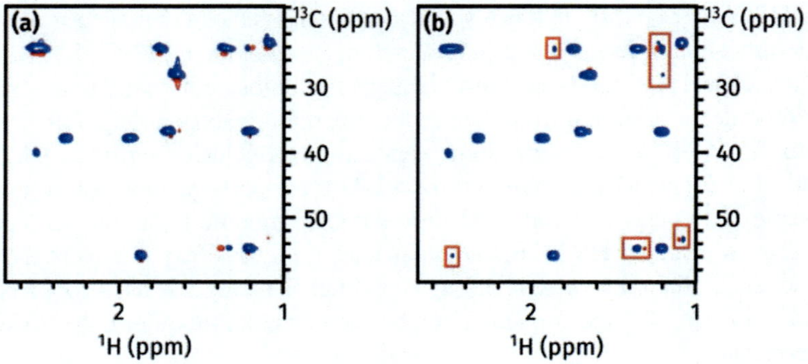

Fig. 4.12 Comparison of COSY-type artefacts in NOAH seHSQC modules. **a** seHSQC1. **b** seHSQC2. The artefacts are highlighted in red boxes. *Data code:* `7A-201115`

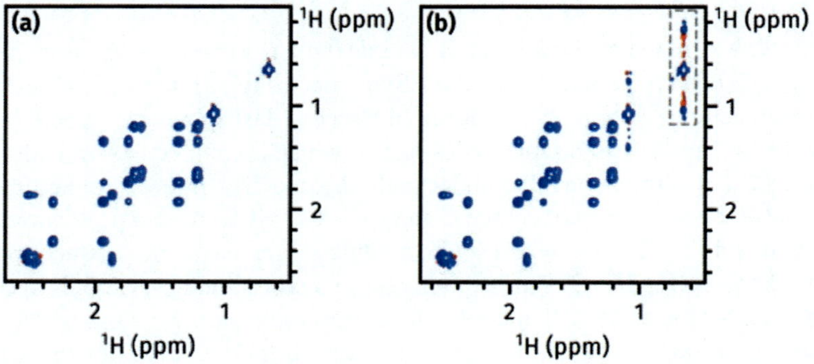

Fig. 4.13 Wing artefacts in CLIP-COSY spectra, taken from NOAH-2 S^+C^c supersequences. **a** Using the seHSQC2 module as shown in Fig. 4.9c. **b** Using a modified seHSQC2 module where the gradient echo just before t_1 is removed. The wing artefacts flanking the two rightmost diagonal peaks can be clearly seen; the boxed artefacts are studied further in Fig. 4.14. *Data code:* `7A-201115`

This gradient can indeed be removed in the seHSQC2 module without compromising the preservation of $^1\mathrm{H}^{!\mathrm{C}}$ magnetisation. However, the omission of this gradient leads to artefacts in the CLIP-COSY spectrum, which I term *'wing artefacts'* because of the way they flank peaks in the CLIP-COSY (Fig. 4.13). These artefacts can most clearly be seen in the diagonal peaks of methyl groups, but are present across the entire spectrum.

A clue as to their origin is provided by the characteristic frequencies of the artefacts:

$$\left(\Omega_1, \Omega_2 = \Omega_I \pm \frac{\Omega_I \cdot [t_{1,\mathrm{HSQC}}/2]}{t_{1,\mathrm{COSY}}}, \Omega_I\right), \tag{4.11}$$

where $t_{1,\mathrm{X}}$ refers to the value of t_1 in experiment X (it does not matter which increment is used, because the ratio of t_1 in the fraction above is constant). In accordance with this, when the seHSQC indirect-dimension spectral width is reduced, the artefacts

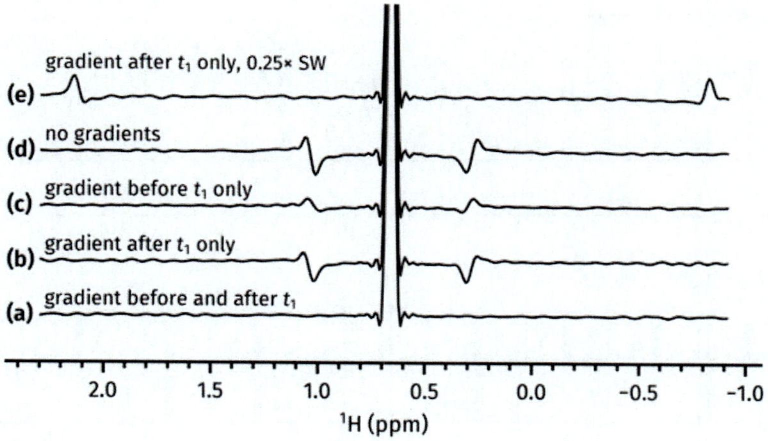

Fig. 4.14 A closer look at wing artefacts in CLIP-COSY spectra, taken from NOAH-2 S^+C^c supersequences with various modifications made to the seHSQC2 module. The spectra being plotted are 1D slices through $F_2 = 0.67\,ppm$ of the 2D CLIP-COSY spectrum; this corresponds to the boxed region in Fig. 4.14b. **a** Using the default seHSQC2 module, with one gradient before and one after t_1. **b** With only the gradient after t_1 (the intensity of the gradient before t_1 is set to 0). **c** With only the gradient before t_1. **d** With neither gradient. **e** With only the gradient after t_1, and additionally with the seHSQC indirect-dimension spectral width reduced by a factor of 4. *Data code:* `7A-201115`

are displaced outwards (Fig. 4.14f). This suggests that these artefacts arise from $^1H^{!C}$ magnetisation which is frequency-labelled in two different t_1 periods: specifically, it evolves once during *half* of the seHSQC t_1 period, and then again during the CLIP-COSY t_1 period. It is thus necessary to apply CTP gradients on both sides of t_1 in order to suppress evolution of this stray magnetisation during both halves: if either or both of the gradients are removed, the artefacts are present (Fig. 4.14a–d). Alternatively, the placement of a purge gradient before the first ^{13}C 90° pulse accomplishes a similar result, as in the original seHSQC2 implementation (Fig. 4.10a).

These wing artefacts are unique to fast-pulsing experiments such as NOAH supersequences: in a typical 2D experiment, these are less likely to arise because of the long(er) recovery delays between data acquisition periods. In the sections which follow, we will see these wing artefacts surface repeatedly. Fortunately, it proves to be relatively easy to suppress them through judicious use of gradients.

4.3.1.4 Multiplicity Editing

Multiplicity editing is reasonably easy to include in all of the seHSQC sequences above. It suffices to add a spin echo of total duration $4\Delta = 1/^1J_{CH}$ just after t_1, while also making sure to change some pulse phases to account for the extra ^1H 180° pulse (Fig. 4.15).

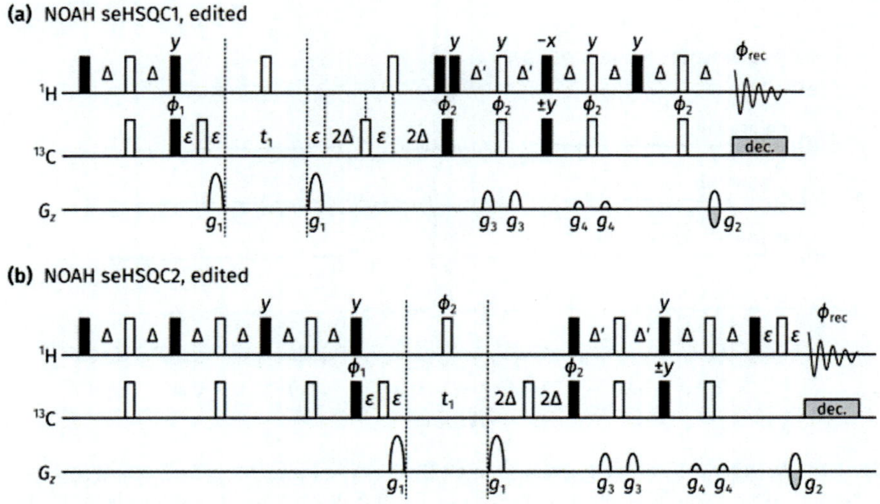

Fig. 4.15 Multiplicity-edited NOAH seHSQC modules. **a** Edited seHSQC1. **b** Edited seHSQC2. All symbols have the same meaning as in Fig. 4.9

The inclusion of editing does not make a substantial difference in the sensitivity comparisons (Fig. 4.16). However, it is interesting to note that the edited seHSQC2 in fact performs *better* than the edited HSQC in terms of preserving bulk magnetisation, as evidenced by the COSY intensities in Fig. 4.16c which are greater than unity. This can be explained by the fact that in the editing period of the HSQC experiment (and seHSQC1), the bulk magnetisation is placed in the transverse plane. The evolution of homonuclear couplings will thus lead to a small loss in the amount of $^{1}\mathrm{H}^{!\mathrm{C}}$ magnetisation preserved. In the seHSQC2 sequence, the bulk magnetisation is longitudinal during the editing period, so does not evolve under J_{HH}.

4.3.1.5 Choice of Δ'

As discussed previously, there are several possible values for the delay Δ' in the seHSQC experiment. I therefore also specifically investigated the possibility of setting $\Delta' = 1/(4 \cdot {}^{1}J_{\mathrm{CH}})$; the corresponding sensitivity comparisons are shown in Fig. 4.17.

As can be seen, the sensitivity boosts obtained for CH groups are higher than in the corresponding spectra with $\Delta' = 1/(8 \cdot {}^{1}J_{\mathrm{CH}})$ (Fig. 4.11). However, there is also a small sensitivity *loss* for CH_2 and CH_3 groups as compared to the original HSQC: this is consistent with previous studies [56]. For CH_3 groups, which typically have large intensities in HSQC-type spectra, this is unlikely to be of any consequence; however, for diastereotopic CH_2 groups this sensitivity loss may not be desirable.

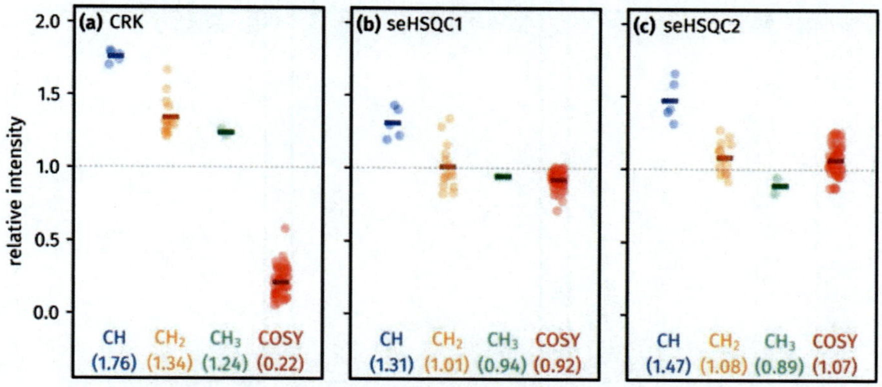

Fig. 4.16 Sensitivity comparisons for NOAH-2 $S^{+}C^{c}$ supersequences with multiplicity editing included; the reference intensities are taken from a likewise edited NOAH-2 SC^{c} supersequence. The delay Δ' was set to $1/(8 \cdot {}^{1}J_{\mathrm{CH}})$. **a** Using the edited CRK seHSQC. **b** Using the edited seHSQC1 module. **c** Using the edited seHSQC2 module. *Data code:* `7A-201115`

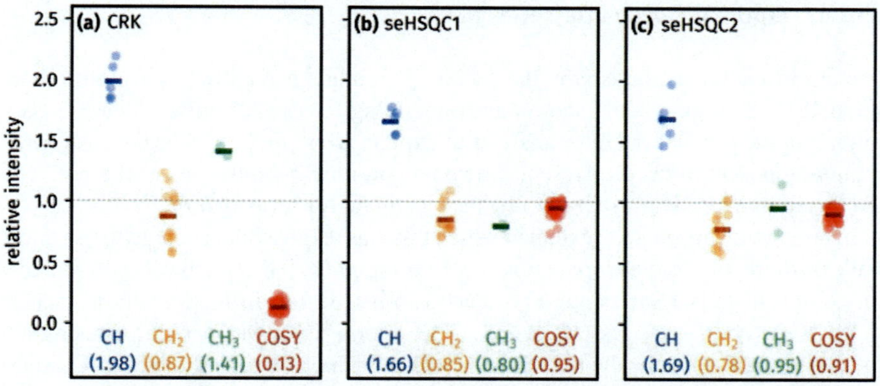

Fig. 4.17 Sensitivity comparisons with Δ' set to $1/(4 \cdot {}^{1}J_{\mathrm{CH}})$. Multiplicity editing was not used. **a** Using the CRK seHSQC. **b** Using the seHSQC1 module. **c** Using the seHSQC2 module. *Data code:* `7A-201115`

It is interesting to note that the performance of the seHSQC1 module is much closer to that of the seHSQC2 here. The fact that seHSQC1 performs worse with the reduced value of $\Delta' = 1/(8 \cdot {}^{1}J_{\mathrm{CH}})$ suggests that there are some inefficiencies in this section of the pulse sequence; however, there is no immediate explanation for this in the product operators (the relevant terms in the seHSQC1 are entirely similar to that in the CRK, save for a minus sign).

The conclusions drawn are entirely similar when multiplicity editing is enabled (Fig. 4.18), so will not be further discussed.

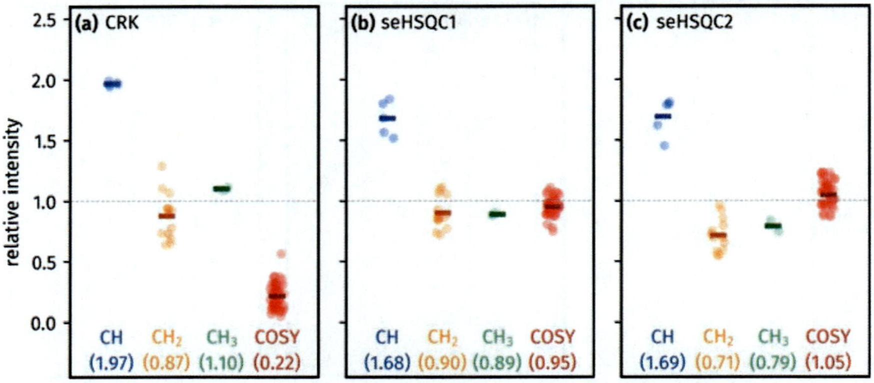

Fig. 4.18 Sensitivity comparisons with Δ' set to $1/(4 \cdot {}^{1}J_{\mathrm{CH}})$ and with multiplicity editing. **a** Using the edited CRK seHSQC. **b** Using the edited seHSQC1 module. **c** Using the edited seHSQC2 module. *Data code:* `7A-201115`

4.3.1.6 Optimal Control for seHSQC1

One unresolved question is why the seHSQC1 module has a lower sensitivity than the seHSQC2, despite being shorter and containing fewer 180° pulses. There is also nothing in the product operator analysis to explain why this should be the case. One remaining possibility is the presence of some non-ideality in the pulses themselves, and the composite ^{1}H pulse is an obvious candidate for investigation.[18]

In order to investigate the extent to which this was responsible, I sought to use optimal control theory to develop a *universal rotation pulse* (URP) which could replace this composite pulse. Such a pulse must accomplish the transformations shown in Eq. (4.10), namely $I_z \to I_y$ and $I_y \to I_x$.[19] This optimisation was performed using an interior-point algorithm (the default in MATLAB `fmincon`) and the ESCALADE method [60] for the calculation of analytic derivatives. Pulse fidelity was calculated as an average over 51 ^{1}H spins, evenly distributed across a frequency range of 11 kHz (corresponding to 15.7 ppm at 700 MHz); the duration of the pulse was fixed at 200 μs and 200 pulse points were used. The maximum B_1 allowed was 15 kHz; the optimisation was run with a 10% variation in B_1 to account for the effects of RF inhomogeneity.

The ^{13}C 90° pulse which is simultaneously applied during the pulse sequence was not optimised together with this. Since the ^{1}H URP has a longer duration than this, the sequence must be modified slightly (Fig. 4.19). In particular, the delays α

[18] The concept of a *composite pulse* [59] is usually associated with greater efficiency/uniformity, but here I have used the term to loosely refer to a consecutive series of pulses.

[19] The effect of this pulse on x-magnetisation is not important for the seHSQC1, but is in fact fully determined by these two constraints: $UI_xU^\dagger = -\mathrm{i}U(I_yI_z - I_zI_y)U^\dagger = -\mathrm{i}(UI_yU^\dagger UI_zU^\dagger - UI_zU^\dagger UI_yU^\dagger) = -\mathrm{i}(I_xI_y - I_yI_x) = I_z$. This reveals a geometric interpretation of this pulse element: it is actually a 120° rotation about the vector (1, 1, 1), which interchanges the three Cartesian axes; it is closely related to the C_3 symmetry operation in the octahedral point group (O_h).

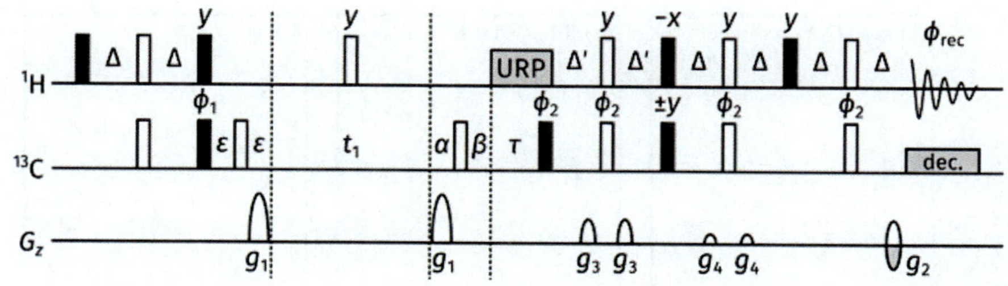

Fig. 4.19 seHSQC1 pulse sequence using a ^{1}H URP in place of the double-90° composite pulse. τ is the difference in duration between the URP and the ^{13}C 90° hard pulse. Delays are set as: $\alpha = \varepsilon + \tau/2$; $\beta = \varepsilon - \tau/2$

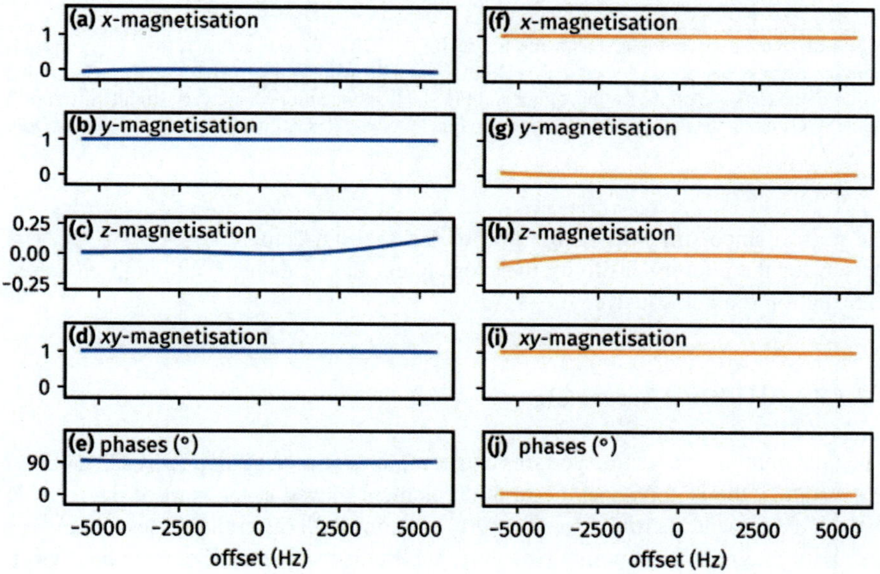

Fig. 4.20 Simulated performance of the optimised URP on z- and y-magnetisation. The pulse fidelity was 99.99%. **a–e** Using z-magnetisation as input: the plots, respectively, show the amount of x-magnetisation, y-magnetisation, z-magnetisation, transverse magnetisation ($M_{xy} = \sqrt{M_x^2 + M_y^2}$) and the phase of the transverse magnetisation generated, as a function of offset frequency. **f–j** The same, but using y-magnetisation as input

and β must be chosen in order to satisfy the relations $2\varepsilon = \alpha + \beta$ (to ensure $^{1}\text{H}^{!\text{C}}$ magnetisation is refocused) and $\alpha = \beta + \tau$ (to ensure that for $^{1}\text{H}^{\text{C}}$ magnetisation, the ^{13}C chemical shift evolves for a total duration of t_1).

This optimisation process yielded a pulse with 99.99% fidelity: the (theoretical) performance of this pulse on z- and y-magnetisation is shown in Fig. 4.20. However, when tested in the actual seHSQC1 experiment, this failed to yield any substantial difference compared to the original double 90° pulse (Fig. 4.21a and b). Importantly,

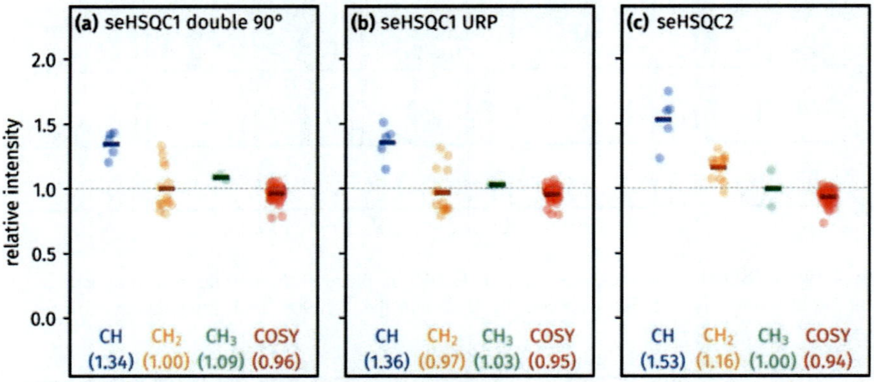

Fig. 4.21 Sensitivity comparisons using the seHSQC1 URP. Δ' was set to $1/(8 \cdot {}^1J_{\mathrm{CH}})$; no multiplicity editing was used. **a** Original seHSQC1 module with double ^{1}H 90° pulse. (Note that different datasets were used for this figure and Fig. 4.11, so the numbers are very slightly different.) **b** seHSQC1 using the optimised URP shown in Fig. 4.20. **c** seHSQC2 module for comparison. *Data code:* `7A-220110`

the performance still falls below that of the seHSQC2 module (Fig. 4.21c). The reason for the poorer sensitivity therefore likely lies elsewhere, although an actual explanation remains elusive.

4.3.1.7 BIG-BIRD Versus ZIP

The final point in this section pertains to the implementation of the seHSQC2 module. As it stands, the isotope-specific rotation element placed at the start of the module is the ZIP element: its role is to effect 90° rotations with different phases on the $^1\mathrm{H}^{\mathrm{C}}$ and $^1\mathrm{H}^{!\mathrm{C}}$ magnetisation pools. However, such pulse sequence elements have been known for a long time: these include TANGO [61], BANGO [62], BIRD [63–65], BIG-BIRD [66] and TIG-BIRD [67]. In particular, the BIG-BIRD element, which allows for independent excitation of $^1\mathrm{H}^{\mathrm{C}}$ and $^1\mathrm{H}^{!\mathrm{C}}$ magnetisation with arbitrary flip angles and phases, can be designed to accomplish the same overall effect as the ZIP element. For the seHSQC2 without multiplicity editing, the relevant BIG-BIRD sequence is

$$45^\circ_{45^\circ}(^1\mathrm{H})\text{–}2\Delta\text{–}180^\circ(^1\mathrm{H},\,^{13}\mathrm{C})\text{–}2\Delta\text{–}45^\circ_{225^\circ}(^1\mathrm{H}), \qquad (4.12)$$

where β_ϕ represents a hard pulse with flip angle β and phase ϕ. For the edited seHSQC2, the phases must be altered slightly:

$$45^\circ_{315^\circ}(^1\mathrm{H})\text{–}2\Delta\text{–}180^\circ(^1\mathrm{H},\,^{13}\mathrm{C})\text{–}2\Delta\text{–}45^\circ_{135^\circ}(^1\mathrm{H}). \qquad (4.13)$$

Figure 4.22 provides sensitivity comparisons between the BIG-BIRD and ZIP versions of the seHSQC2. The fact that similar numbers were achieved with both

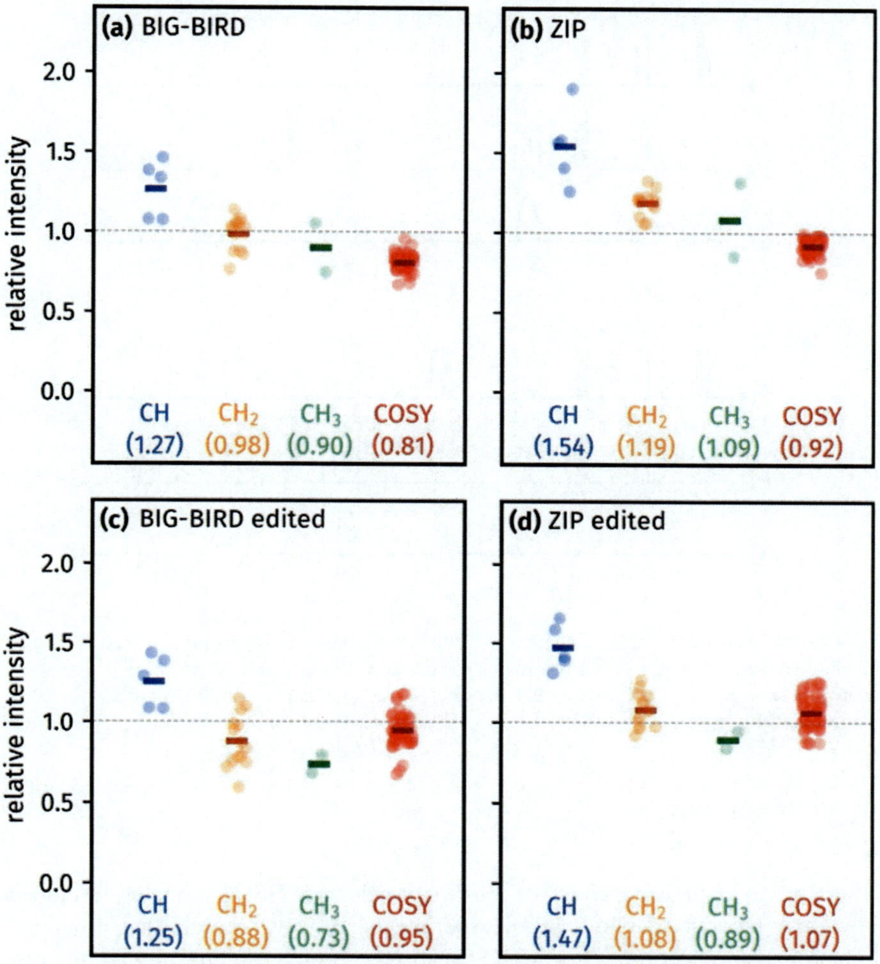

Fig. 4.22 Sensitivity comparisons of BIG-BIRD and ZIP pulse elements in seHSQC2 module. The reference dataset being compared against is still the NOAH-2 SC^c (but with editing in **c** and **d**. The delay Δ' was set to $1/(8 \cdot {}^1J_{CH})$. **a** Unedited seHSQC2 using BIG-BIRD. **b** Unedited seHSQC2 using ZIP. **c** Edited seHSQC2 using BIG-BIRD. **d** Edited seHSQC2 using ZIP. *Data code:* `7A-201115`

elements indicates that the BIG-BIRD element is (largely) effecting the desired rotations on the different magnetisation pools. However, in all respects, the performance of the BIG-BIRD sequence was not as good as the ZIP version: both the seHSQC sensitivity itself, as well as the sensitivity of the later CLIP-COSY, were slightly decreased when the BIG-BIRD element was used. This is true regardless of whether multiplicity editing was used.

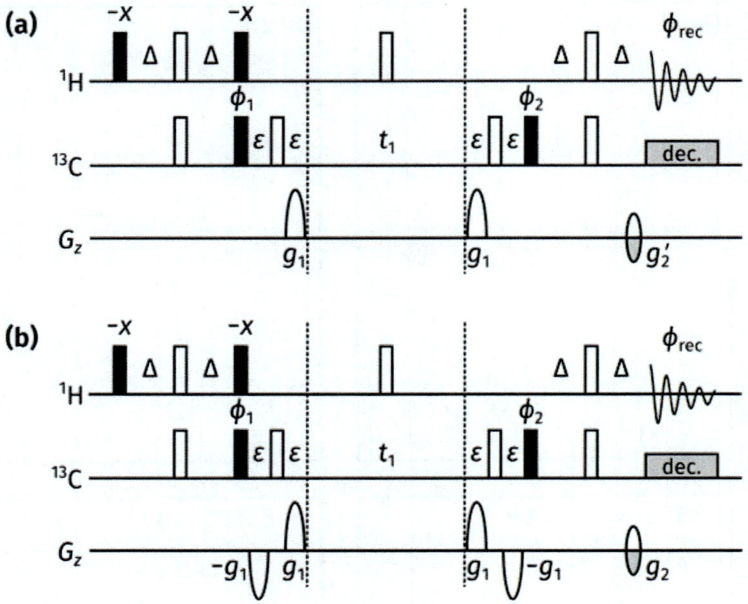

Fig. 4.23 NOAH HMQC pulse sequences. **a** With two encoding gradients around t_1. **b** With four encoding gradients around t_1. Phase cycling is performed with $\phi_1 = (x, -x)$, $\phi_2 = (x, x, -x, -x)$ and $\phi_{rec} = (x, -x, -x, x)$. The delay Δ is set to $1/(4 \cdot {}^1J_{NH})$. Gradient amplitudes are: $g_1 = 80\%$; $g_2 = \pm 32.4\%$; $g_2' = g_2/2$

4.3.2 ^{15}N HMQC

In Sect. 4.3.3, I will discuss how the sensitivity-enhanced HSQC modules developed above may be adapted into ^{1}H–^{15}N experiments. However, before that, I make a slight detour to cover the ^{15}N HMQC experiment, which (up until my DPhil) was the experiment of choice for detecting one-bond ^{1}H–^{15}N correlations.

4.3.2.1 CTP Gradient Scheme

The HMQC module is based on the ASAP-HMQC reported by Kupče and Freeman [13]. In their original report, the ASAP-HMQC used a symmetric gradient scheme similar to that in the seHSQC modules previously described (Fig. 4.23a). However, in the NOAH HMQC module [32], this was expanded into a four-gradient scheme where bipolar gradient *pairs* were placed before and after t_1 (Fig. 4.23b). This was likely implemented in order to allow the final gradient, g_2, to have as large an amplitude as possible. In heteronuclear experiments, this final gradient is particularly important for dephasing bulk magnetisation which is transverse just prior to detection (due to pulse imperfections or relaxation). If this gradient is too weak, this unwanted magnetisation will be incompletely dephased, leading to artefacts in the resulting spectrum.

Strategies to maximise the amplitude of this gradient are particularly crucial in ^{1}H–^{15}N experiments (as compared to ^{1}H–^{13}C experiments) for two reasons. Firstly, the natural abundance of ^{15}N (0.36%) is even smaller than ^{13}C (1.1%), which translates into a weaker signal: better artefact suppression must be achieved in order for the signal to not be obscured. Secondly, the gyromagnetic ratio of ^{15}N is also smaller: since $g_2/g_1 \propto |\gamma_N/\gamma_H|$, g_2 will naturally be smaller if no modifications are made to the pulse sequence.

In this respect, the four-gradient scheme in Fig. 4.23b is superior to the two-gradient scheme, because the gradient g_2 will have an amplitude of $4\gamma_N g_1/\gamma_H$. However, when used in a NOAH supersequence, this leads to wing artefacts in downstream modules, since bulk $^{1}H^{!N}$ magnetisation effectively does not experience any coherence order selection during t_1. This motivates a return to the two-gradient scheme of Fig. 4.23a, but to compensate for the fact that the decoding gradient g_2' has half of the amplitude of g_2, I instead *lengthened* all CTP gradients from their usual duration of 1 ms to 2.5 ms. This ensures that any stray transverse bulk magnetisation at the end of the HMQC module is effectively dephased by the final gradient.

The HMQC spectra thus obtained are shown in Fig. 4.24. These, and all other ^{1}H–^{15}N spectra in this chapter, were acquired assuming a $^{1}J_{NH}$ value of 90 Hz. The first column shows the spectra obtained with the original four-gradient scheme: although the artefacts at 2.2 ppm are reasonably well-suppressed in the HMQC module (Fig. 4.24b), the CLIP-COSY spectrum clearly has a set of wing artefacts (Fig. 4.24c). (Note that the wing artefacts occur at different frequencies compared to the ^{13}C seHSQC case, because the t_1 increment in the ^{15}N HMQC module is different.)

The second column shows what happens when the two-gradient scheme is adopted without changing the gradient duration. Although the CLIP-COSY wing artefacts disappear (Fig. 4.24f), the HMQC artefacts are over twice as intense (Fig. 4.24e) and (in this case) have comparable intensity to the desired peaks.

By using the two-gradient scheme and increasing the gradient duration to 2.5 ms, we obtain the best of both worlds: the HMQC artefacts are well-suppressed (Fig. 4.24h, in fact even better than in the original spectrum), and the CLIP-COSY is free of wing artefacts (Fig. 4.24i). The only drawback is a slight loss in signal intensity in the HMQC, which arises due to diffusion and relaxation during the longer pulse sequence. However, this decrease is only on the order of 5% (for this example), which is a totally acceptable price to pay in return for the improved spectral quality.

The signal loss does suggest that we should check whether such a long gradient is truly needed. I therefore ran the two-gradient HMQC experiment with gradient durations ranging from 1 ms to 2.5 ms; the signal and artefact intensities in each of these experiments are plotted in Fig. 4.25. Generally, there is little variation in the intensities of the two desired peaks. The artefact intensity is more erratic, which possibly reflects the fact that gradient dephasing varies sinusoidally with the gradient duration τ due to the spatial integral

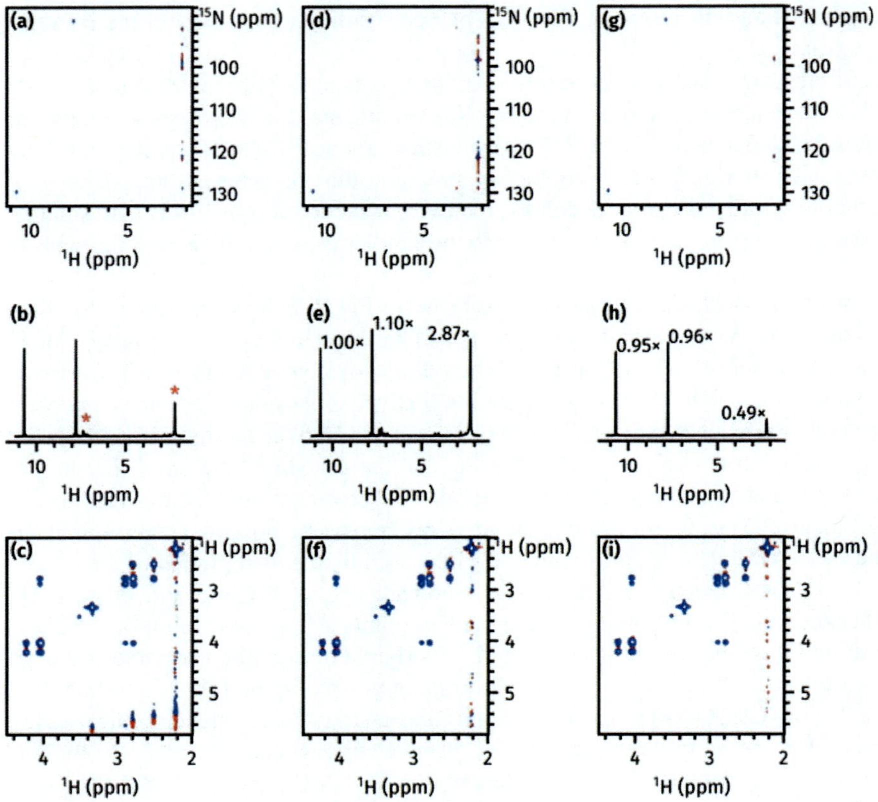

Fig. 4.24 Comparison of HMQC and CLIP-COSY spectra obtained from NOAH-3 $M_N S^+ C^c$ supersequences, acquired using different HMQC gradient schemes. In the first row, the HMQC spectrum itself is shown. In the second row, the positive projection of the HMQC spectrum onto the F_2 axis is shown; the numbers indicate peak intensities with respect to the reference dataset (the left column). The asterisks indicate artefacts caused by bulk magnetisation not sufficiently dephased by the final gradient. In the third row, (an inset of) the CLIP-COSY spectrum is shown. **a–c** Using the four-gradient scheme of Fig. 4.23b, with 1 ms gradients. **d–f** Using the two-gradient scheme of Fig. 4.23a, with 1 ms gradients. **g–i** Using the two-gradient scheme of Fig. 4.23a, with 2.5 ms gradients. *Data code:* `7Z-200926`

$$\frac{1}{L}\int_{-L/2}^{L/2} \exp(\mathrm{i}\gamma G z \tau)\,\mathrm{d}z = \frac{\sin(\gamma G L \tau/2)}{\gamma G L \tau/2}, \tag{4.14}$$

where L is the sample length and G the gradient amplitude (see also Eq. (1.93)). Nevertheless, increased gradient durations clearly lead to better artefact suppression, which (at least in these datasets) is greatest with 2.5 ms gradients. This value of 2.5 ms was therefore fixed on as a suitable choice for ^{15}N experiments, including the seHSQC which is discussed in Sect. 4.3.3.

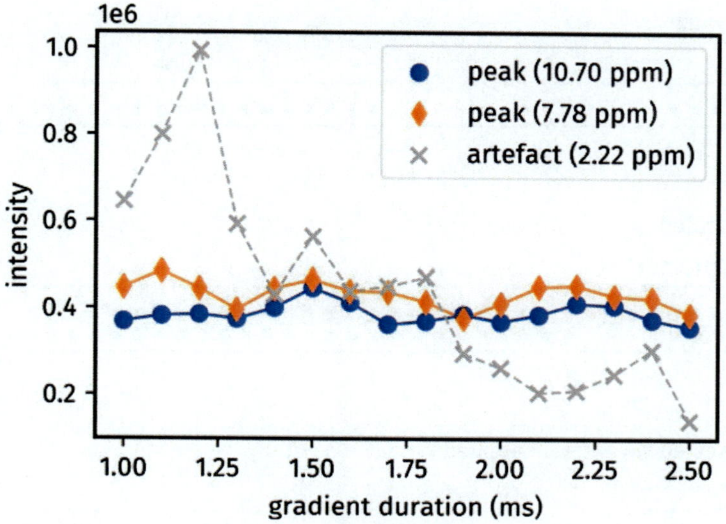

Fig. 4.25 Effect of CTP gradient duration on ^{15}N HMQC signal and artefact intensities, as measured by (absolute) peak heights in the positive F_2 projection. *Data code:* `7Z-200926`

4.3.2.2 *k*- and SW-Scaling

^{1}H–^{15}N spectra of small molecules are often relatively sparse in the indirect dimension and do not typically need to be acquired with the same resolution as a ^{1}H–^{13}C experiment. However, if these two different classes of experiments were to be combined together within a supersequence, it would not ordinarily be possible to individually toggle the number of t_1 increments recorded for each module.

One way of changing this would be to decrease the frequency with which the ^{15}N t_1 duration is incremented in the pulse program. This practice was first proposed by Parella et al. in the context of time-shared NMR [27, 68], and in this thesis is called 'k-scaling'. This means that each t_1 increment in the ^{15}N experiment is acquired several times before t_1 is incremented, which effectively leads to k times fewer t_1 increments, but with each increment having k times the number of scans (after the data have been combined). An alternative to this, which leads to the same resolution in the indirect dimension, would be to increase the spectral width of the ^{15}N module (which is encoded as the `CNST40` parameter in GENESIS) by the same factor of k: this leads to very similar results, and is referred to as 'SW-scaling' here.

Figure 4.26 illustrates all of these possibilities in detail, focusing only on the t_1 sampling schedule of the scaled ^{15}N experiments. In a 'typical' NOAH supersequence (Fig. 4.26a), 256 t_1 increments would be recorded (corresponding to the eight blue dots); for the gramicidin sample used in this section, the largest value of t_1 (denoted as AQ_1) is around 60 ms. The default TopSpin processing routine for 2D NMR applies forward linear prediction (LP) [69–72] in the indirect dimension, such as to double the number of data points: the data obtained through LP are indicated by orange dots.

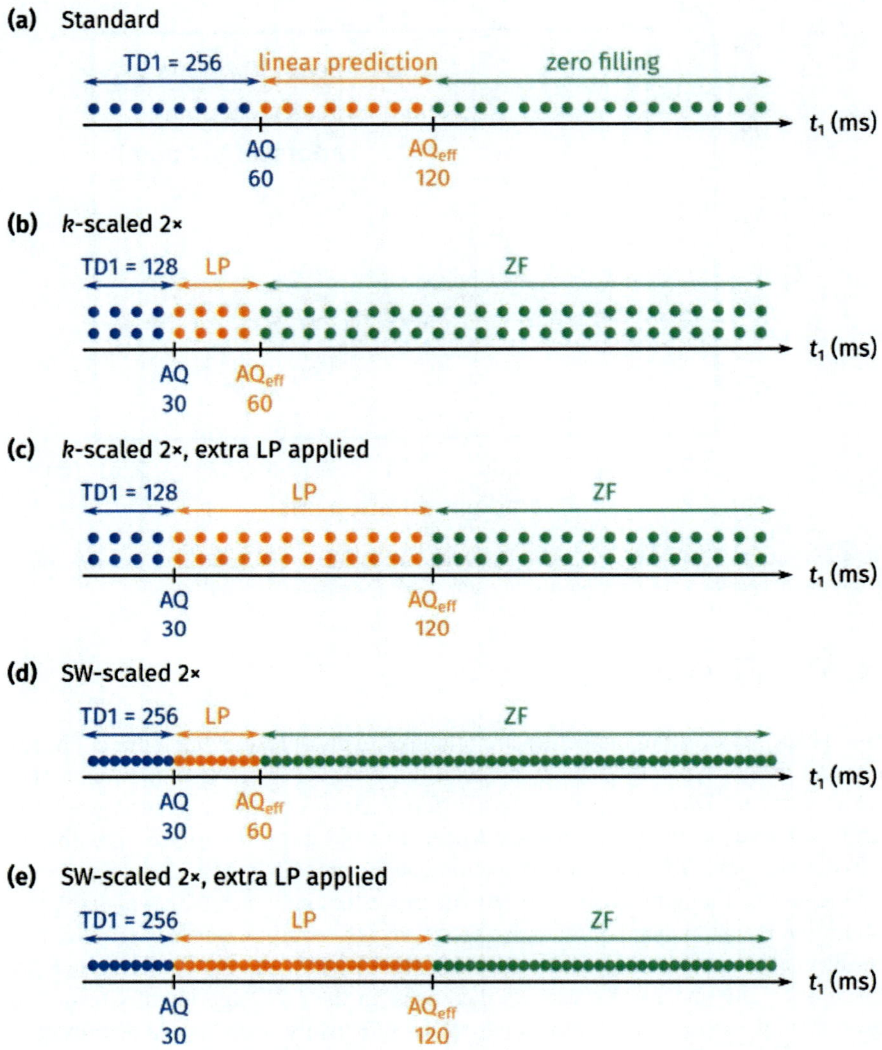

Fig. 4.26 Pictorial representation of k- and SW-scaling in NOAH ^{15}N modules. Each blue dot represents 32 t_1 increments physically acquired as part of the experiment; the experimental time is proportional to the number of blue dots, and is constant across all of these experiments. Orange dots represent 32 t_1 increments obtained through forward linear prediction, and green dots represent zeroes. **a** The default experiment. **b–e** k-scaled experiment, without and with extra linear processing to restore the original value of $AQ_{1,\text{eff}}$. **d–e** SW-scaled experiment, without and with extra linear processing to restore the original value of $AQ_{1,\text{eff}}$

This leads to an *effective* AQ_1 which is twice as large, denoted as $AQ_{1,\text{eff}}$. Finally, zero filling (ZF) is applied to increase the digital resolution to a given value, as represented by the green dots.

In the k-scaled spectrum, both AQ_1 (and thus $AQ_{1,\text{eff}}$) are halved, in return for a doubling of the number of scans (Fig. 4.26b). The resolution in the indirect dimension is proportional to $AQ_{1,\text{eff}}$; thus, this also corresponds to a $2\times$ decrease in resolution. However, this can be counteracted through the use of extra linear prediction (Fig. 4.26c) to extend $AQ_{1,\text{eff}}$ back to its original value. The SW-scaled spectra (Fig. 4.26d and e) are almost entirely equivalent to the k-scaled spectra in terms of their effect on AQ_1 and $AQ_{1,\text{eff}}$, except that the sampling schedule is *very* slightly different.

Figure 4.27 shows the effects of performing k- or SW-scaling on the ^{15}N HMQC module, without any extra linear prediction beyond the default. In this case, these scaling procedures alone can in fact lead to significant gains in SNR (as measured by peak heights), with up to $2.6\times$ improvements being observed at the extreme of 8-fold k- or SW-scaling. This likely arises because the J_{HH} modulation in the indirect dimension (visible in the standard spectrum, Fig. 4.27b) is no longer being resolved: the largest gains are attained for the peak at 123 ppm, where this modulation is especially prominent.

The effect of scaling *and* adding extra linear prediction is shown in Fig. 4.28. In general, this combination of scaling plus LP leads to improvements in spectral SNR of up to $6\times$. However, this is accompanied by distortions in the F_1 multiplet structure, especially for the k-scaled spectra (Fig. 4.28c–h).

However, caution should be exercised when interpreting these results. Although this improvement in SNR is genuine, it is not necessarily the case that this represents a true improvement in *detection sensitivity*: in other words, processing techniques such as LP do not always allow for better discrimination between signal and noise [73, 74].[20] Further tests would need to be done on more dilute samples[21] to ascertain the benefits of the scaling-plus-LP routine on spectra with lower intrinsic SNRs.

4.3.3 ^{15}N Sensitivity-Enhanced HSQC

The implementation of the ^{15}N seHSQC modules is entirely identical to the ^{13}C versions previously discussed (Sect. 4.3.1), except for the following points:

- CTP gradient amplitudes were set using $g_1 = 80\%$ and $g_2 = \mp 16.2\%$ (note the sign change because of the negative magnetogyric ratio of ^{15}N);
- CTP gradient durations were lengthened to 2.5 ms (as discussed for the ^{15}N HMQC) in order to suppress artefacts in the seHSQC itself;
- the delays Δ and Δ' were both set to $1/(4 \cdot {}^{1}J_{\text{NH}})$ to maximise sensitivity for NH groups (as before, the value of ${}^{1}J_{\text{NH}}$ was set to 90 Hz);
- ^{13}C adiabatic pulses were replaced with hard pulses on ^{15}N.

[20] In recent years, this issue has been investigated more thoroughly in the context of NUS [75].

[21] Or, perhaps, less sensitive instruments—although the 700 MHz cryoprobe used here was the only option for these supersequences.

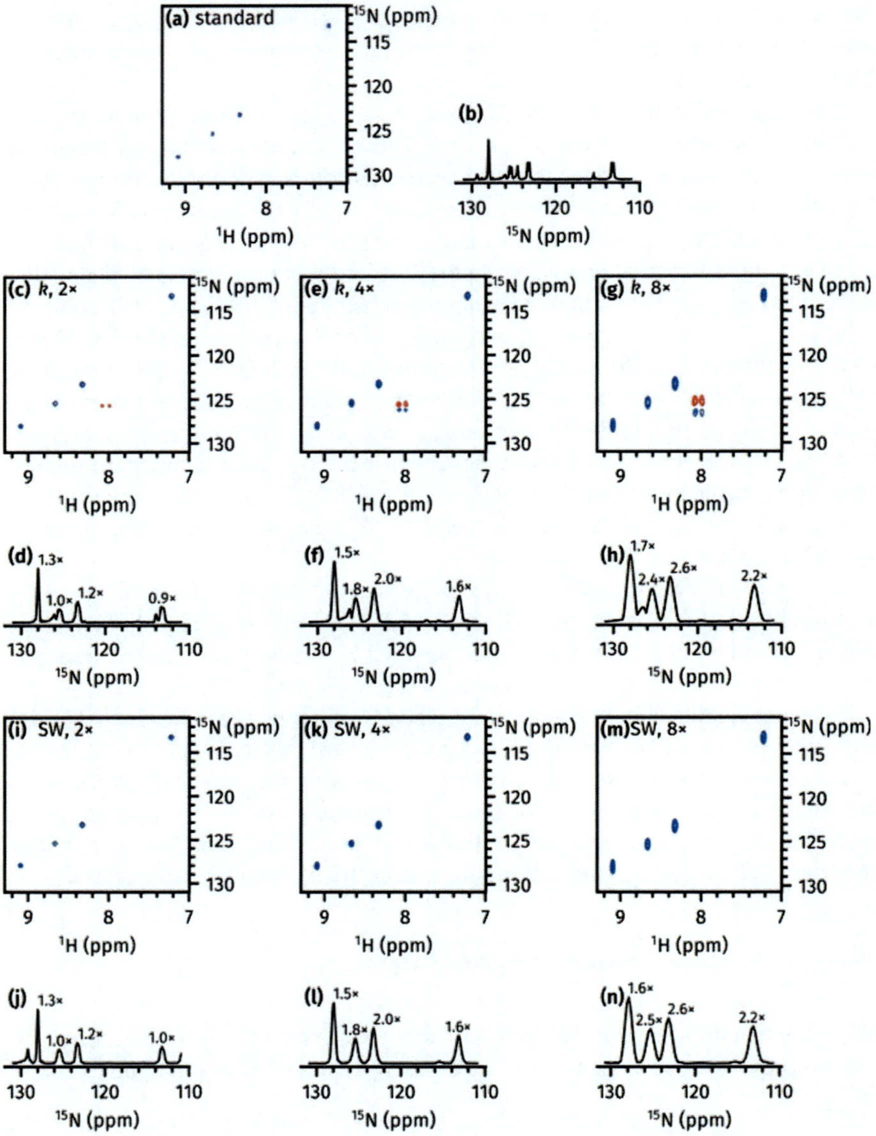

Fig. 4.27 Effects of *k*- and SW-scaling on NOAH HMQC spectrum (taken from NOAH-3 $M_N S^+ C^c$ supersequences). Each HMQC spectrum is shown together with a positive projection onto the F_1 axis. The relative SNR of each peak, with respect to the standard spectrum, is indicated on each of the other projections. **a–b** Standard spectrum. **c–h** *k*-scaled spectra. **i–n** SW-scaled spectra. *Data code:* `7G-210310`

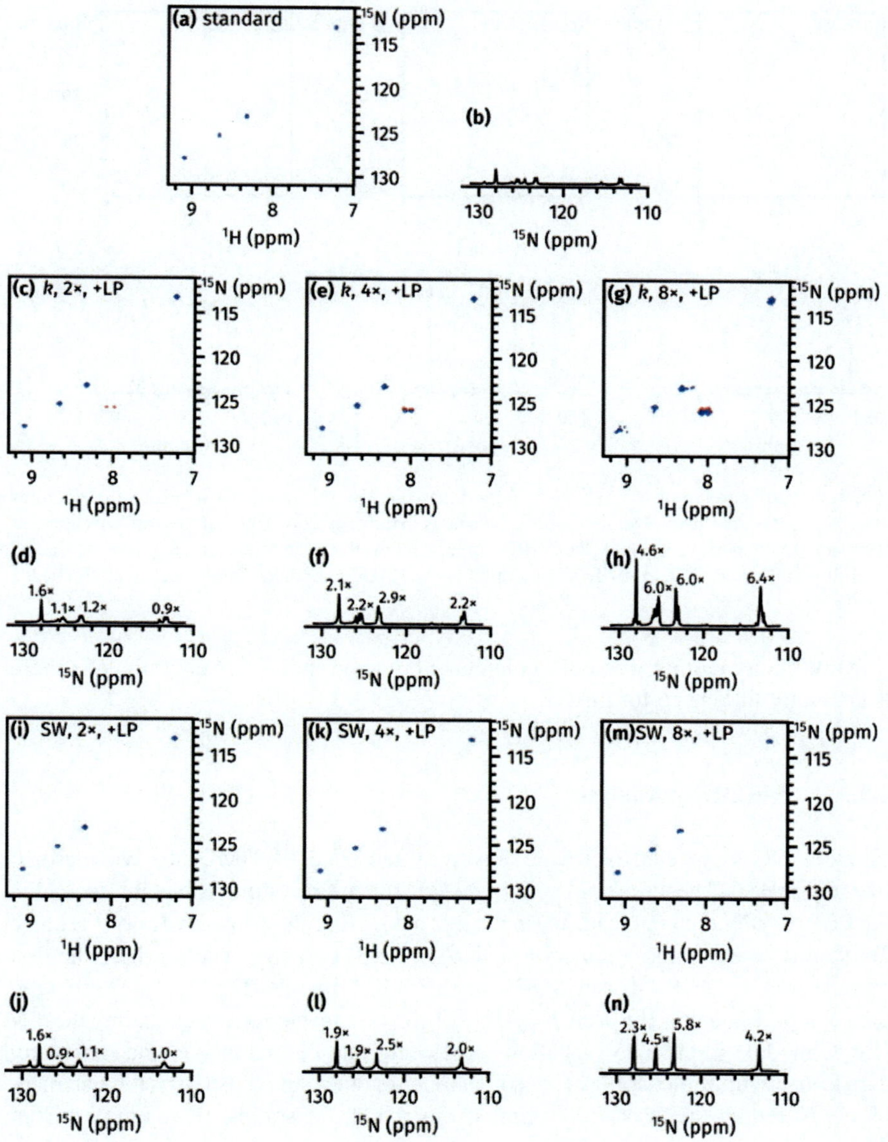

Fig. 4.28 The same as in Fig. 4.27, but with extra linear prediction applied to all scaled spectra to bring $AQ_{1,\text{eff}}$ up to its original value in the standard spectrum. Linear prediction of k times more points leads to a $\sqrt{k}$ increase in noise; to account for this, all spectra are plotted with the same noise level. **a–b** Standard spectrum. **c–h** k-scaled spectra. **i–n** SW-scaled spectra. *Data code:* `7G-210310`

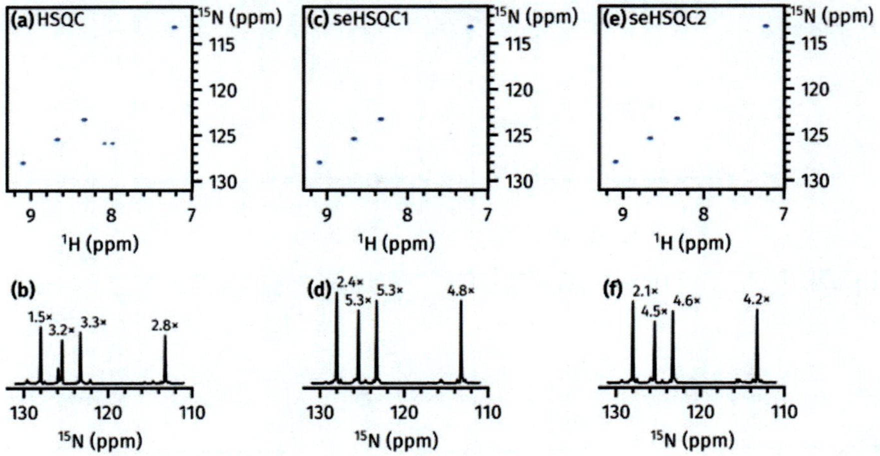

Fig. 4.29 Comparison of NOAH ^{1}H–^{15}N modules; the 2D spectra and their positive projections onto the F_1 axis are shown. Numbers above each peak in the projections indicate sensitivity increases compared to the HMQC module (the HMQC spectra themselves can be seen in, e.g., Fig. 4.27a and b). **a–b** HSQC module. **c, d** seHSQC1 module. **e–f** seHSQC2 module. *Data code:* `7G-201017`

Multiplicity editing was not implemented or used in the ^{15}N experiments; there is generally little need for this.

4.3.3.1 Sensitivity Analysis

The seHSQC is expected to provide sensitivity gains over the HMQC for two reasons: first, the seHSQC peaks are not split by $^{n}J_{\mathrm{HH}}$ in the indirect dimension, and secondly, the PEP transfer element leads to (in theory) a doubling in sensitivity for NH groups. To quantify these gains, I ran several NOAH-3 $\mathrm{XS^{+}C^{c}}$ experiments, where the first module was a ^{15}N HSQC, seHSQC1, or seHSQC2 module (Fig. 4.29). These were then compared against the basic NOAH-3 $\mathrm{M_{N}S^{+}C^{c}}$ supersequence. Including the ^{15}N HSQC module in this analysis allows us to determine how much of the sensitivity gain is due to multiplet collapse, and how much is due to the sensitivity enhancement block. Note that the peak at 8.0 ppm is folded and, for simplicity, is omitted from this analysis.

As can be seen, the (non-enhanced) HSQC alone allows for substantial sensitivity increases over the HMQC, with over $3\times$ gains being accomplished simply due to multiplet collapse. The seHSQC experiments push this even further: the PEP transfer block provides an approximate $1.5\times$ sensitivity improvement for all peaks, which is on top of the boost already provided by the HSQC. In contrast to the ^{13}C case, the ^{15}N seHSQC1 module outperforms the seHSQC2 in terms of raw sensitivity, though (as before) there is no clear explanation for this subtle difference.

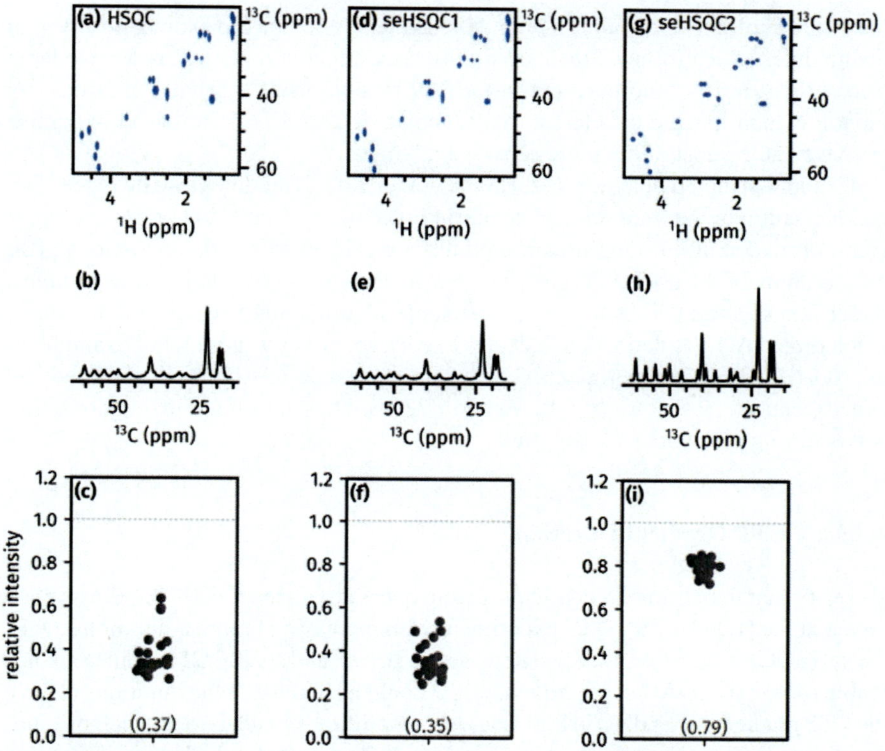

Fig. 4.30 Comparison of the ^{13}C seHSQC module from NOAH-3 XS$^+$C^c supersequences, where X is a ^{1}H–^{15}N module. The 2D spectra, their positive projections onto the F_1 axis, as well as the relative intensities of all peaks (as compared to the corresponding module in the NOAH-3 M$_N$S$^+$C^c supersequence), are shown. Numbers in parentheses are averages over all peaks. **a–c** X = ^{15}N HSQC. **d–f** X = ^{15}N seHSQC1. **g–i** X = ^{15}N seHSQC2. *Data code:* `7G-201017`

However, this apparent advantage of the seHSQC1 vanishes when looking at the ^{13}C seHSQC module which comes later in the supersequence. When preceded by the ^{15}N HSQC or seHSQC1 modules, this module suffers from severely decreased sensitivity, compared to when the seHSQC2 module is used (Fig. 4.30). As the spectra and their F_1 projections show, this reduction in sensitivity actually stems from broadening in the indirect dimension. This in turn arises because ^{1}H$^{!N}$ magnetisation (which includes the ^{1}H^C component sampled in this module—by definition, protons coupled to ^{13}C are not coupled to ^{15}N) precesses in the transverse plane during the t_1 period of the ^{15}N HSQC and seHSQC1. Evolution of $^nJ_{HH}$, as well as T_2 relaxation, lead to a t_1-dependent signal loss which (after Fourier transformation) is manifested as line broadening and significantly reduced peak heights (Fig. 4.30c and f).

While the ^{15}N seHSQC2 module also does not preserve ^{1}H$^{!N}$ magnetisation as well as the HMQC module (with an approximately 20% loss, see Fig. 4.30i), this is not anywhere as drastic as with the other modules. Even more importantly, these losses are constant across all t_1 increments, so there is no line broadening observed in the modules which follow. Small sensitivity losses in later modules are a reasonable price to pay for the improvement in the ^{15}N module itself: typically, it is the ^{15}N module which has the smallest intrinsic sensitivity, so it is desirable to maximise this, even at the cost of the more sensitive modules.

It should be noted that the ^{13}C seHSQC1 (and HSQC) modules also do place $^{1}H^{!C}$ magnetisation in the transverse plane during t_1. However, they do not cause (appreciable) broadening in downstream modules, as evidenced by the almost complete preservation of CLIP-COSY peak heights in Fig. 4.11. This is because indirect-dimension spectral widths for ^{13}C experiments are often much wider, and the acquisition times (AQ_1, as defined in Sect. 4.3.2) correspondingly shorter. For example, in the NOAH-3 $XS^{+}C^{c}$ experiments shown here, the values of AQ_1 for the ^{15}N and ^{13}C experiments are, respectively, 60.1 ms and 5.38 ms. Thus, the effects of magnetisation losses during t_1 are not so keenly felt.

4.3.3.2 CTP Gradient Duration

To verify that the 2.5 ms CTP gradients were necessary in the ^{15}N seHSQC as well, I repeated the NOAH-3 $S_N^{+}S^{+}C^{c}$ experiment with increasing gradient durations. Only the seHSQC2 module was studied here, as the HSQC and seHSQC1 modules are not viable for use in NOAH supersequences. It should be noted that the durations of *only* the CTP gradients g_1 and g_2 (in Fig. 4.9c) were modified; g_3 and g_4 retain their original durations of 1 ms. The results, shown in Fig. 4.31, are qualitatively very similar to that obtained with the HMQC (in Fig. 4.25): increasing the gradient duration does not lead to substantial differences in signal intensity, but does provide much improved artefact suppression. In this case, equivalent results can be obtained using gradient durations of 2.2 ms to 2.3 ms; however, 2.5 ms still appears to be a reliable 'default' value which can be used without having to perform further optimisations.[22]

4.3.3.3 *k*- and SW-Scaling

Finally, the effects of k- and SW-scaling were also investigated for the seHSQC2 module. The effects of just performing scaling are very minimal (Fig. 4.32): unlike the HMQC module where lowering the resolution avoids losses in peak height due to multiplet structure, there are no such gains for the seHSQC module. The use of indirect-dimension linear prediction, however, is rather more successful for the seHSQC module (Fig. 4.33): again, this likely stems from the lack of multiplet structure in F_1. As before, the application of LP to the SW-scaled spectra yields better peak shapes than the k-scaled spectra (compare, e.g., Fig. 4.33g and m). The same caveats regarding detection sensitivity are also applicable in this instance.

[22] I am aware that this statement somewhat undermines my arguments for the use of POISE optimisations in Chap. 3.

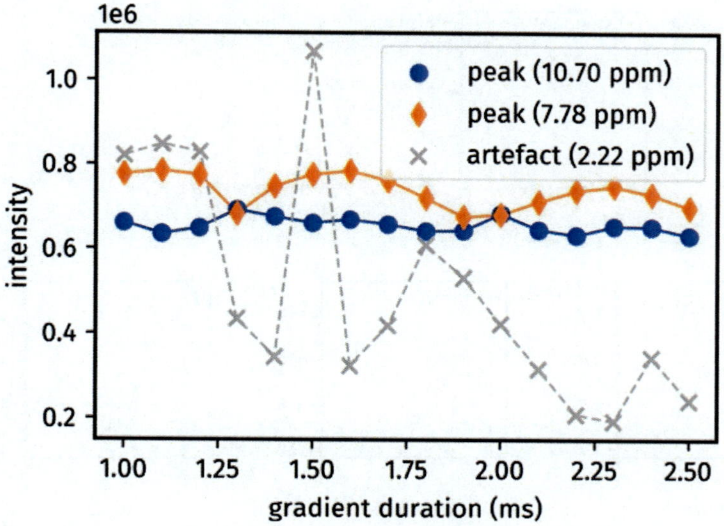

Fig. 4.31 Effect of CTP gradient duration on ^{15}N seHSQC2 signal and artefact intensities, as measured by (absolute) peak heights in the positive F_2 projection. *Data code:* `7Z-200926`

4.3.4 Dual HSQC and HSQC-TOCSY

Having completed our survey of ^{15}N modules, we now return to ^{13}C modules. In particular, this section explores how *two* (or more) modules drawing on $^{1}H^{C}$ magnetisation can be combined in the same supersequence. All of the ^{13}C modules described so far fully consume $^{1}H^{C}$ magnetisation: if we were to crudely concatenate two such modules, the second of these modules will have greatly reduced sensitivity. This is acceptable only if the second module has a far greater intrinsic sensitivity. More ideally, we should like to have a way to *partition* the magnetisation pool between the two different modules as needed: this would allow us to balance their sensitivities.

This can be accomplished by modifying the first module to only excite a portion of $^{1}H^{C}$ magnetisation, storing the remainder along $+z$. The NOAH HSQC module—which is derived from the ASAP-HSQC experiment [11, 45], previously described in Sect. 3.4.5—does in fact allow for this, through a modification to the INEPT delay. I therefore begin by describing SSX-type supersequences, where the first HSQC only partially excites $^{1}H^{C}$ magnetisation, and the second consumes the remainder. Here, X represents a homonuclear module, which in this section is the CLIP-COSY; however, the conclusions drawn here are applicable to any homonuclear module.

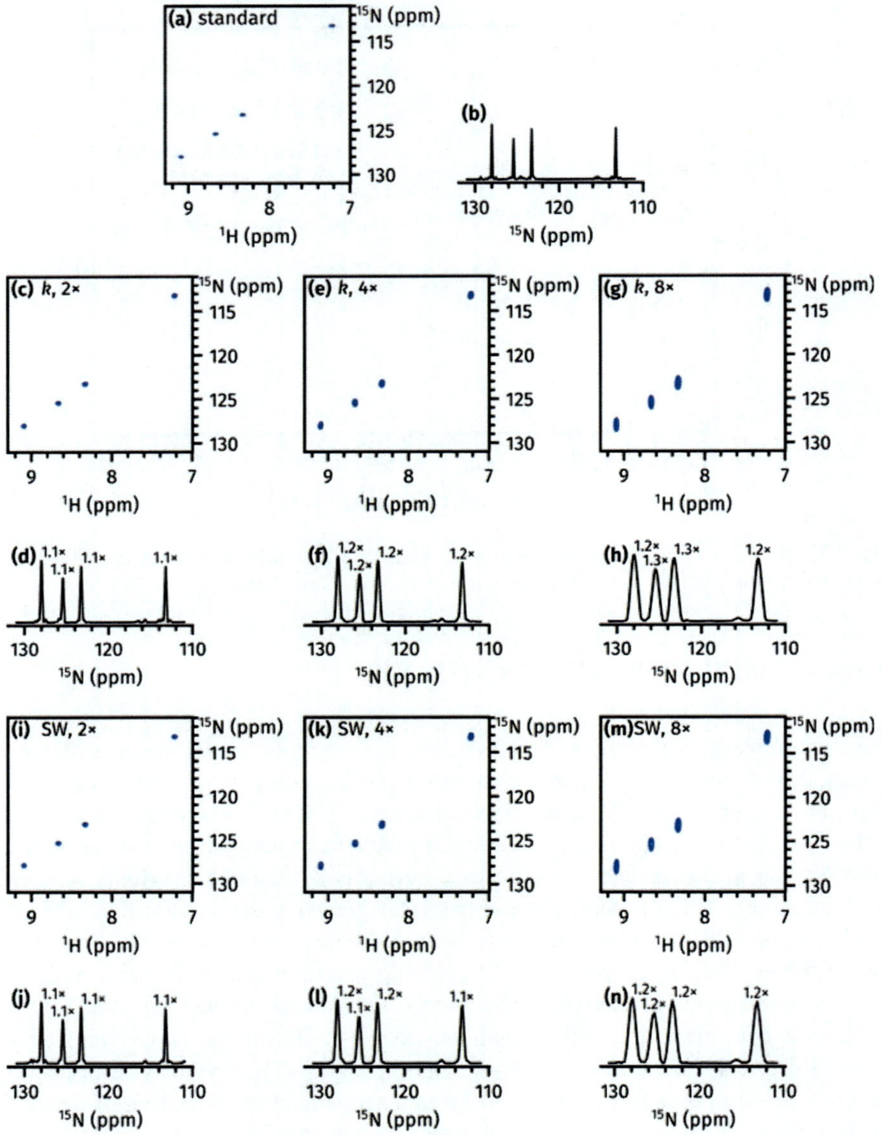

Fig. 4.32 Effects of *k*- and SW-scaling on NOAH seHSQC2 spectrum (taken from NOAH-3 $S_N^+S^+C^c$ supersequences). Each seHSQC spectrum is shown together with a positive projection onto the F_1 axis. The relative SNR of each peak, with respect to the standard spectrum, is indicated on each of the other projections. **a–b** Standard spectrum. **c–h** *k*-scaled spectra. **i–n** SW-scaled spectra. *Data code:* `7G-210310`

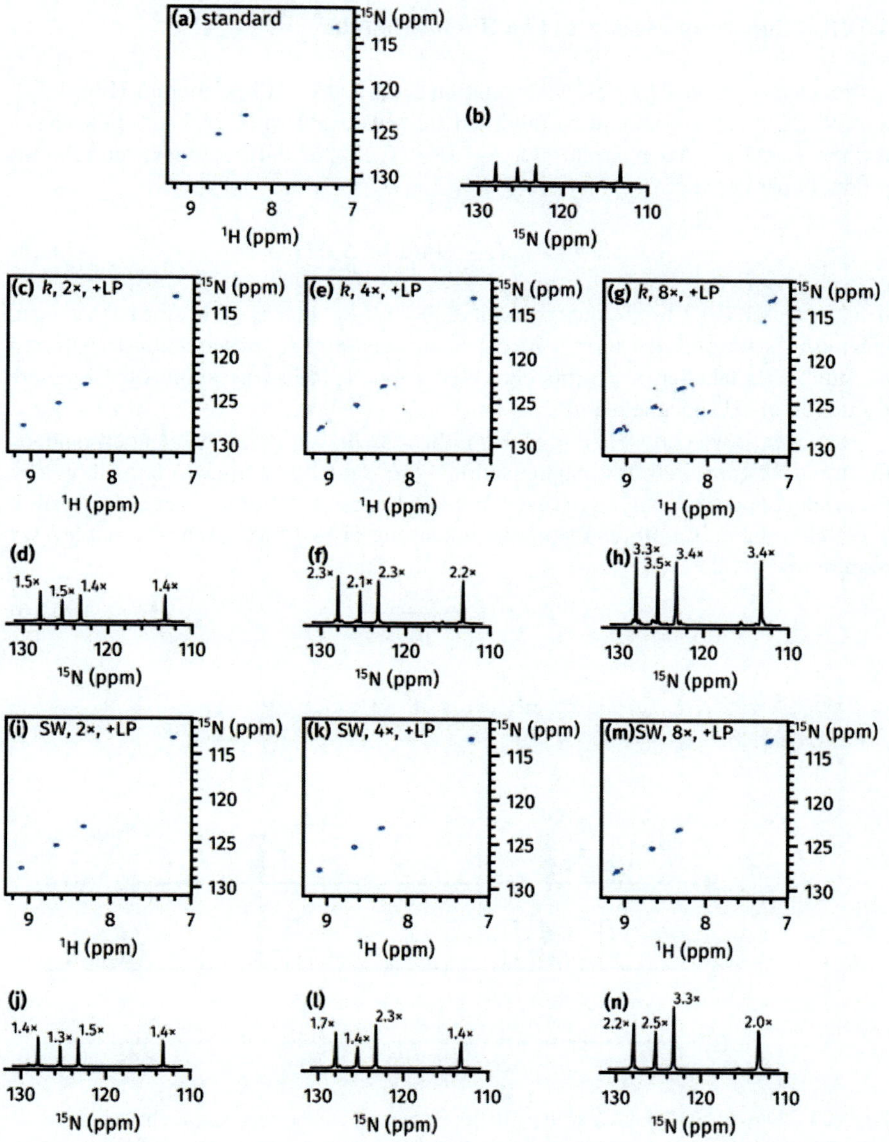

Fig. 4.33 The same as in Fig. 4.32, but with extra linear prediction applied to all scaled spectra to bring $AQ_{1,\mathrm{eff}}$ up to its original value in the standard spectrum. Linear prediction of k times more points leads to a $\sqrt{k}$ increase in noise; to account for this, all spectra are plotted with the same noise level. **a–b** Standard spectrum. **c–h** k-scaled spectra. **i–n** SW-scaled spectra. *Data code:* `7G-210310`

4.3.4.1 Supersequences with two HSQC Modules

In order to accomplish partial $^{1}H^{C}$ excitation in the NOAH HSQC module (Fig. 4.34), the INEPT delay Δ_E should be modified from its usual value of $1/(4J)$ (where J is short for $^{1}J_{CH}$). After the $90°(I)$–Δ_E–$180°(I, S)$–Δ_E INEPT block, the relevant product operators are

$$\cos(2\pi J\Delta_E)I_y - \sin(2\pi J\Delta_E)2I_xS_z. \tag{4.15}$$

In a 'normal' INEPT block, the choice of $\Delta_E = 1/(4J)$ makes the cosine term vanish, leaving us with only the term $-2I_xS_z$. Since this term is subsequently transferred to spin S and labelled in t_1, this choice of $\Delta_E = 1/(4J)$ corresponds to *complete* excitation of $^{1}H^{C}$ magnetisation.

However, if we choose $\Delta_E < 1/(4J)$, then the first I_y term—which corresponds to 'stored' or unexcited $^{1}H^{C}$ magnetisation—is eventually returned to I_z, as shown by the product operator analysis in Fig. 4.34.[23] To be precise, in order to excite a fraction f of $^{1}H^{C}$ magnetisation (and store the remaining $(1 - f)$ for the next module), we require that

$$\Delta_E = \frac{2\Delta \arcsin f}{\pi}, \tag{4.16}$$

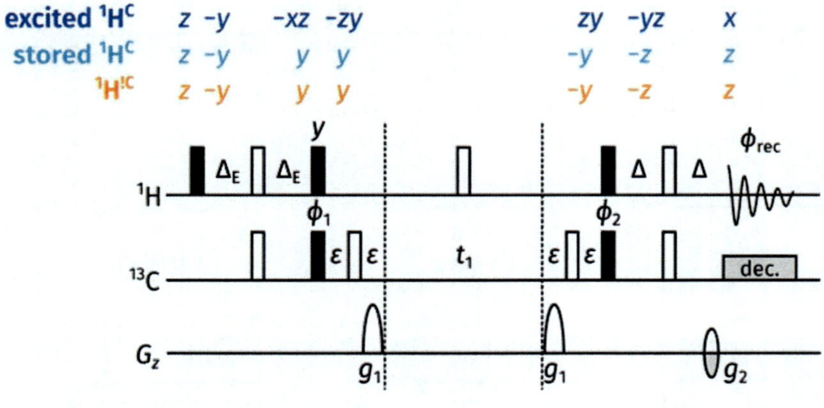

Fig. 4.34 NOAH HSQC module with partial excitation, where the delay Δ_E is changed from the usual value of $1/(4J)$. As shown by the product operator analysis, a proportion f of $^{1}H^{C}$ magnetisation is stored along $+z$ at the end of the sequence and can be sampled again in a later module. Gradient amplitudes are $(g_1, g_2) = (75\%, \pm 18.9\%)$; all other symbols have the same meaning as in Fig. 4.9. (The different CTP gradient amplitudes are chosen to avoid accidental refocusing between different modules.)

[23] In the context of the ASAP experiments (both HMQC and HSQC), this is termed *Ernst angle* excitation because the unexcited magnetisation is stored for later repetitions of the *same* experiment, which mirrors the original Ernst angle for the 1D pulse–acquire experiment. However, for NOAH supersequences, the magnetisation is stored for other modules, so it is more correct to refer to this as a *partial* excitation.

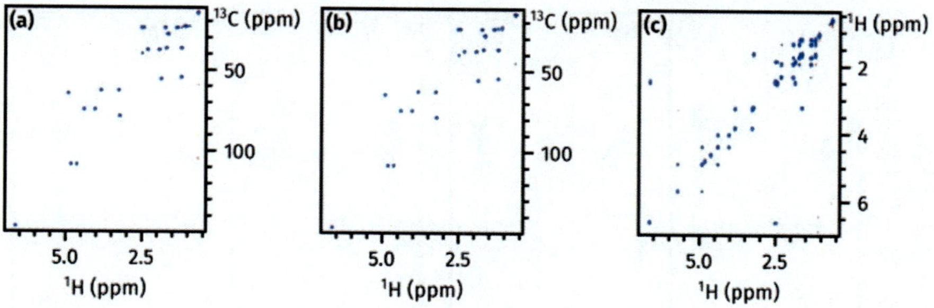

Fig. 4.35 Spectra from a NOAH-3 SSC^c supersequence, where the INEPT delay of the first HSQC module was modified to only excite a fraction $f = 0.8$ of $^1H^C$ magnetisation. **a** First HSQC. **b** Second HSQC. **c** CLIP-COSY. *Data code:* `7A-201010`

where Δ is the usual value of $1/(4J)$. The spectra obtained from a NOAH-3 SSC^c supersequence, using this partial $^1H^C$ excitation in the first HSQC module (with $f = 0.8$), are shown in Fig. 4.35.

In Fig. 4.36a, the intensities of these spectra are compared against the HSQC and CLIP-COSY in a NOAH-2 SC^c supersequence. As expected, the first HSQC retains (around) 80% of its 'base' sensitivity, corresponding to the fraction of excited magnetisation. The second HSQC spectrum *nominally* only has access to the remaining 20% of magnetisation: however, due to recovery of $^1H^C$ magnetisation during the FID of the first HSQC, this number is boosted to 65%. The CLIP-COSY sensitivity is decreased compared to a NOAH-2 SC^c supersequence, but only marginally: this is because of cumulative losses from the two HSQC modules.

Figure 4.36b and d detail several possible modifications to this basic scheme. In the first of these (Fig. 4.36b), we further increase the sensitivity of the second HSQC module by converting it into a seHSQC2 module.[24] The CLIP-COSY sensitivity is slightly decreased due to poorer $^1H^{!C}$ preservation, which is consistent with Sect. 4.3.1. Note, however, that it is not possible to change the *first* HSQC into a seHSQC experiment: none of the seHSQC versions discussed in Sect. 4.3.1 preserve unused $^1H^C$ magnetisation in the same way as the HSQC module above.

Alternatively, it is also possible to include a period of isotropic mixing between the two HSQC modules: here, the DIPSI-2 sequence [76] was chosen. Since the $^1H^C$ magnetisation pool has been (partially) depleted, and the $^1H^{!C}$ magnetisation pool is (almost) full, the application of DIPSI mixing should in theory lead to transfer of polarisation from the $^1H^{!C}$ pool to $^1H^C$. However, when tested, this was not found to have a beneficial impact: in fact, small *decreases* in sensitivity were observed for both the second HSQC module and the final CLIP-COSY (Fig. 4.36c).

Lastly, it should also be noted that the acquisition of two HSQC spectra in one experiment has also been accomplished in a different way by Parella and coworkers [38, 39]. This relies on 'splitting' up the CRK seHSQC experiment, such that the

---

[24] The sensitivity increases are not uniform because they depend on multiplicity.

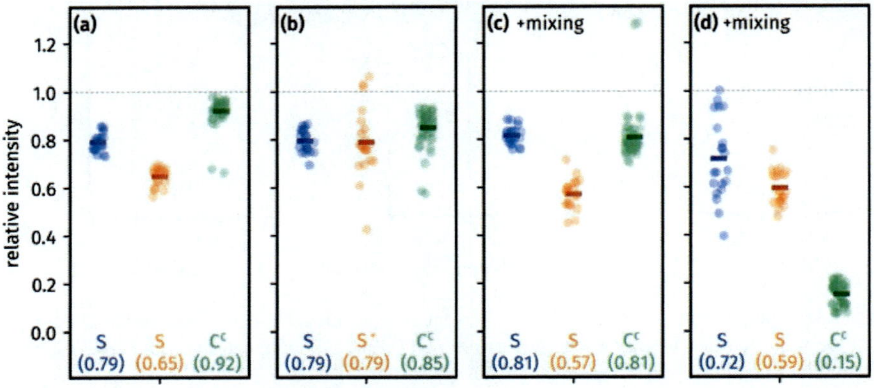

Fig. 4.36 Comparisons of (se)HSQC and CLIP-COSY sensitivities of NOAH-3 SSC^c and SS$^+$C^c supersequences. The three groups of dots in each subplots refer to the three respective modules in the supersequence. Peak intensities are normalised against the HSQC and CLIP-COSY experiments in a NOAH-2 SC^c supersequence; numbers in parentheses indicate averages over all peaks. **a** NOAH-3 SSC^c with $f = 0.8$ (the same spectra as shown in Fig. 4.35). **b** NOAH-3 SS$^+$C^c with $f = 0.8$. **c** NOAH-3 SSC^c with $f = 0.8$, plus 35 ms DIPSI-2 mixing after the first HSQC module. **d** A NOAH-3 SSC^c supersequence, but using the split-seHSQC implementation of Nolis et al. [38] (as opposed to the ASAP-HSQC module) for the double HSQC. *Data code:* `7A-201010`

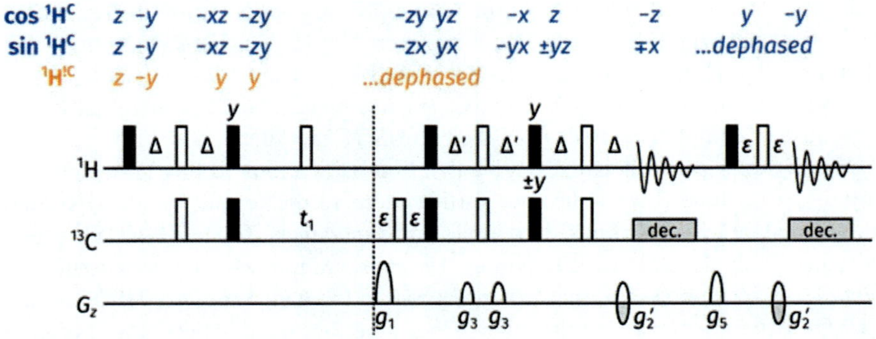

Fig. 4.37 Split-seHSQC experiment used for the collection of two HSQC spectra [38]. All symbols have the same meaning as in Fig. 4.9a; g_5 is a purge gradient with arbitrary amplitude

cosine- and sine-modulated components are separately detected (Fig. 4.37).[25] Specifically, an additional detection period can be inserted before the final ^{1}H 90° pulse, where the cosine-modulated component is stored as longitudinal magnetisation and the sine-modulated component is transverse (and in-phase).

However, this implementation does not fare as well in the context of the NOAH-3 SSC^c supersequences investigated here. The sensitivities of the three modules when using this are shown in Fig. 4.36e: the two HSQC spectra have comparable,

[25] In my paper [1], this was referred to as an 'MFA' (multiple-FID acquisition) implementation. I refrain from using this term here because it is ambiguous—after all, NOAH experiments are also multiple-FID experiments.

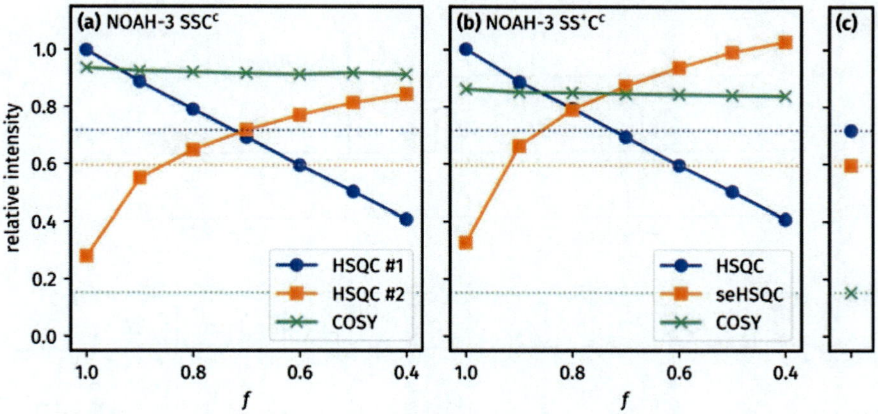

Fig. 4.38 Sensitivity comparisons for NOAH-3 SSC^c and NOAH-3 SS$^+$C^c supersequences. Dotted lines corresponding to the split-seHSQC sensitivities are drawn to guide the eye. Intensities are compared against the HSQC and CLIP-COSY modules in a NOAH-2 SC^c experiment, and averaged over all peaks. **a** NOAH-3 SSC^c supersequences acquired with different values of f. **b** NOAH-3 SS$^+$C^c supersequences acquired with different values of f. **c** Using the split-seHSQC implementation. *Data code:* `7A-201010`

but slightly poorer, sensitivity than the $f = 0.8$ partial excitation scheme used in Fig. 4.36a. Two other drawbacks of the split-seHSQC implementation are inherited from the parent CRK seHSQC experiment. Firstly, there is a slightly larger spread in peak sensitivities, because the CRK seHSQC does not uniformly boost peak intensities. Furthermore, since ^{1}H$^{!C}$ magnetisation is dephased, the CLIP-COSY which follows suffers from a large drop in intensity. Prepending the experiment with the ZIP element to make a 'split seHSQC2' does not help, because this would place the bulk magnetisation in the transverse plane during the first FID. Finally, the split-seHSQC implementation does not easily allow for features such as multiplicity editing to be independently applied to one of the two signals.[26]

Figure 4.38 summarises all of these factors, and further shows how sensitivity varies as a function of the parameter f. From Fig. 4.38a and b, it is clear that the sensitivity of the first HSQC is directly proportional to f. The sensitivity of the second HSQC module in Fig. 4.38a stems from both the unexcited $(1 - f)$ proportion of ^{1}H^C magnetisation, plus any that has recovered during the first FID: this latter component decreases as f increases. The CLIP-COSY intensities generally do not vary with f. In Fig. 4.38b, the second HSQC module is replaced with the seHSQC2: this allows f to be increased in order to balance the sensitivities of the two HSQC spectra. Finally, for completeness, the split-seHSQC implementation is shown in Fig. 4.38c; the numbers here are the same as in Fig. 4.36d. It is clear that there are several

[26] It should at least be *possible* to apply editing to one signal using a modification similar to that done by Nolis et al. [39] However, that is substantially more complicated than in the NOAH/ASAP implementation, where each module is separate.

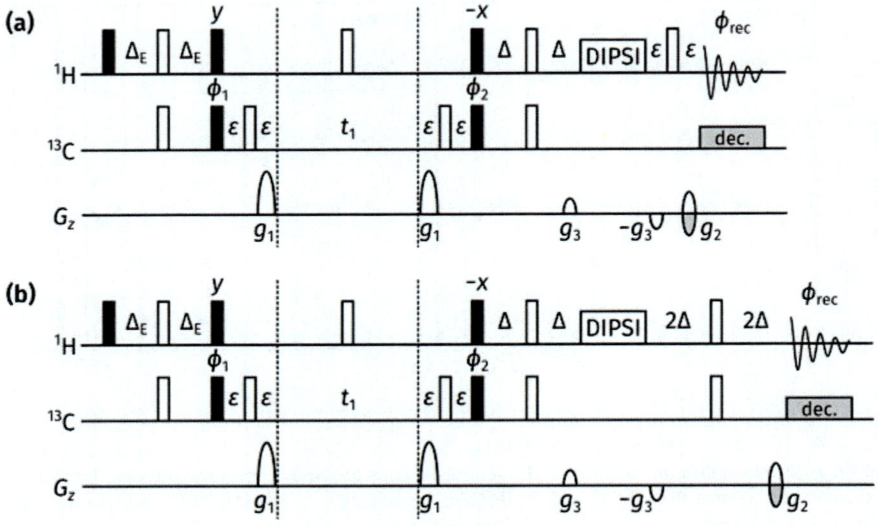

Fig. 4.39 NOAH HSQC-TOCSY module. **a** Without inversion of HSQC signals ('direct' responses). **b** With inversion of HSQC signals

values of f which allow the NOAH/ASAP-type implementation to outperform the split-seHSQC experiment in sensitivity.

On its own, the acquisition of two HSQC spectra—as has so far been shown—is not particularly groundbreaking. However, it is possible to differentiate the two HSQC signals and thereby extract more information. For example, one spectrum may be run without decoupling in order to measure one-bond coupling constants [38, 77]; multiplicity editing may be implemented [45]; or the indirect-dimension spectral width of one of the HSQC spectra can be changed in order to make use of spectral aliasing techniques [25, 39]. Most interesting of all, though, would be to modify the pulse sequence itself to obtain different *types* of ^{1}H–^{13}C spectra: for example, the addition of an isotropic mixing block to the first HSQC yields an HSQC-TOCSY + HSQC combination [38], which I now discuss.

4.3.4.2 HSQC-TOCSY

The addition of DIPSI mixing inside the partial-excitation HSQC module leads to an HSQC-TOCSY module (Fig. 4.39a), which is conceptually identical to the ASAP-HSQC-TOCSY experiment also previously reported by Luy and coworkers [78]. Note that the ^{1}H 90° pulse immediately after t_1 has a different phase from that in the HSQC module: this ensures that $^1\mathrm{H}^{!C}$ (and unused $^1\mathrm{H}^{C}$) magnetisation is returned to $+z$ after accounting for the extra 180° pulse at the end.

The HSQC-TOCSY module may optionally be implemented with inversion of 'direct', HSQC-type responses, which correlate ^{13}C nuclei with directly bonded

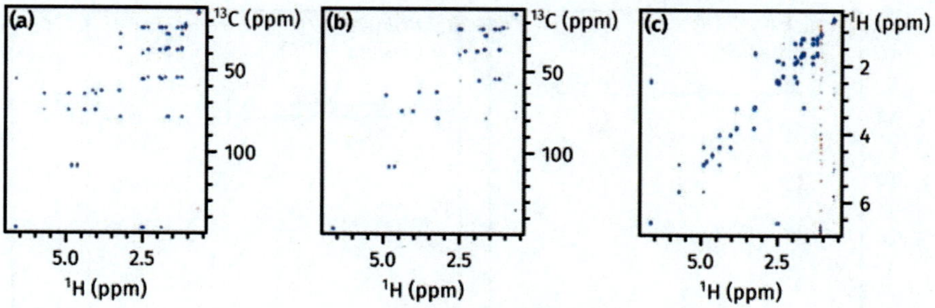

Fig. 4.40 Spectra from a NOAH-3 $S^TS^+C^c$ supersequence. **a** HSQC-TOCSY ($f = 0.9$). **b** seHSQC2. **c** CLIP-COSY. *Data code:* `7A-210126`

protons. This is to be contrasted with the 'indirect' responses, which correlate ^{13}C nuclei with other remote protons in the same spin system, and represent magnetisation which has been transferred during the isotropic mixing period. This inversion can be accomplished by expanding the final gradient echo to a duration of $4\Delta = 1/J$ (Fig. 4.39b); ^{13}C-bound protons will evolve under $^1J_{\mathrm{CH}}$ and acquire an additional 180° phase shift compared to remote protons.[27]

The use of such editing in the HSQC-TOCSY is perhaps slightly risky, because it may lead to unwanted peak cancellation through accidental overlap. Furthermore, if used together with a second HSQC module (as discussed here), this information does not need to be encoded in the HSQC-TOCSY. Thus, it is not enabled by default; the user must explicitly turn it on through an acquisition flag. The concept of inverting remote peaks is more useful in the HSQC-COSY, which will be discussed in Sect. 4.3.5.

With this module in hand, it is possible to implement either NOAH-3 S^TSC^c or NOAH-3 SS^TC^c supersequences which yield both HSQC and HSQC-TOCSY data. In both cases, the first of the two ^{13}C modules can be adjusted to only partially excite $^1H^C$ magnetisation. However, the former is preferred as the HSQC-TOCSY has a lower intrinsic sensitivity compared to the HSQC: thus, it is better to place this first in the supersequence, and to also use a larger value of f in order to balance the relative sensitivities. Figure 4.40 shows an example of the spectra obtained with a setting of $f = 0.9$.

Finally, I performed a series of experiments to establish whether changing the first module from an HSQC (i.e. SSC^c) to an HSQC-TOCSY (i.e. NOAH-3 S^TSC^c) would affect the sensitivities of the two latter modules. The results are shown in Fig. 4.41a and b, where the solid lines represent the supersequences starting with the HSQC-TOCSY, and the dashed lines those starting with the HSQC. At large values of f, corresponding to depletion of the $^1H^C$ pool, the subsequent ^{13}C module experiences sensitivity *gains* when the HSQC-TOCSY is used (i.e. the solid orange line lies above the dashed orange line). However, this changes at smaller values of

[27] Of course, the remote protons may also happen to be bonded to ^{13}C, but the chance of this is only roughly 1%.

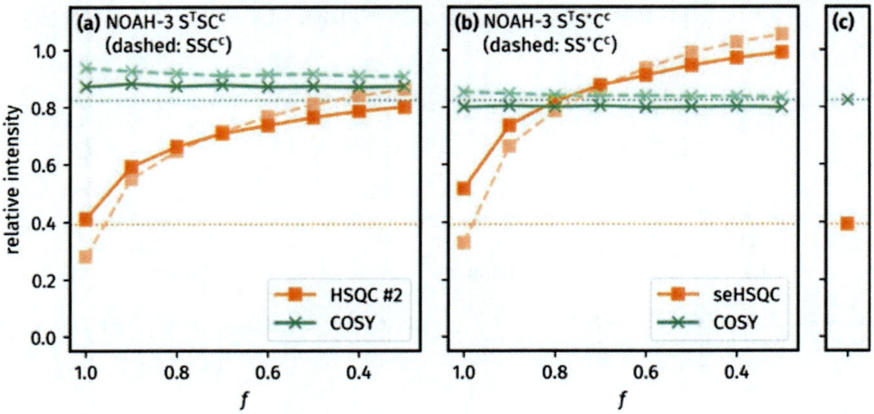

Fig. 4.41 Sensitivity comparisons for the latter two modules in NOAH-3 SSC^c and NOAH-3 S^{T}SC^c supersequences. Intensities are compared against the HSQC and CLIP-COSY modules in a NOAH-2 SC^c experiment, and averaged over all peaks. **a** NOAH-3 S^{T}SC^c supersequences acquired with different values of f; the dashed lines are the corresponding intensities in the NOAH-3 SSC^c supersequence. **b** The same, but with the seHSQC2 as the second module instead. **c** The corresponding intensities of the second and third modules in a seHSQC-TOCSY + HSQC + CLIP-COSY supersequence: this is to be compared with the first data point ($f = 1$) in (a). Dotted lines corresponding to these sensitivities are drawn to guide the eye. *Data code:* `7A-210126`

f, where a drop in intensity is instead observed. For the CLIP-COSY module, the HSQC-TOCSY leads to slightly lower sensitivities across the board. All of these changes are, however, extremely minor: the supersequences shown here still work entirely as expected.

Figure 4.41c shows one final alternative which has not yet been considered: the use of the seHSQC-TOCSY [57] as the first module. This experiment is similar to the seHSQC2 module, but with the addition of isotropic mixing. Although this fully maximises the intensity of the HSQC-TOCSY component (not shown in Fig. 4.41), and also preserves $^1\mathrm{H}^{!\mathrm{C}}$ magnetisation for the CLIP-COSY (as evidenced by the relatively high CLIP-COSY intensity in Fig. 4.41c), it cannot retain any portion of $^1\mathrm{H}^{\mathrm{C}}$ magnetisation. Thus, any later ^{13}C module which draws on the same magnetisation pool will only be able to use any signal which has recovered during the FID, just like in an experiment run with $f = 1$. In this case, the second module is a plain HSQC (not seHSQC): thus, its sensitivity is similar to the leftmost point in Fig. 4.41a.

4.3.5 HSQC-COSY

One downside of the HSQC-TOCSY experiment discussed in the previous section is the (largely) indiscriminate transfer of magnetisation effected by the DIPSI mixing.

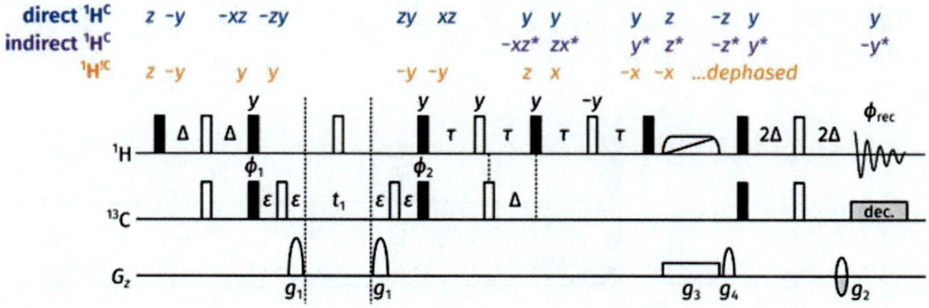

Fig. 4.42 HSQC-CLIP-COSY experiment with product operator analysis for the HSQC-COSY signal derived from the $^1H^C$ magnetisation, as well as the bulk $^1H^{!C}$ magnetisation. Both the 'direct', HSQC-type peaks, as well as the 'indirect' responses arising from coherence transfer in the perfect echo block, are analysed. The shorthand notation for product operators is expanded here to deal with a three-spin IKS system, where I and S are a mutually bonded ^{1}H–^{13}C pair as usual, and K is a 'remote' proton coupled to I. Terms with asterisks are on spin K; thus, for example, $zx*$ refers to an I_zK_x term. The delay τ is chosen to be $1/(4 \cdot \sum {}^nJ_{\mathrm{HH}})$; typically, this sum of couplings is set as 30 Hz, leading to a value of $\tau = 8.33\,ms$. The ZQF gradient g_3 should be calibrated as per Thrippleton et al. [83]; g_4 is a purge gradient with arbitrary amplitude. All other symbols have the same meaning as in Fig. 4.9

It is difficult to determine how many transfer steps give rise to each peak, unless several HSQC-TOCSY spectra with different mixing times are recorded, which is a time-consuming process.[28]

For unambiguous spectral assignment, it can be preferable to have a mixing process which only effects coherence transfer across a single coupling: this can be described as an HSQC-COSY experiment. Similar experiments, including the H2BC [79, 80], 2BOB/H2OBC [81] and HMQC-COSY [82] have previously been reported: the 2BOB experiment in particular has previously been incorporated in NOAH supersequences [51]. These often use constant-time techniques in order to remove $^nJ_{\mathrm{HH}}$ splittings and minimise linewidths in the indirect dimension; however, the drawback of this approach is that peak *amplitudes* are modulated by $^nJ_{\mathrm{HH}}$. Furthermore, it is not generally possible to obtain absorption-mode lineshapes for all peaks in the spectrum: typically the 'direct' responses are in-phase absorption, and 'indirect' responses antiphase dispersion. (Here, the terms in-phase and antiphase are used with respect to $^nJ_{\mathrm{HH}}$; specifically, for the indirect peaks, this refers to the 'active' coupling over which the coherence was transferred).

---

[28] Of course, as per the previous section, this could be done in a NOAH supersequence where each HSQC-TOCSY spectrum is assigned a fraction of the $^1H^C$ magnetisation. However, one must be wary of stretching the same magnetisation pool too thinly between many different modules: the resulting decrease in signal intensity may lead to a crossing into the sensitivity-limited regime.

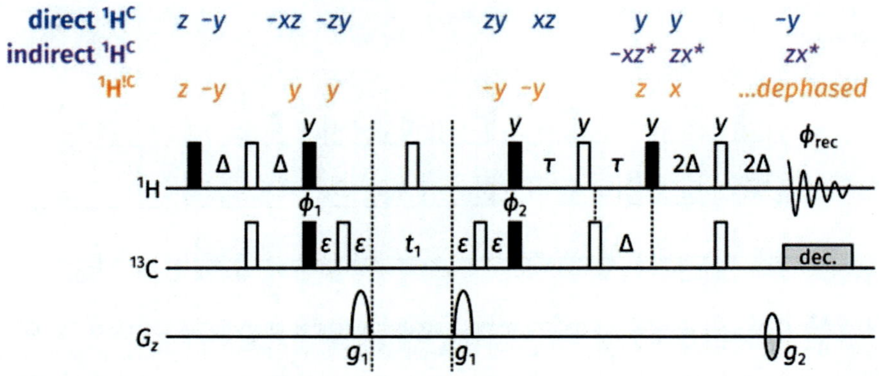

Fig. 4.43 'Double spin echo' HSQC-COSY experiment; all symbols have the same meaning as in Fig. 4.42

4.3.5.1 HSQC-CLIP-COSY

These problems are circumvented by the HSQC-CLIP-COSY experiment [84, 85], where the basic HSQC experiment is combined with clean in-phase (CLIP) coherence transfer using a perfect echo [86–88] and ZQF [83] (Fig. 4.42). The use of an HSQC-type experiment means that ${}^{n}J_{\mathrm{HH}}$ does not evolve during t_1, and the CLIP transfer ensures that all peaks have in-phase absorption lineshapes. Much like the HSQC-TOCSY experiment before it, this experiment may be implemented with direct/indirect response editing: it is this version of the experiment which is shown in Fig. 4.42.[29] If this editing is not desired, then the final 2Δ–$180°(I, S)$–2Δ spin echo can simply be reduced to a minimal gradient echo, i.e. ε–$180°(I)$–ε.

Unfortunately, the HSQC-CLIP-COSY experiment does not preserve ${}^{1}\mathrm{H}^{!\mathrm{C}}$ magnetisation; thus, it cannot be directly used in a NOAH supersequence without sacrificing the sensitivity of terminal homonuclear modules. In particular, the ZQF used in the HSQC-CLIP-COSY experiment dephases all magnetisation that is not along z. If we wished to retain ${}^{1}\mathrm{H}^{!\mathrm{C}}$ magnetisation, it would therefore have to (somehow) be placed along z during this period, and would have to be differentiated from the HSQC-COSY signal *after* the ZQF, which is not possible.[30] On top of that, the experiment also cannot be modified to provide partial excitation of ${}^{1}\mathrm{H}^{\mathrm{C}}$ magnetisation, which limits the ways in which it can be combined with other ${}^{13}\mathrm{C}$ modules.

---

[29] Note that this direct/indirect editing is orthogonal to *multiplicity editing*, which labels both direct and indirect responses with a sign that depends on the multiplicity of the ${}^{13}\mathrm{C}$ nucleus detected in t_1. The addition of multiplicity editing leads to highly confusing spectra, so is ignored here—although the option to do so *is* provided in the GENESIS pulse programs if the user so desires.

[30] To be precise, although the ${}^{1}J_{\mathrm{CH}}$ Hamiltonian could be used to separate bulk magnetisation and the HSQC-type 'direct' responses, the 'indirect' responses cannot be differentiated.

4.3.5.2 Double Spin Echo HSQC-COSY

Before tackling these problems directly, I discuss a simpler version of the HSQC-COSY experiment, which uses a simple spin echo for $^{n}J_{HH}$ evolution instead of the perfect echo of the CLIP version. Together with the final 4Δ spin echo, this forms a 'double spin echo' (DSE) version of the HSQC-COSY experiment (Fig. 4.43); the 'double' refers only to the mixing section after t_1. The removal of the CLIP coherence transfer element leads to mixed lineshapes in this experiment, where the direct responses are (mostly) in-phase absorption, and the indirect responses (mostly) antiphase dispersion.[31]

Like the CLIP version, this DSE experiment does not preserve bulk magnetisation; it is also incompatible with partial $^{1}H^{C}$ excitation. However, it does provide more raw sensitivity than the CLIP version: in the CLIP version, any antiphase signal at the end of the perfect echo is destroyed by the ZQF. All of this available signal is sampled in the DSE version, but this comes at the cost of not having pure absorption lineshapes.

It should be mentioned that prepending this DSE HSQC-COSY with the ZIP element would in fact return the bulk $^{1}H^{!C}$ magnetisation to $+z$ at the end of the sequence. However, I did not manage to test this experimentally, as my focus was on the development of the *triple* spin echo HSQC-COSY below.

4.3.5.3 Triple Spin Echo HSQC-COSY

In the DSE HSQC-COSY, the first of the two spin echoes after t_1 serves a dual purpose: $^{1}J_{CH}$ is allowed to evolve for a duration of $\tau - (\tau - \Delta) + \Delta = 2\Delta$ (thus generating peaks which are in-phase with respect to $^{1}J_{CH}$), and $^{n}J_{HH}$ evolves for the total duration of 2τ (allowing coherence transfer to remote spins). The triple spin echo (TSE) HSQC-COSY is derived from the DSE version by separating the first spin echo into two distinct parts: one for $^{1}J_{CH}$ refocusing, and one for $^{n}J_{HH}$ evolution. As shown by the product operator analysis in Fig. 4.44a, this not only preserves the bulk $^{1}H^{!C}$ magnetisation, but is also compatible with partial $^{1}H^{C}$ excitation.

The problem with the TSE version is the spectral quality: consider an even larger spin system of the form $S - I - K - L$, where $\{I, K, L\}$ are ^{1}H and S is ^{13}C. At the end of t_1, single-quantum magnetisation on spin I is present. Any evolution of J_{IK} in the first spin echo (of duration 2Δ) leads to terms of the form $2I_yK_z$, and can then be transferred onto spin K by the subsequent 90°_x pulse. This magnetisation can then further evolve under J_{KL} during the 2τ spin echo, and then be transferred a *second* time by the 90°_{-x} pulse to spin L. Since the intensity of this transfer pathway is proportional to $\sin(2\pi J_{IK}\Delta)$, these 'relay' artefacts are especially prominent for large J_{IK}.

[31] The qualifier *mostly* is required because evolution of $^{n}J_{HH}$ during the final spin echo leads to mixtures of in-phase absorption and antiphase dispersion. This evolution is generally small, though, because 2Δ is smaller than τ.

In order to suppress these, the experiment must be run a second time but with an additional 180° ^{13}C pulse inserted during the second spin echo (Fig. 4.44b).[32] Any magnetisation still on spin I will therefore evolve under $^{1}J_{CH}$ for a total period of 4Δ, leading to net inversion. However, magnetisation which has already been transferred to spin K at this point will not be inverted: it is this magnetisation which is responsible for the relay artefacts. Therefore, by subtracting the two datasets (or equivalently, inverting the receiver phase and adding the two datasets), the desired peaks will be added up and the artefacts cancelled through subtraction.

The insertion of this 180° pulse also causes any unexcited ^{1}H^{C} magnetisation to be inverted. In order to ensure that this is returned to $+z$ at the end of the sequence

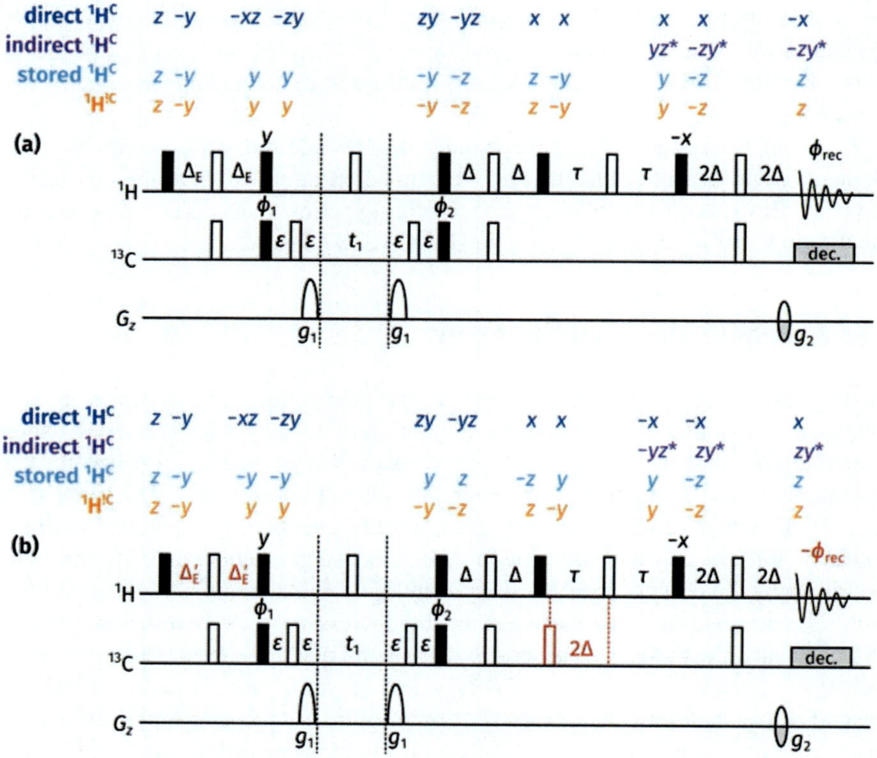

Fig. 4.44 'Triple spin echo' (TSE) HSQC-COSY experiment. **a** The first part of the experiment; this can be used on its own, but leads to spurious 'relayed' peaks arising via coherence transfer over two scalar couplings. **b** The second part of the experiment; co-adding this dataset with the first part leads to suppression of the relayed peaks. All symbols have the same meaning as in Fig. 4.42

---

[32] A very similar strategy is used in the Bruker seHSQC pulse sequence `hsqcedetgpsisp2.4` to suppress the COSY-type artefacts. However, to the best of my knowledge, this sequence has not been published anywhere.

(and not $-z$), the initial INEPT delay must be *lengthened* instead of shortened, as per:

$$\Delta'_{\mathrm{E}} = \frac{2\Delta(\pi - \arcsin f)}{\pi}, \tag{4.17}$$

where, as before, f is the proportion of magnetisation to be excited.

4.3.5.4 Supersequences

The spectral quality, and sensitivity, of all of these modules is captured in Figs. 4.45 and 4.46. The CLIP version (Figs. 4.45a and 4.46a) has the best lineshapes, but its sensitivity is slightly lower. The DSE version (Figs. 4.45b and 4.46b) provides greater sensitivity, but at the cost of impure lineshapes.

For the TSE version, we first look at the importance of the relay artefact suppression procedure described above. The extra relay artefacts are clearly visible in Fig. 4.45c, acquired using the 'basic' sequence in Fig. 4.44a only. (As previously mentioned, these arise from large $^{n}J_{\mathrm{HH}}$ which evolve during the 2Δ spin echo; for this specific compound, the offending couplings are ^{3}J between two axial protons in a six-membered ring, and ^{2}J in diastereotopic methylenes). The presence of these peaks largely defeats the purpose of using an HSQC-COSY experiment; in fact, this unoptimised TSE HSQC-COSY is qualitatively very similar to an HSQC-TOCSY acquired with a short mixing time of 10 ms (Fig. 4.45d).[33] However, these artefacts can be efficiently removed using the suppression technique described in the text above; the result in Fig. 4.45e is qualitatively similar to the two other HSQC-COSY experiments. Using this suppression technique, the sensitivity of the TSE HSQC-COSY (when acquired with $f = 1$) falls between that of the CLIP and DSE versions.

Figure 4.46 also shows the relative sensitivities of the later modules in NOAH-3 S^{c}SC^{c} supersequences. In all of the first three cases (CLIP, DSE, and TSE HSQC-COSY with $f = 1$, Fig. 4.46a and c), the HSQC sensitivity is low (ca. 20%) because no $^{1}\mathrm{H}^{\mathrm{C}}$ magnetisation is retained for it to use. Thus, the signal derives only from $^{1}\mathrm{H}^{\mathrm{C}}$ magnetisation which has recovered during the HSQC-COSY FID. However, when the TSE version is used, partial $^{1}\mathrm{H}^{\mathrm{C}}$ excitation can be used to control this sensitivity: for example, when $f = 0.8$ (Fig. 4.46d), the HSQC-COSY sensitivity is decreased (on average equalling that of the HSQC-CLIP-COSY), but the sensitivity of the HSQC module is almost doubled. Generally, choosing a value of $f < 1$ allows for the HSQC-COSY and HSQC sensitivities to be better balanced.

Finally, the CLIP-COSY module suffers when the HSQC-CLIP-COSY or the DSE HSQC-COSY are used, because both of these dephase $^{1}\mathrm{H}^{!\mathrm{C}}$ magnetisation; however, the TSE version successfully preserves around 70% of this magnetisation for it, regardless of the value of f. (This value is slightly lower than the approximately 90% magnetisation preserved by the HSQC module, because the bulk magnetisation

---

[33] The HSQC-TOCSY even gives better peak shapes, since the DIPSI mixing transfers in-phase magnetisation to in-phase magnetisation.

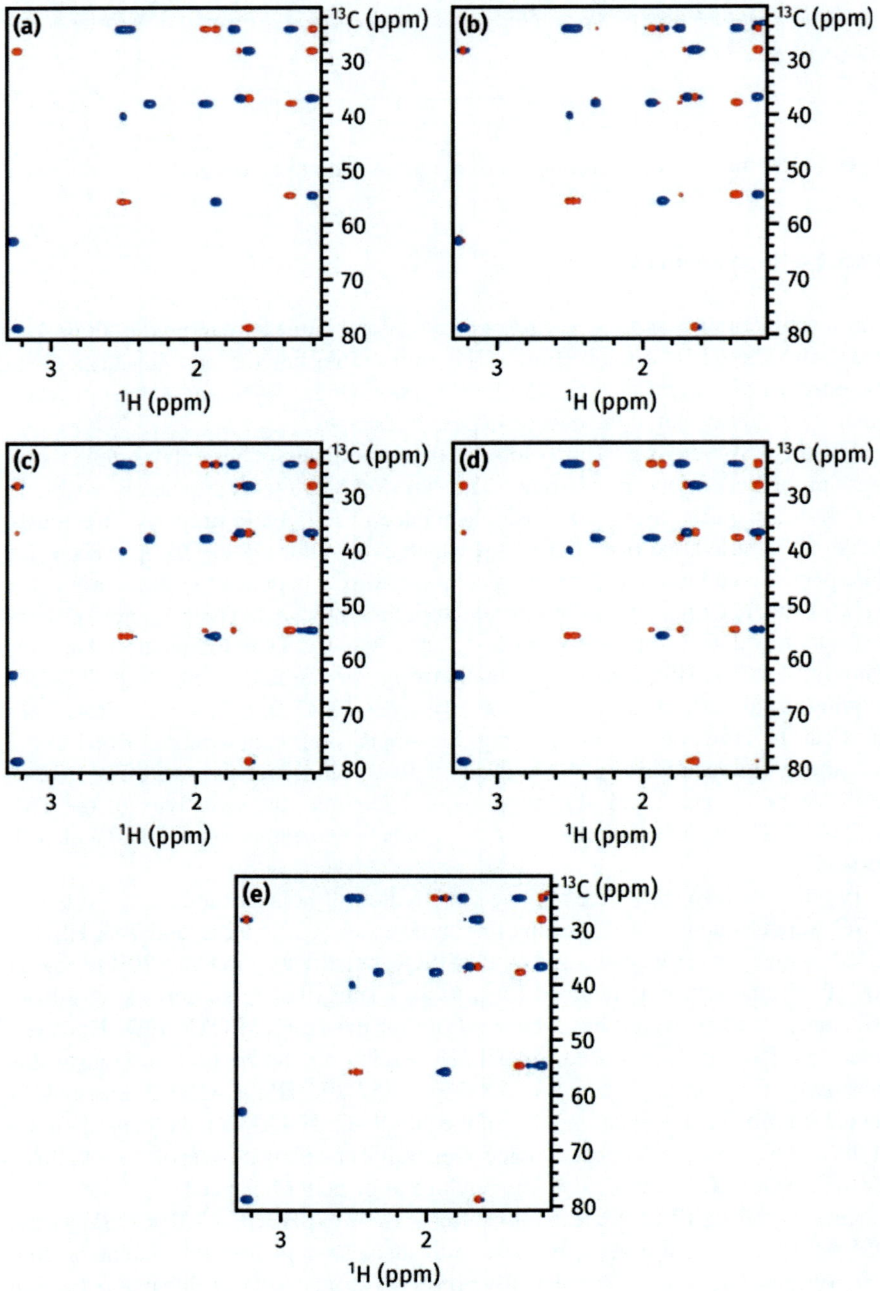

Fig. 4.45 HSQC-COSY and HSQC-TOCSY spectra, taken from (respectively) NOAH-3 S^cSC^c and NOAH-3 S^TSC^c experiments. **a** HSQC-CLIP-COSY. **b** DSE HSQC-COSY. **c** TSE HSQC-COSY without suppression of relay artefacts (described in the text). **d** HSQC-TOCSY with 10 ms mixing time. **e** TSE HSQC-COSY with suppression of relay artefacts. *Data code:* `7A-210303`

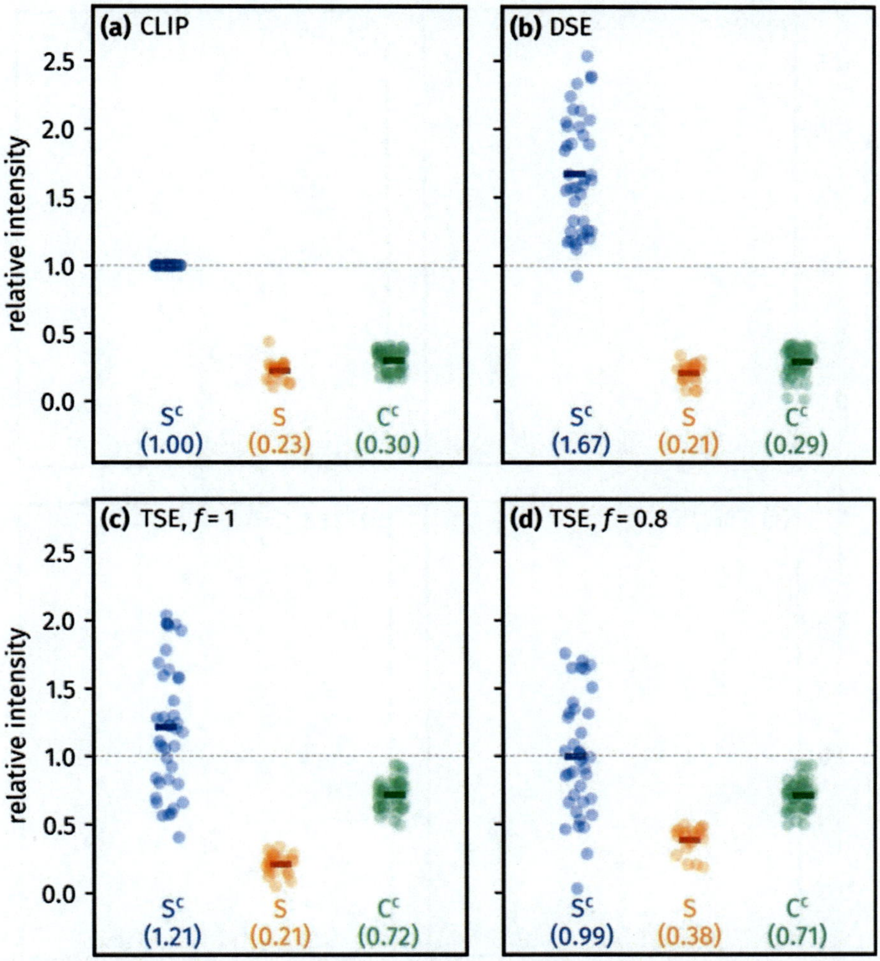

Fig. 4.46 Sensitivity comparisons for all three modules in NOAH-3 S^{c}SC^{c} supersequences. Peak intensities are relative to the HSQC-CLIP-COSY module itself (the leftmost column in (a)), as well as HSQC and CLIP-COSY spectra from a NOAH-2 SC^{c} experiment. Numbers in parentheses indicate averages over all peaks. **a** HSQC-CLIP-COSY. **b** DSE HSQC-COSY. **c** TSE HSQC-COSY, acquired with $f = 1$. **d** TSE HSQC-COSY, acquired with $f = 0.8$. *Data code:* `7A-210723`

is placed in the transverse plane during the 2τ spin echo (Fig. 4.44a) and experiences losses due to $^{n}J_{\mathrm{HH}}$ evolution.)

4.3.5.5 HSQC-COSY in Context

In NOAH-3 S^{c}SC^{c}-type supersequences, using the TSE HSQC-COSY module here appears to be a sensible option as it is capable of preserving some ^{1}H^{C} magnetisation

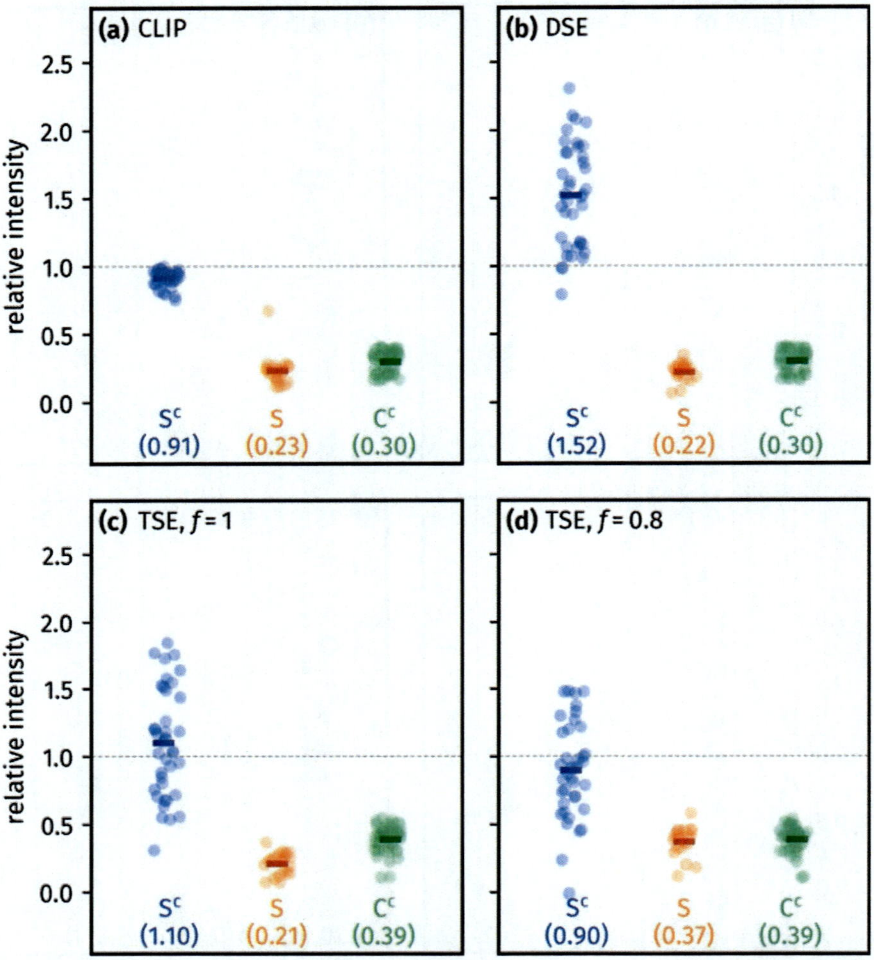

Fig. 4.47 Sensitivity comparisons for the three last modules in NOAH-4 BS^cSC^c supersequences. Peak intensities are relative to the HSQC-CLIP-COSY module from a NOAH-3 S^cSC^c supersequence, and HSQC and CLIP-COSY spectra from a NOAH-2 SC^c experiment (these are the same reference spectra as used in Fig. 4.46). Numbers in parentheses indicate averages over all peaks. **a** HSQC-CLIP-COSY. **b** DSE HSQC-COSY. **c** TSE HSQC-COSY, acquired with $f = 1$. **d** TSE HSQC-COSY, acquired with $f = 0.8$. *Data code:* `7A-210723`

for the HSQC, as well as all $^1H^{!C}$ magnetisation for the CLIP-COSY. However, this may not necessarily be so important in the context of a larger supersequence—particularly one which begins with the HMBC module, which *already* dephases $^1H^{!C}$ magnetisation (meaning that there is not much of it to preserve).

Figure 4.47 provides the same sensitivity comparisons as in Fig. 4.46, but in the context of a NOAH-4 BS^cSC^c supersequence instead. The HSQC-COSY and HSQC modules largely follow the same pattern as before, but with an approximate 10% loss across the board: this reflects the imperfect preservation of $^1H^C$ magnetisation by the *zz*-HMBC module. The CLIP-COSY module, however, has a substantially lower sensitivity regardless of which HSQC-COSY module is chosen. When the CLIP or DSE HSQC-COSY modules are used, the CLIP-COSY retains only roughly 30% of its original intensity: this is the same as in Fig. 4.46. With the TSE HSQC-COSY, this is boosted to around 40% because there is one extra FID in which the $^1H^{!C}$ polarisation can be recovered. However, the use of the HMBC module at the beginning effectively places an upper limit on the amount of signal available to this module.

For virtually all homonuclear modules (including the CLIP-COSY), this small difference in sensitivity will not make a real difference in the interpretability of the spectrum. This is especially so considering that the HMBC module—which has a far lower sensitivity—is also present in the supersequence: a CLIP-COSY with 30% of its original sensitivity is still more intense than the HMBC experiment. This argument was used in justifying the NOAH-3 BSC^c experiment, and logically, should be equally applicable to the NOAH-4 BS^{c}SC^c experiment. In this case, the only compelling reason to use the TSE HSQC-COSY would be to preserve a portion of $^1H^C$ magnetisation for a later ^{13}C module. Thus, in this context, the decision of which HSQC-COSY module to use is slightly more nuanced: the cleaner lineshapes provided by the CLIP version, or the sensitivity of the DSE version, may be more preferable.

4.3.6 2DJ and PSYCHE

So far, I have exclusively used the CLIP-COSY as the final homonuclear module in NOAH supersequences. In general, since homonuclear modules are placed at the end of supersequences, there is rarely any need to adapt them for NOAH supersequences, as they do not need to preserve any magnetisation. One exception to this is the family of 2DJ and pseudo-2D pure shift experiments, where the spectral width in the indirect dimension is extremely small, on the order of 50 Hz. In such cases, the number of t_1 increments required is far smaller than for a typical 2D experiment. Since—by default—each module in a NOAH supersequence is acquired with the same number of increments, directly concatenating such modules to the end of a supersequence would therefore prove suboptimal.

However, as described previously in Sects. 4.3.2 and 4.3.3, it is possible to reduce the number of t_1 increments for a particular module, and in exchange, increase the number of scans recorded for that module. In the context of ^{15}N modules, this was a 'special' procedure referred to as k-scaling; however, for these homonuclear modules, it is natural and necessary. One difference in the implementation is that, instead of specifying a value k by which `TD1` is scaled down and `NS` scaled up, the user is allowed to directly specify the number of t_1 increments as an integer (`CNST37` in TopSpin). This value must be a divisor of the *normal* number of t_1 increments for all other modules (or equivalently, k must be an integer). The 'indirect-dimension

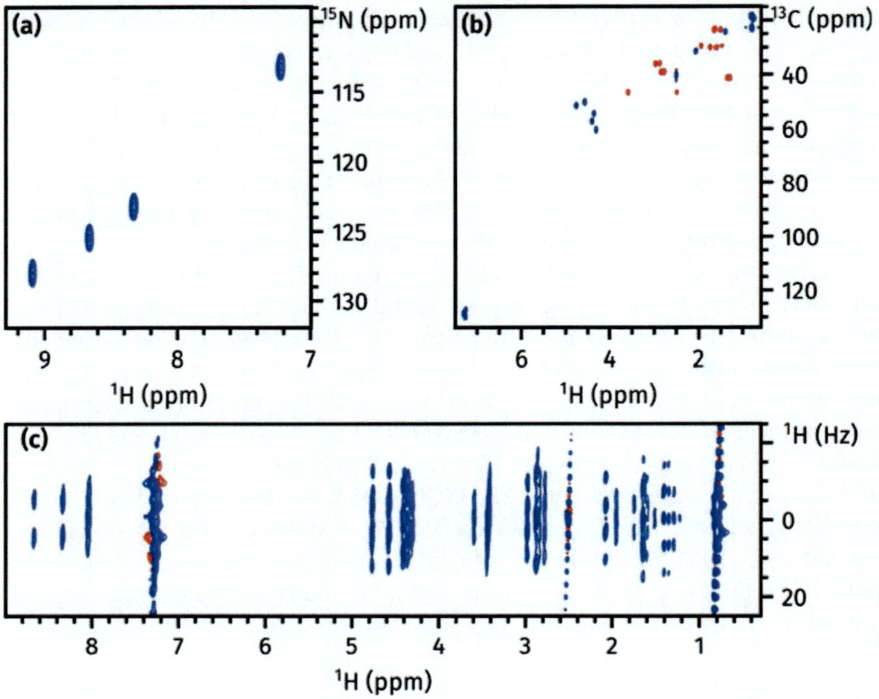

Fig. 4.48 Spectra obtained from a NOAH-3 $\mathrm{S_N^+S^+J}$ supersequence. **a** ^{15}N seHSQC (128 t_1 increments and 2 scans per increment). **b** ^{13}C seHSQC (128 t_1 increments and 2 scans per increment). **c** PSYCHE 2DJ (32 t_1 increments and 8 scans per increment). *Data code:* `7G-201028`

spectral width' is specified as `CNST38`: for pure shift modules this quantity is more properly referred to as the reciprocal of the chunk size, i.e. $1/T_{\text{chunk}}$.

The modules thus implemented include the standard magnitude-mode 2DJ experiment, the PSYCHE 2DJ experiment [89], the original 1D PSYCHE [90] and the 1D TSE-PSYCHE [89]. An example of the data thus obtained (with the PSYCHE 2DJ) is shown in Fig. 4.48. One particular advantage of including PSYCHE-type modules in NOAH supersequences is that sensitivity is not likely to be at a premium: this is partly because of the increased number of scans but also partly because other 2D experiments have comparably low sensitivity (meaning that in the time needed to acquire an HSQC, for example, the PSYCHE experiment will also have sufficient sensitivity). This allows the user to choose a relatively small flip angle (ca. 10°) for the PSYCHE saltire pulses in order to minimise artefacts from imperfect decoupling (see also Sect. 2.2.3).

On top of this, for the 1D (TSE-)PSYCHE sequences, the extra transients can be used to perform SAPPHIRE averaging [91]. Ordinarily, the PSYCHE pulse sequence seeks to refocus J-couplings in the middle of each chunk (of duration T_{chunk}). This is accomplished through the *prefocusing* of J-couplings in a spin echo of total duration

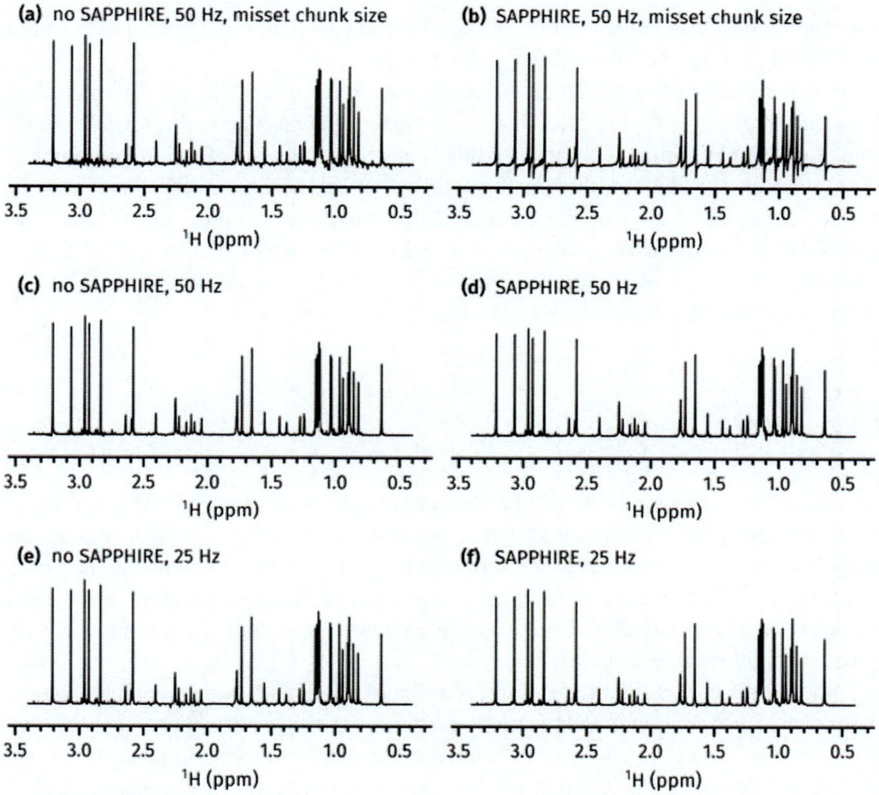

Fig. 4.49 A series of 1D PSYCHE spectra obtained from the NOAH-3 $S_N^+S^+P$ supersequence (saltire flip angle of 15°). **a** 1D PSYCHE spectrum acquired without SAPPHIRE averaging, and an incorrect chunk size which is not an integral number of complex data points. The 'indirect-dimension spectral width' (more precisely, $1/T_{\text{chunk}}$) is 50 Hz. **b** 1D PSYCHE spectrum with 8-step SAPPHIRE averaging and an incorrect chunk size; this manifests as phase errors which cannot be corrected. **c, d** The same as **a** and **b**, but with the chunk size automatically rounded in the pulse program. **e–f** The same as **c** and **d**, but with double the chunk size. This leads to a more striking difference between the data acquired without and with SAPPHIRE, as discussed in the original paper [91]. *Data code:* `7C-211123`

$T_{\text{chunk}}/2$ [92]. In the SAPPHIRE procedure, each chunk of the pure shift interferogram is instead collected several times, all while varying the exact point in time where J-coupling is perfectly refocused. Summation of these data leads to the suppression of artefacts which arise due to the periodic J-modulation in the interferogram. This averaging is somewhat analogous to a phase cycle, and performing an 8-step SAPPHIRE averaging procedure (for example) would often require the experiment to be lengthened beyond the duration which is truly necessary. However, in the context of NOAH, since each chunk is acquired at least k times, a k-step SAPPHIRE procedure can be incorporated 'for free'. In the GENESIS pulse programs, this feature is enabled by default, although it can be turned off in developer mode where non-SAPPHIRE pure shift modules are available.

A final and more prosaic implementation detail is that in the 1D pure shift modules, the chunk size T_{chunk} is automatically rounded to the nearest even multiple of the dwell time τ_{dw} (`DW` in TopSpin). This ensures that each chunk consists of an integral number of complex data points. Although this can be set by the user manually, it is very easy to forget, especially for someone not fully acquainted with the experiment; the results can be quite different, as illustrated in Fig. 4.49. When an n-step SAPPHIRE averaging is used, the requirement is even stricter: T_{chunk} must be a multiple of $2n\tau_{\text{dw}}$. This is also encoded in the pulse programs.

4.3.7 *HMBC*

The HMBC module is one which in fact does not fit perfectly into the NOAH principle of only exciting magnetisation which is needed. As described in Sect. 4.1.3, the HMBC module should only require magnetisation of protons which have long-range couplings to ^{13}C; however, it ends up exciting *all* ^{1}H$^{!C}$ magnetisation. This leads to sharply reduced, and also unbalanced, intensities of homonuclear modules which come later in the supersequences.

I made some early (and brief) attempts at devising a pulse sequence which sought to discriminate these two magnetisation components using a perfect echo [87]. However, this was quickly abandoned as it proved very difficult to *also* retain ^{1}H^{C} magnetisation. I do not claim here that it is impossible to come up with a pulse sequence which does this, but it is certainly not easy, and ultimately I turned my focus to improving (rather than replacing) the HMBC module.

4.3.7.1 Suppression of One-Bond Artefacts

One of the issues with the NOAH HMBC module was that there were an unusual amount of one-bond artefacts, which arise from ^{1}H^{C} magnetisation that is allowed to evolve during the pulse sequence. Generally, HMBC experiments seek to suppress this through the use of a low-pass J-filter (LPJF, see also Sect. 3.4.7). In addition to a standard second-order LPJF, the NOAH HMBC module *also* contains a zz-filter, which stores ^{1}H^{C} magnetisation along $+z$ before the LPJF. Thus, in theory, one-bond artefacts should be suppressed in the NOAH HMBC to an even greater extent.

However, this expectation is not borne out: in some cases, the NOAH HMBC has *more intense* one-bond artefacts when compared against a standard HMBC experiment (Fig. 4.50a and c). Even performing a POISE optimisation of the LPJF delays, as described in Sect. 3.4.7, did not lead to any reduction in artefacts. I hypothesised instead that these artefacts arose from imperfect manipulation of ^{1}H^{C} magnetisation by the zz-filter. In particular, any *antiphase* magnetisation (of the form I_xS_z or I_yS_z) generated after the zz-filter would be reconverted into in-phase I_x or I_y terms, which would not be destroyed by the LPJF.[34] (The LPJF works based on the assumption that it needs to destroy in-phase magnetisation, which is true if the excitation element is just a ^{1}H 90° pulse, but is not necessarily applicable in the NOAH HMBC.)

Such antiphase terms can, however, be easily removed by the addition of a ^{13}C 90° pulse, which transforms them into a mixture of double- and zero-quantum terms: these are either unobservable, or can be efficiently dephased by CTP gradients. This technique is used in the CLIP-HSQC family of experiments [77, 85], as well as the LPJF itself. In this case, we simply need to add an additional ^{13}C 90° pulse between the zz-filter and the LPJF: this pulse is highlighted in Fig. 4.3b. This results in a striking reduction of the one-bond artefacts, as shown in Fig. 4.50b. The suppression accomplished with the addition of this 90° pulse is superior to that in the standard HMBC (Fig. 4.50c), and is comparable to a standard HMBC with a third-order LPJF (Fig. 4.50d). Of course, the NOAH module can also be 'upgraded' to use a third-order LPJF. In GENESIS pulse programs, this can be done using the `-DLP3` acquisition flag (although the results were not evaluated here).

4.3.7.2 Gradient Selection Schemes

Another point which was investigated (but bore less fruit) was the gradient scheme used for CTP selection. The NOAH module, as shown in Fig. 4.3b, uses a 'symmetric' scheme where two gradients of equal amplitude surround the t_1 period: this encoding is later decoded by a third gradient just prior to acquisition. However, other choices exist: for example, the Bruker standard library HMBC (derived from the work of Cicero et al. [93]) uses only two gradients which have unequal amplitudes. This gradient scheme cannot be directly used in a NOAH HMBC module, though. This is because the zz-filter element places $^{1}\mathrm{H}^{\mathrm{C}}$ magnetisation along the $+z$ axis just before the HMBC J-evolution delay (see Fig. 4.3b). This magnetisation later experiences the ^{1}H 180° pulse in the middle of t_1, which means that an extra 180° pulse must be added at the very end of the sequence to finally return it to $+z$. It is necessary to place at least one gradient after this 180° pulse to ensure proper CTP selection (in the 'symmetric' scheme of Fig. 4.3b, this is fulfilled by the gradient g_2).

Using this knowledge, it is possible to construct several 'asymmetric' gradient schemes:

1. 'Scheme A' (Fig. 4.51a) is modified from the Bruker standard library to include a 180° pulse and gradient at the end. The presence of an additional gradient means that there is a free parameter, here denoted as α, which can be used to control the relative amplitudes of these three CTP gradients. The gradient amplitudes are chosen as follows:

$$\text{echo:} \qquad g_1 = gc_1; \qquad g_2 = g; \qquad g_3 = gc_2; \tag{4.18}$$

$$\text{antiecho:} \qquad g_1 = g; \qquad g_2 = gc_1; \qquad g_3 = gc_2, \tag{4.19}$$

[34] It would be nice to back this up with simulations, but I did not have the time to run these.

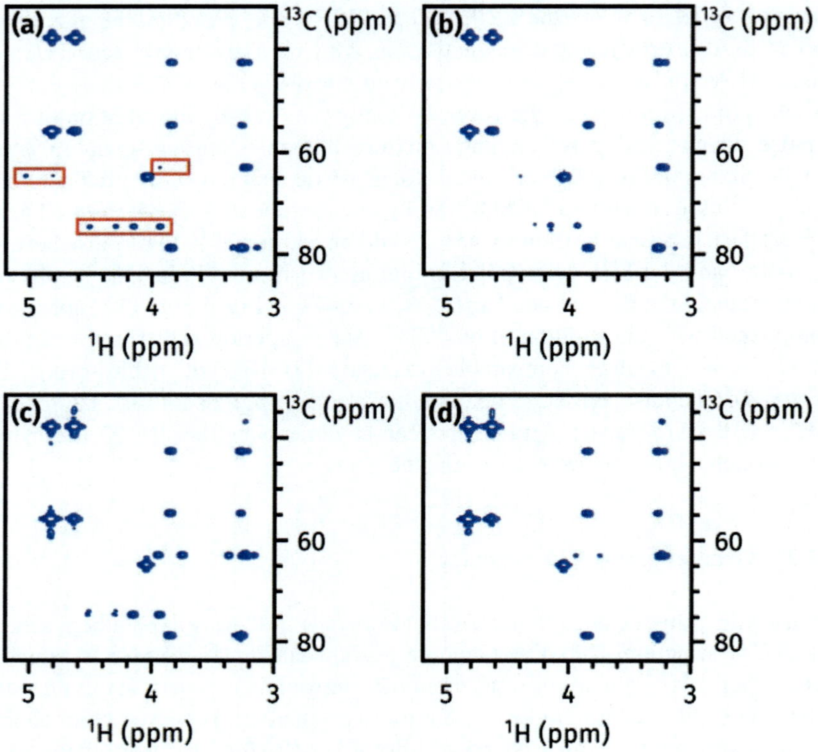

Fig. 4.50 **a** NOAH *zz*-HMBC module without the additional 90° pulse. One-bond artefacts are highlighted in red. **b** NOAH *zz*-HMBC with the 90° pulse. **c** Standard library HMBC with a second-order LPJF. **d** Standard library HMBC with a third-order LPJF. *Data code:* `7A-210916`

where $c_1 = -\alpha(\gamma_H - \gamma_C)/(\gamma_H + \gamma_C)$ and $c_2 = (1 - \alpha)(\gamma_H - \gamma_C)/\gamma_H$. In principle g is also a free parameter; for maximum suppression of artefacts I chose a relatively large value of 80%.

2. In 'Scheme B' (Fig. 4.51b), the 180° pulse is shifted to immediately after t_1, before any of the CTP gradients have been applied. This means that there is no need for a third gradient, and the CTP gradient amplitudes can be directly taken from the standard library sequence:

$$\text{echo:} \qquad g_1 = g; \qquad g_2 = gc; \tag{4.20}$$

$$\text{antiecho:} \qquad g_1 = gc; \qquad g_2 = g, \tag{4.21}$$

where $c = -(\gamma_H - \gamma_C)/(\gamma_H + \gamma_C)$ and $g = 80\%$.

3. 'Scheme C' (Fig. 4.51c) simply does not add a 180° pulse, but instead modifies the phases of the *zz*-filter in order to place $^1H^C$ magnetisation along the $-z$ axis during the HMBC J-evolution delay. Here, the gradient amplitudes are the same as those in the standard library sequence as well as in scheme B.

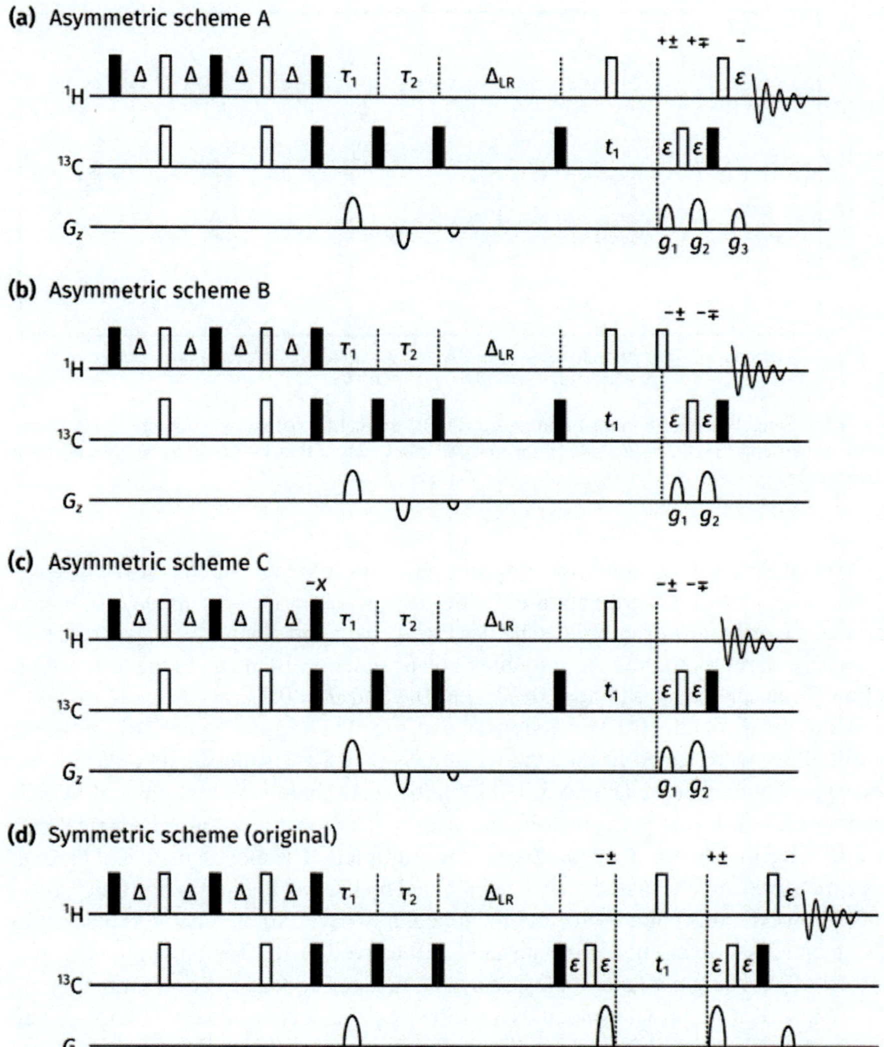

Fig. 4.51 Alternative CTP gradient schemes investigated for the NOAH HMBC module. The coherences selected for during each gradient are indicated above each gradient, using the same notation for product operators as described in the *Preface*: the 'upper' term (e.g. + in ±) refers to the echo experiment and the 'lower' term to the antiecho experiment. So, for example, +± refers to selection of I_+S_+ during the echo experiment and I_+S_- during the antiecho experiment. Gradient amplitudes are as described in the text. **a** 'Asymmetric scheme A', modified from the standard library sequence to include an additional 180° pulse and gradient. **b** 'Asymmetric scheme B', where the 180° pulse is shifted forward to the end of t_1. **c** 'Asymmetric scheme C', which modifies the *zz*-filter instead of using an extra 180° pulse. **d** The original 'symmetric' scheme (the same as in Fig. 4.3b), placed here for convenience

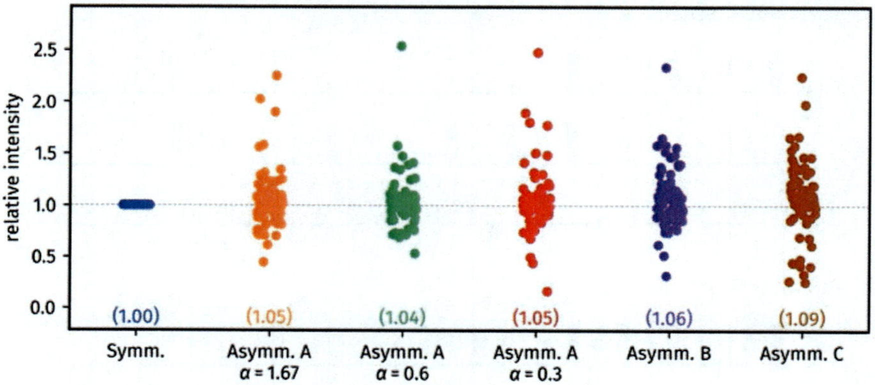

Fig. 4.52 Sensitivities of various asymmetric HMBC gradient schemes, as compared to the symmetric scheme in Fig. 4.3b. Each dot indicates one crosspeak in the HMBC spectrum; numbers in parentheses represent averages over all peaks. *Data code:* `7A-211226`

It is of interest to note two limiting cases of scheme A. When $\alpha = (\gamma_H + \gamma_C)/(\gamma_H - \gamma_C) \approx 1.67$, we obtain the ratio $g_1 : g_2 : g_3 = 1 : -1 : \pm 2\gamma_C/\gamma_H$, which mimics the original 'symmetric' scheme (Fig. 4.51d); and when $\alpha = 1$, we have that $g_3 = 0$, i.e. a return to the two-gradient tactic of schemes B and C. In the tests which follow, I ran scheme A with $\alpha = 1.67$, $\alpha = 0.6$ and $\alpha = 0.3$.

All of the different HMBC versions above, plus the original 'symmetric' scheme in Fig. 4.3b, were tested in the context of a NOAH-2 BS supersequence using the andrographolide sample (Fig. 4.52). As can be seen, there is not much at all which separates the different versions (outliers with $> 2\times$ 'sensitivity improvements' can be attributed to different J-modulation in the multiplet). The most sensitive of these is asymmetric scheme C, which may be explained by the fact that it has one fewer 180° pulse: however, this comes with an immediate drawback. Since scheme C places the $^1H^C$ magnetisation along $-z$ during the LPJF as well as the J-evolution delay Δ_{LR} (a total of ca. 70 ms), relaxation losses during this period lead to poorer retention of $^1H^C$ magnetisation for later modules, as shown by the decreased HSQC sensitivities in Fig. 4.53. In contrast, all the other gradient schemes retain $^1H^C$ magnetisation equally well.

The final point worth studying is the quality of the HMBC spectrum itself. To do this, we need to look at the actual spectra (Fig. 4.54). For the most part, the spectra are all the same; however, there is a notable set of artefacts present in Fig. 4.54b (scheme A with $\alpha = 1.67$) as well as Fig. 4.54e (scheme B), highlighted in red boxes. These artefacts occur at the frequencies

$$(\Omega_1, \Omega_2) = \left(\Omega_S \pm \frac{\Omega_I}{2}, \Omega_I\right), \tag{4.22}$$

and are in fact 'wing' artefacts similar to that observed in other modules (Sect. 4.3.1 to 4.3.3). In this case, they arise due to imperfect refocusing of the ^1H chemical shift

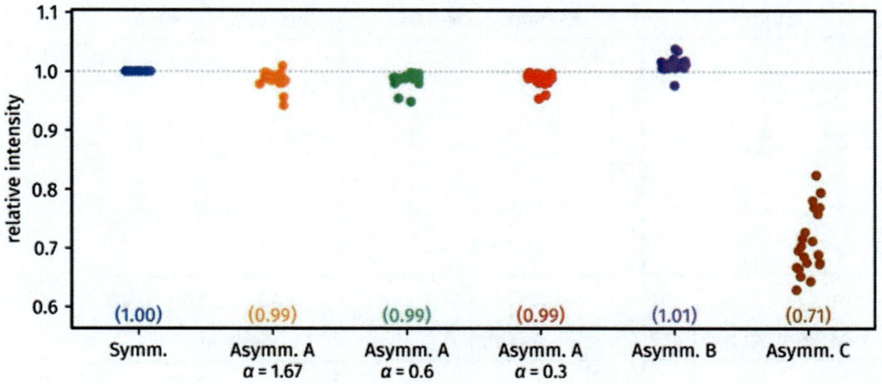

Fig. 4.53 Sensitivities of the HSQC module in a NOAH-2 BS supersequence, where the HMBC module is implemented using the gradient schemes of Fig. 4.52. Each dot indicates one crosspeak in the HSQC spectrum; numbers in parentheses represent averages over all peaks. *Data code:* `7A-211226`

during t_1: specifically, whenever $I_zS_\pm$ terms are present during the second half of t_1. Extra evidence for the origin of these artefacts comes from the observation that when the 180° pulse in the middle of t_1 is phase cycled, the artefacts are removed. In a standard HMBC, these terms would not be detected in the final FID; however, in this case, the addition of an extra 180° pulse after t_1 provides a mechanism by which they may be converted back into observable spin-I magnetisation (through off-resonance effects or miscalibration).

The poorer performance may therefore be understood as follows: when scheme A is acquired with $\alpha = 1.67$, the gradients g_1 and g_2 have equal and opposite amplitudes, and so do not enforce any coherence selection on spin I during the second half of t_1. Likewise, scheme B contains no gradients during the second half of t_1.

The characteristics of these gradient schemes are summarised in Table 4.4. As can be seen, the 'best' schemes are either the original symmetric scheme, or asymmetric scheme A with $\alpha \neq 1.67$. However, there is not much difference between these: it is not clear whether the improvement in sensitivity is reproducible across a wide range of samples, and in any case, the gains are extremely marginal.

4.3.7.3 Other Artefacts

It has been established that the HMBC module is not fully ideal in terms of magnetisation preservation. However, there are also some other curious phenomena in supersequences containing the HMBC module which have not been fully described in the literature. One of these is the presence of *inverted peaks* in the homonuclear X module(s) in a NOAH-3 BSX supersequence: this is illustrated in Fig. 4.55a with the CLIP-COSY module (X = C^c). It is not clear why this occurs, because the HMBC module (and the gradients which follow) should dephase all $^{1}H^{!C}$ magnetisation.

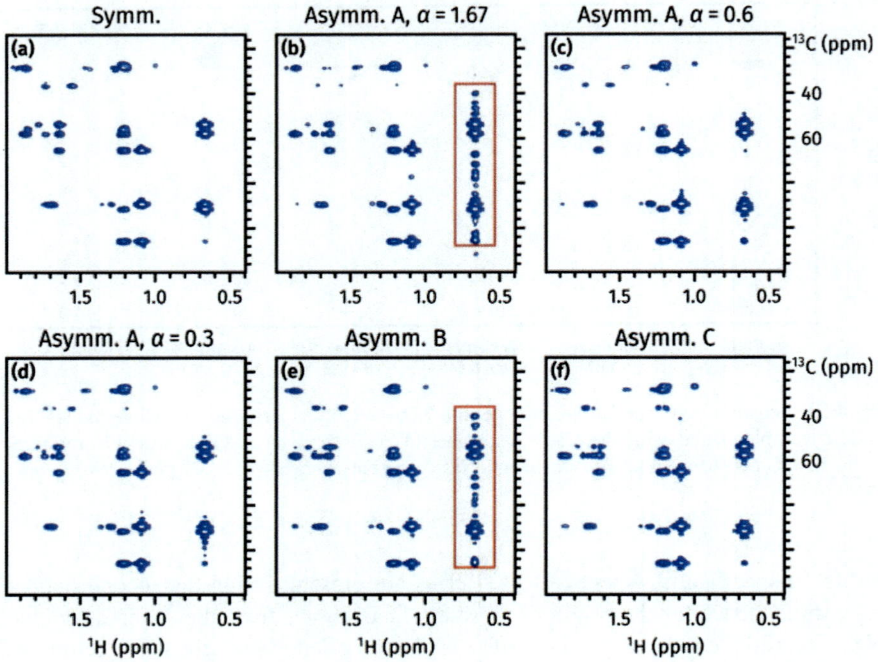

Fig. 4.54 HMBC spectra acquired with the gradient schemes of Fig. 4.52. Extra 'wing' artefacts present in two of the spectra (asymmetric scheme A with $\alpha = 1.67$, **b**, and asymmetric scheme B, **e**) are highlighted in red boxes. *Data code:* `7A-211226`

Table 4.4 Comparison of HMBC gradient schemes discussed in this section: the data are a summary of Figs. 4.52, 4.53 and 4.54. *Data code:* `7A-211226`

Gradient scheme	HMBC sensitivity	HSQC sensitivity	Wing artefacts
Symmetric	1	1	No
Asymmetric A, $\alpha = 1.67$	1.05	0.99	Yes
Asymmetric A, $\alpha = 0.6$	1.04	0.99	No
Asymmetric A, $\alpha = 0.3$	1.05	0.99	No
Asymmetric B	1.06	1.01	Yes
Asymmetric C	1.09	0.71	No

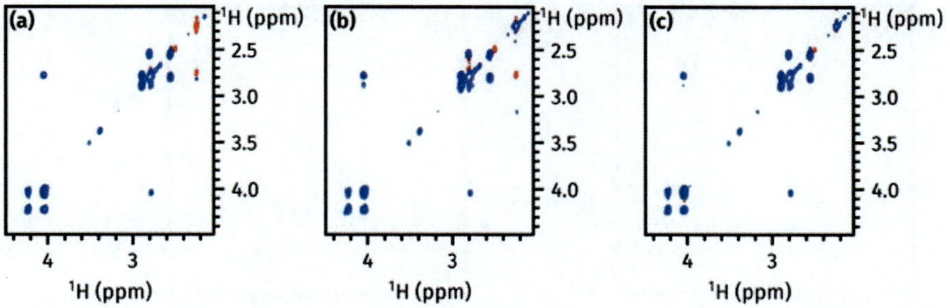

Fig. 4.55 **a** CLIP-COSY from a NOAH-3 BSC^c supersequence, acquired with a ^{1}H 90° pulse width of 11.28 μs (this value was obtained using the POISE calibration described in Sect. 3.4.1). An inverted peak is visible at 2.25 ppm. **b** The same, but acquired using a 90° pulse width of 11.78 μs. **c** CLIP-COSY from a NOAH-4 BSSC^c supersequence. The 90° pulse width was 11.28 μs, the same as in **a**. *Data code:* `7Z-220214`

Although this leads to reduced sensitivity in the homonuclear module, in that the signal derives from polarisation which has recovered during the preceding FIDs, it is not clear why this polarisation should be *negative*. One clue lies in the fact that these peaks are very sensitive to the ^{1}H 90° pulse width: simply changing this by 0.5 μs is sufficient to restore the correct signal sign (Fig. 4.55b).

The modules placed between the HMBC and the homonuclear module also play an important role. When *two* HSQC modules are used, i.e. a NOAH-4 BSSC^c supersequence (using $f = 0.7$ as described in Sect. 4.3.4—although this is unlikely to matter), the negative peaks are no longer observed (Fig. 4.55c). In fact, having *no modules* between the HMBC and the homonuclear module is also (at least sometimes) acceptable: a separate set of data shows that the inverted peaks in an NOAH-3 BS$^+$C^c experiment are not seen in a NOAH-3 S$^+$BC^c supersequence (Fig. 4.56a and b). The use of isotropic mixing (implemented as a sequence of adiabatic pulses [94], and referred to as 'ASAP' mixing) just before the homonuclear module does not remedy this (Fig. 4.56c and d). Unfortunately, a good explanation for these artefacts has remained elusive.

4.3.7.4 ^{15}N HMBC Module

The entirety of this section has—until now—been devoted to the ^{13}C HMBC module. However, the techniques used in constructing this, including the implementation of the *zz*-filter, are equally applicable to a ^{15}N HMBC. For simplicity, the NOAH ^{15}N HMBC module uses a first-order LPJF (since directly bonded NH pairs are less common); this may be omitted if desired. To minimise the number of pulses, a simple magnitude-mode version of the HMBC is used (Fig. 4.57). The implementation of this module within supersequences is discussed in greater detail in Sect. 4.5, within the context of *generalised supersequences*.

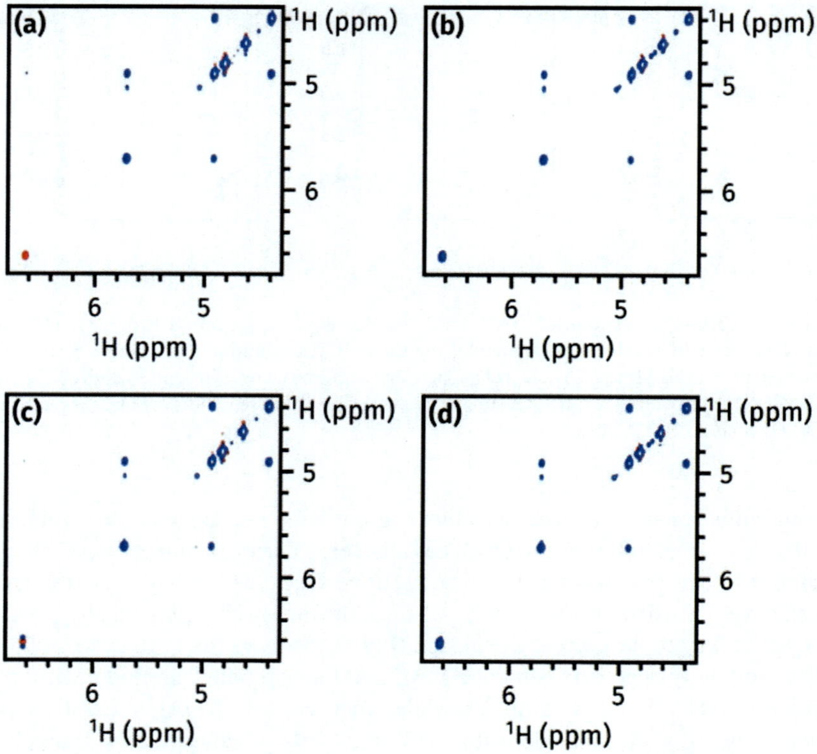

Fig. 4.56 **a** CLIP-COSY from a NOAH-3 $BS^{+}C^{c}$ supersequence. An inverted diagonal peak can be seen at 6.6 ppm. **b** From a NOAH-3 $S^{+}BC^{c}$ supersequence. **c, d** The same as **a** and **b**, but with 40 ms ASAP mixing placed just before the CLIP-COSY module. *Data code:* `7A-211227`

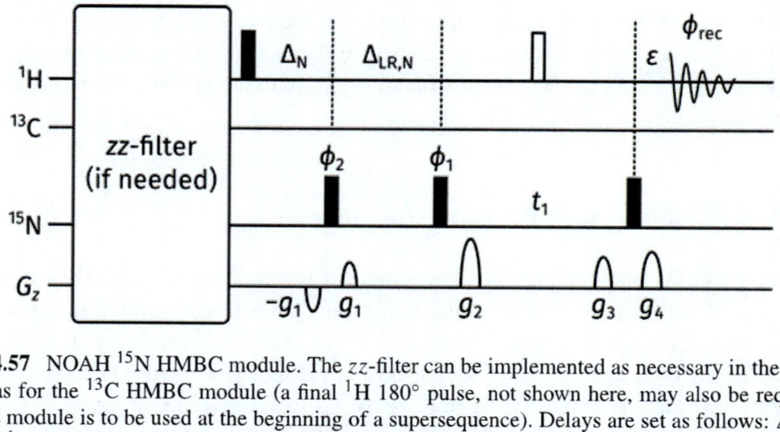

Fig. 4.57 NOAH ^{15}N HMBC module. The *zz*-filter can be implemented as necessary in the same way as for the ^{13}C HMBC module (a final ^{1}H 180° pulse, not shown here, may also be required if this module is to be used at the beginning of a supersequence). Delays are set as follows: $\Delta_N = 1/(2 \cdot {}^{1}J_{NH})$; $\Delta_{LR,N} = 1/(2 \cdot {}^{n}J_{NH})$. Phase cycling is performed using $\phi_1 = \phi_{rec} = (x, -x)$ and $\phi_2 = (x, x, -x, -x)$. Gradient amplitudes are $(g_1, g_2, g_3, g_4) = (5\%, 70\%, 30\%, 50.1\%)$

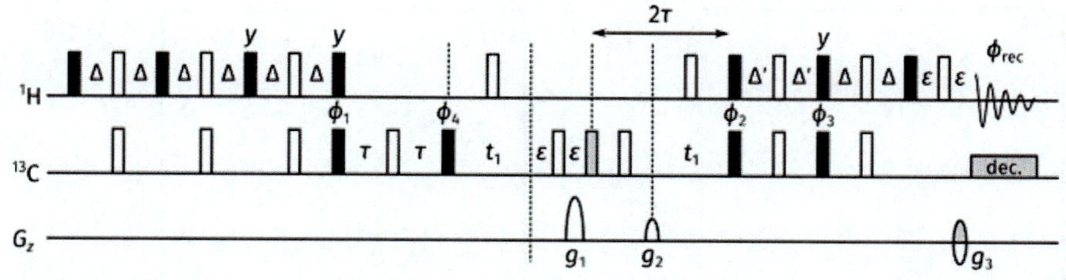

Fig. 4.58 NOAH 1,1-ADEQUATE module. The grey filled bar is a 120° pulse, optimised for ^{13}C double-quantum to single-quantum coherence transfer [97]. Delays are set as follows: $\Delta = 1/(4 \cdot {}^1J_{CH})$, $\Delta' = 1/(8 \cdot {}^1J_{CH})$, and $\tau = 1/(4 \cdot {}^1J_{CC})$. Phase cycling is performed using $\phi_1 = (x, -x)$, $\phi_2 = (x, x, -x, -x)$, $\phi_3 = (y, y, -y, -y)$, $\phi_4 = (x, x, x, x, -x, -x, -x, -x)$ and $\phi_{rec} = (x, -x, -x, x, -x, x, x, -x)$. Gradient amplitudes are $(g_1, g_2, g_3) = (78.5\%, 77.6\%, \mp 59\%)$. Echo–antiecho selection is achieved by inverting the sign of g_3 as well as the pulse phase ϕ_3

4.3.8 ADEQUATE

The final 'new' module to be described in this section is the 1,1-ADEQUATE experiment [95], which provides two-bond C–H correlations through a combination of $^1J_{CH}$ and $^1J_{CC}$. Because of the requirement for two adjacent ^{13}C nuclei, the sensitivity of the ADEQUATE experiment is considerably lower than almost all other NOAH modules considered so far. As established in Sect. 4.1.1, sensitivity imbalances between different modules lead to less ideal supersequences with smaller effective time savings $\rho_{t,\mathrm{eff}}$. Thus, the ADEQUATE module has to date not been implemented in NOAH supersequences.[35]

In this section, I only consider the design of the NOAH ADEQUATE module itself, which turns out to be extraordinarily simple. The ^{1}H pulses in the original ADEQUATE experiment mirror that of the CRK seHSQC almost perfectly, save for one extra 180° pulse. Thus, the ADEQUATE module can be modified to preserve $^1\mathrm{H}^{!\mathrm{C}}$ magnetisation using exactly the same strategy in seHSQC2: namely, replacing the first ^{1}H excitation pulse with the ZIP element (Fig. 4.58). In order to compensate for the extra 180° pulse (which arises in the constant-time period of the ADEQUATE), the pulse phases in the ZIP element are modified accordingly: thus, the third ^{1}H 90° pulse has a phase of y (instead of $-y$ in seHSQC2).

An example of a NOAH-2 AB supersequence is shown in Fig. 4.59. This is not the best setting to use the ADEQUATE module in, because the ^{13}C HMBC module has roughly 100 times the sensitivity of the ADEQUATE; however, it demonstrates that the ADEQUATE module does indeed work as intended. More constructive uses of the ADEQUATE module are deferred to Sect. 4.5: in there, I discuss combinations

[35] There is, in fact, one exception to my claim [96]. Here, the ADEQUATE and seHSQC modules were not modified to preserve any magnetisation for later modules. Since the ADEQUATE itself has such low sensitivity, it is likely that the sensitivity losses in the later modules were tolerable; however, no sensitivity comparisons against standalone experiments were provided.

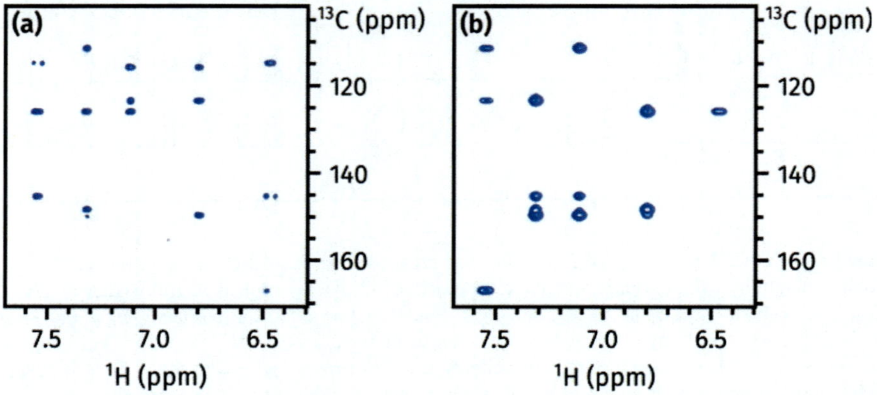

Fig. 4.59 Spectra from a NOAH-2 AB supersequence. **a** 1,1-ADEQUATE module. **b** HMBC module. *Data code:* `7T-211231`

with the (lower-sensitivity) ^{15}N HMBC module, and show how other modules may be added to form so-called 'generalised' supersequences.

4.4 Solvent Suppression in NOAH

Solvent suppression is one of the most important aspects of modern NMR, and has been incorporated into a large number of experiments. It is not a particularly difficult task to implement some basic solvent suppression techniques in NOAH supersequences, as will be described in this section. All suppression techniques shown here can be enabled via acquisition flags (the TopSpin `ZGOPTNS` parameter); this means that there is no need to switch pulse sequences.

4.4.1 Presaturation

A simple first step is to implement simple presaturation of the solvent resonance during the recovery delay of the supersequence. In GENESIS experiments, this can be enabled using the `-DPRESAT` acquisition flag. This often provides excellent suppression in the first module in a supersequence, and to a lesser extent, later modules (as the solvent magnetisation recovers during acquisition periods).

With the same acquisition flag, presaturation is also applied during long delays, in particular mixing times in NOESY experiments. The use of presaturation during the HMBC J-evolution delay was also tested, but was found to be unnecessary, especially since the HMBC module is typically placed first in supersequences and thus comes right after the recovery delay.

4.4.2 Intrinsic Suppression

A number of NOAH modules in fact come with an *intrinsic* form of solvent suppression, in that they return $^{1}\mathrm{H}^{!\mathrm{X}}$ magnetisation (including that of solvents) to the $+z$ axis at the end of the module. This is generally true of many of the HSQC-based NOAH modules, which seek to only sample $^{1}\mathrm{H}^{\mathrm{X}}$ magnetisation pools (X = C or N). Thus, these modules already have far better solvent suppression properties compared to standard library sequences: there is no need for further modification.

For these sequences, it is possible to go one step further and to see how the solvent suppression varies with transmitter offset. This is less relevant if the solvent is just water (where it suffices to put the water peak on resonance), but may be important for samples in other protonated solvents or mixtures thereof. I tested this by recording the first increment of the NOAH HSQC and seHSQC modules on an aqueous sample of sucrose and integrating the water peak. The results (Fig. 4.60) indicate that the HSQC module provides the best suppression across a range of offsets; of the two seHSQC modules, seHSQC1 provides more uniform suppression, whereas seHSQC2 has some regions ('spikes') where suppression is poorer than average.

It is not surprising that the more complicated seHSQC sequences have slightly poorer suppression: there are substantially more pulses involved which lead to cumulative errors, especially for peaks which are further away from resonance. However, it is not entirely clear why seHSQC2 has such large spikes. Some further investigation showed that the poorer suppression partly arises from imperfect cancellation of the water signal by phase cycling. This is, however, not a full explanation; it merely replaces one mystery with another. Nevertheless, it should be reiterated that all of these sequences still provide excellent water suppression. The standard library seHSQC sequences, for example, do not even come close to this level of suppression: although these do dephase water magnetisation prior to acquisition, radiation damping during the FID still leads to a very large water signal.

4.4.3 Excitation Sculpting

Since solvent suppression in HMBC and HSQC-type modules can be adequately accomplished through presaturation and intrinsic suppression, it remains to consider the suppression in homonuclear modules. These modules invariably occur at the end of supersequences, and thus almost any form of solvent suppression can be used: there is no need to consider how magnetisation needs to be preserved for other modules.

In practice, I chose to implement excitation sculpting [98] (ES) just prior to acquisition: the refocusing element chosen was a combination of a selective 180° sinc pulse and a hard 180° pulse. This worked perfectly for almost all of the homonuclear modules used in NOAH supersequences, with only a few adjustments needed, such as for the PSYCHE modules where it was necessary to ensure that chemical shifts and J-couplings evolved for the correct amount of time.

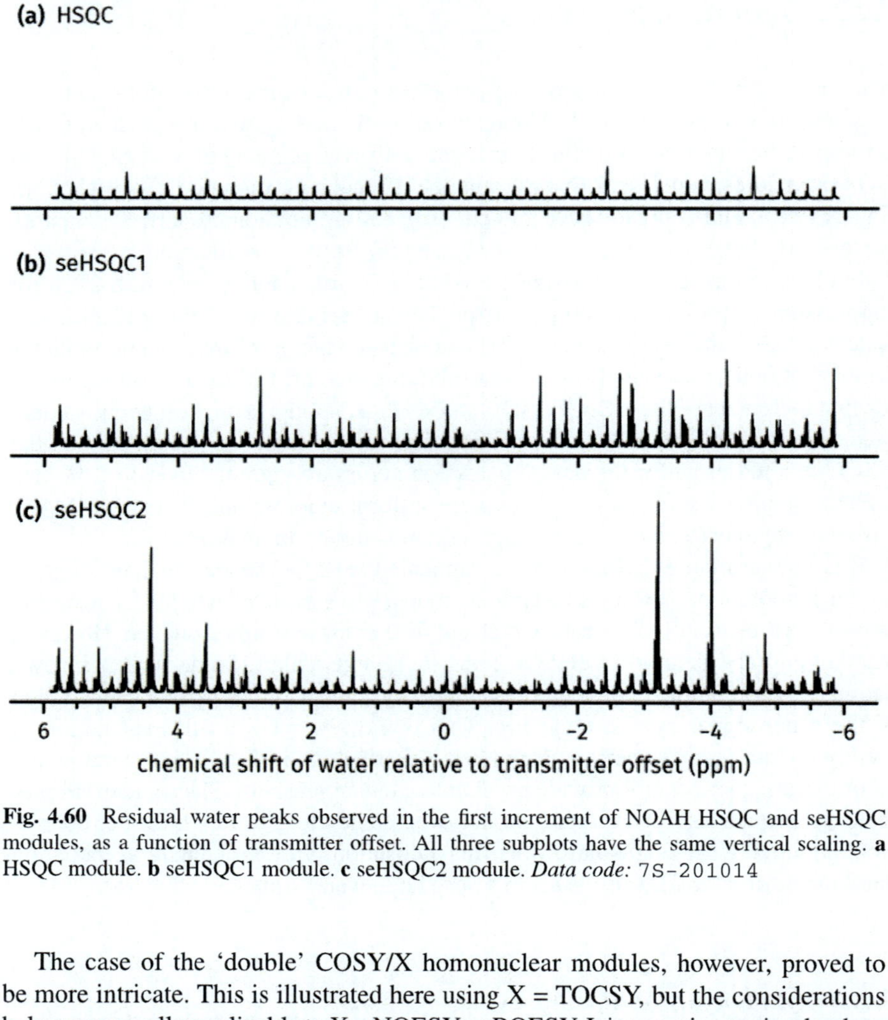

Fig. 4.60 Residual water peaks observed in the first increment of NOAH HSQC and seHSQC modules, as a function of transmitter offset. All three subplots have the same vertical scaling. **a** HSQC module. **b** seHSQC1 module. **c** seHSQC2 module. *Data code:* `7S-201014`

The case of the 'double' COSY/X homonuclear modules, however, proved to be more intricate. This is illustrated here using X = TOCSY, but the considerations below are equally applicable to X = NOESY or ROESY. It is tempting to simply place ES blocks prior to both FIDs: in the COSY module, this dephases transverse solvent magnetisation as desired, and (in principle) should leave longitudinal magnetisation untouched, meaning that the TOCSY module which later consumes this should be unaffected.

However, this is not totally true. *Inside* the bandwidth of the selective 180° pulse, any longitudinal magnetisation experiences an 720° rotation; and *outside* of the bandwidth, it experiences a 360° rotation. However, *between* these two extremes, there is a crossover point where longitudinal magnetisation is sent into the transverse plane and subsequently dephased by gradients. This is reflected in simulations of the ES element, specifically the nulls in Fig. 4.61b: peaks which fall within this range will be lost in the TOCSY module. This is visible in the spectra of Fig. 4.62: although the water suppression obtained using this 'double ES' scheme is better, several peaks in

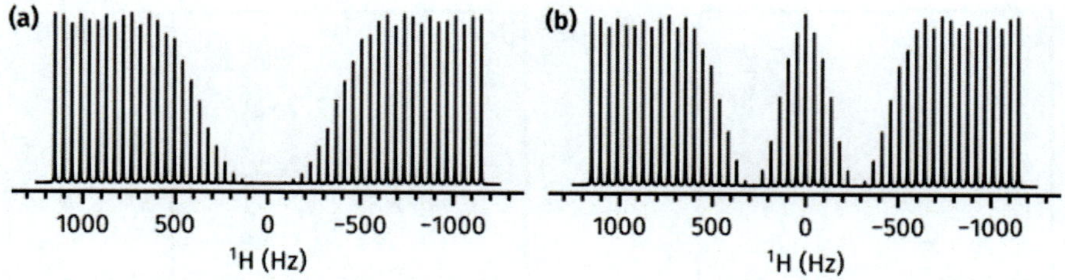

Fig. 4.61 Simulations showing the proportion of magnetisation retained by the excitation sculpting block (using a 2 ms sinc pulse). **a** Retention of transverse magnetisation: this was obtained by simulating the spectrum of a 90°–ES–detect pulse sequence. **b** Retention of longitudinal magnetisation: this was obtained using an ES–90°–detect pulse sequence

the TOCSY spectrum have disappeared, because they fall precisely into these nulls. It proves better to omit the ES in the COSY module: an adequate degree of water suppression in the COSY can still be attained thanks to the use of presaturation.

Of course, it may well be that *no* peaks fall within this null, and thus the better solvent suppression may be obtained at no cost. This was the case when the experiments in Fig. 4.62 were reacquired on a 700 MHz spectrometer; or when the sinc pulse was lengthened to 4 ms. However, such a coincidence cannot always be relied on.

4.5 Parallel and Generalised NOAH Supersequences

To conclude this chapter, I discuss how *multiple* NOAH supersequences may be run in parallel in order to obtain yet more data from a single experiment. As of the time of writing, current software limitations in TopSpin limit the `NBL` parameter (the number of memory blocks) to 5. This makes it presently impossible to linearly extend a supersequence beyond five modules. However, it is possible to *interleave* different supersequences, such that one t_1 increment of supersequence A is acquired, followed by one t_1 increment of supersequence B, before the value of t_1 is increased for both supersequences. In this text, I refer to this as a 'vertical' stacking of modules (as opposed to the traditional NOAH concept, which focuses on 'horizontal' concatenation of modules).

4.5.1 Parallel NOAH Supersequences

This concept is more clearly illustrated in Fig. 4.63, which contains a more explicit depiction of how t_1 is incremented for each module. In Fig. 4.63a, a 'traditional' NOAH-2 ST supersequence is shown: here, t_1 is incremented for both the HSQC and

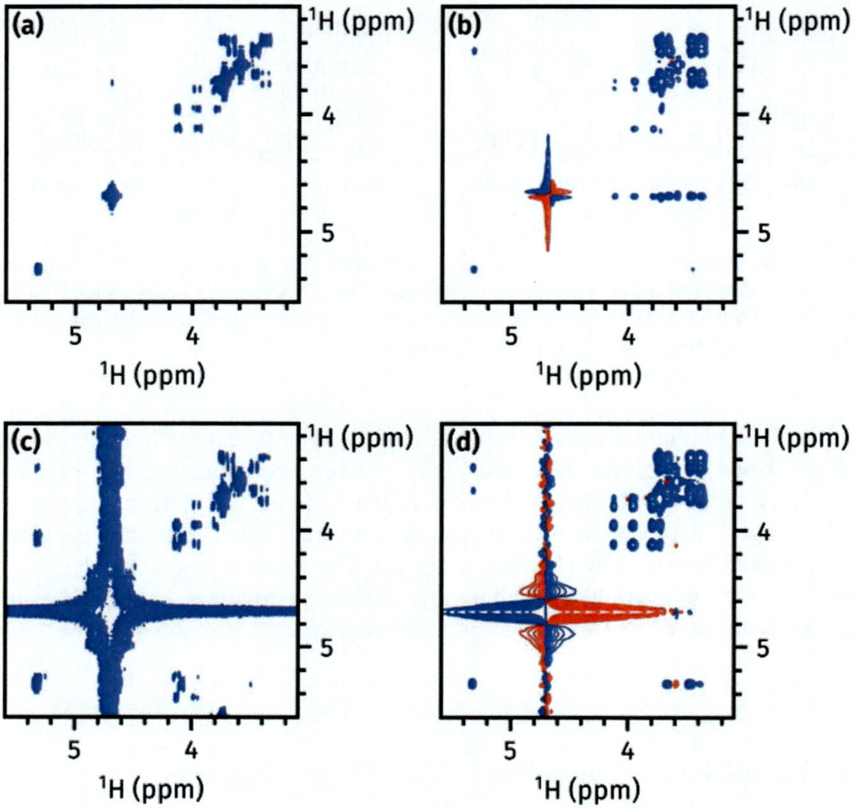

Fig. 4.62 **a–b** COSY and TOCSY spectra obtained from a NOAH-3 SCT supersequence where excitation sculpting was placed before both the COSY and TOCSY FIDs. Although the water suppression is better, some peaks in the TOCSY module are lost. **c, d** The same, except that excitation sculpting was applied only in the TOCSY module. For all spectra shown here, presaturation of the water resonance was applied during the recovery delay. *Data code:* `4S-211105`

TOCSY module at the same time, and on each t_1 increment, the sequence of modules being acquired is always the same.[36] In Fig. 4.63b, this is different: the second module is alternated between a TOCSY and a NOESY. If the total experimental duration (which is largely proportional to the number of recovery delays, d_1) is to be kept the same as before, this means that the TOCSY and NOESY modules will be recorded with half the number of t_1 increments each. Although this loss of resolution may be considered a drawback, this scheme does allow us greater *flexibility* in the design of NOAH supersequences: it is not ideal to acquire both a TOCSY and a NOESY

[36] This description is slightly simplified: it omits the fact that all three modules must be acquired twice per t_1 increment for echo–antiecho or States–TPPI quadrature detection to be carried out. This is mostly an implementation detail, however; it does not materially affect the concepts described in this section.

using a traditional 'linear' supersequence, since both of these modules depend on $^{1}H^{!X}$ magnetisation.

Figure 4.63c shows how the F_1 resolutions for all three modules can be equalised, by performing the equivalent of k-scaling on the HSQC module (described in Sect. 4.3.2). This entails reducing the number of t_1 increments by a factor of 2 (in this case)

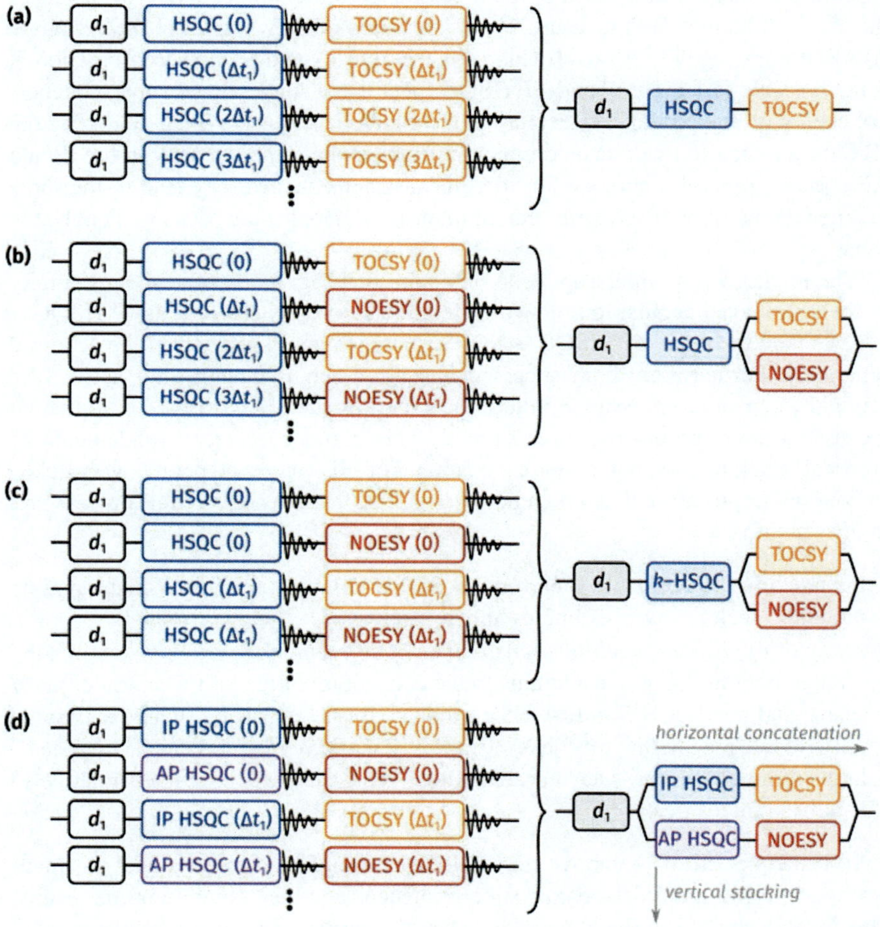

Fig. 4.63 An overview of 'parallel' NOAH supersequences. The left side of each diagram explicitly spells out how t_1 is incremented for each module: the value in parentheses indicates the current value of t_1, which begins at 0 and is incremented by Δt_1 each time. The right side is a 'condensed' depiction of the entire supersequence. **a** A 'standard' NOAH supersequence, where t_1 for every module is incremented at the same time. **b** The first example of a 'parallel' supersequence: the HSQC module is acquired as normal, but the TOCSY and NOESY modules are interleaved in the second slot. **c** The same as in (b), but the HSQC module is subjected to k-scaling (see also Sect. 4.3.2). **d** A fully parallel supersequence, where both the first and second module slots are varied systematically. This amounts to the interleaved acquisition of two different 'standard' supersequences

and acquiring each increment twice instead, which corresponds to a doubling of the number of scans after the data have been appropriately combined. This generally has very little impact on the spectra (as was previously shown), but conveniently leads us into the final example of Fig. 4.63d. Here, instead of acquiring each increment of the HSQC module twice, this slot is used to acquire two different modules, an in-phase (IP) and antiphase (AP) HSQC (both run without ^{13}C decoupling). Through appropriate linear combination of the data, it is possible to isolate the two peaks of the ^{1}H–^{13}C doublets and measure $^{1}J_{CH}$ values: this is known as in-phase/antiphase (IPAP) processing [77, 99–101]. Much like the TOCSY and NOESY combination, it is not possible to acquire IP and AP HSQC spectra in a single, linear supersequence, not even with the partial $^{1}H^{C}$ excitation technique described in Sect. 4.3.4. For the IPAP processing to work, each doublet has to have the same intensity in the IP and AP spectra. A careful choice of f in a linear supersequence may lead to the same *average* intensity, but the distribution of intensities amongst the peaks will not be the same.

The reader will no doubt notice at this point that Fig. 4.63d corresponds simply to the interleaved acquisition of two different supersequences (one is IP-HSQC + TOCSY and the other AP-HSQC + NOESY), which could just as well be acquired separately. Furthermore, both supersequences are acquired with only half of the original F_1 resolution (assuming the same experimental time as Fig. 4.63a), which means that there are *no real time savings*. This is indeed correct: fundamentally, 'vertical' stacking does not increase the number of FIDs recorded per recovery delay, so does not improve the time efficiency (or ρ_t). The main benefits of this process are, in my opinion:

- the *flexibility* to construct supersequences that combine previously incompatible modules (such as in the examples above); and
- a way of *adjusting for relative sensitivities of different modules*. For example, the arrangement in Fig. 4.63c amounts to the acquisition of the HSQC module for $2n$ scans, and the TOCSY and NOESY modules for n scans each (where n is some positive integer). In this way, less intrinsically sensitive modules can be assigned a larger number of scans, such that all modules in the supersequence have (relatively) equalised intensities.

In principle, all of the above may still be accomplished in a roundabout manner by acquiring multiple separate supersequences and then combining the results. This is most obvious in Fig. 4.63d. However, the current implementation of parallel supersequences provides *convenience* for the user, as all experiments may be simultaneously acquired and processed. This factor should not be overlooked, especially considering that all the different constructions (Fig. 4.63b–d) must be processed in a slightly different manner: it is easier to do this from within a single experiment, rather than attempting to combine FIDs from separate datasets after the fact. There is also another argument for acquiring increments in an interleaved manner: this helps to minimise drifts in the spectra which arise from environmental instabilities such as temperature fluctuations.

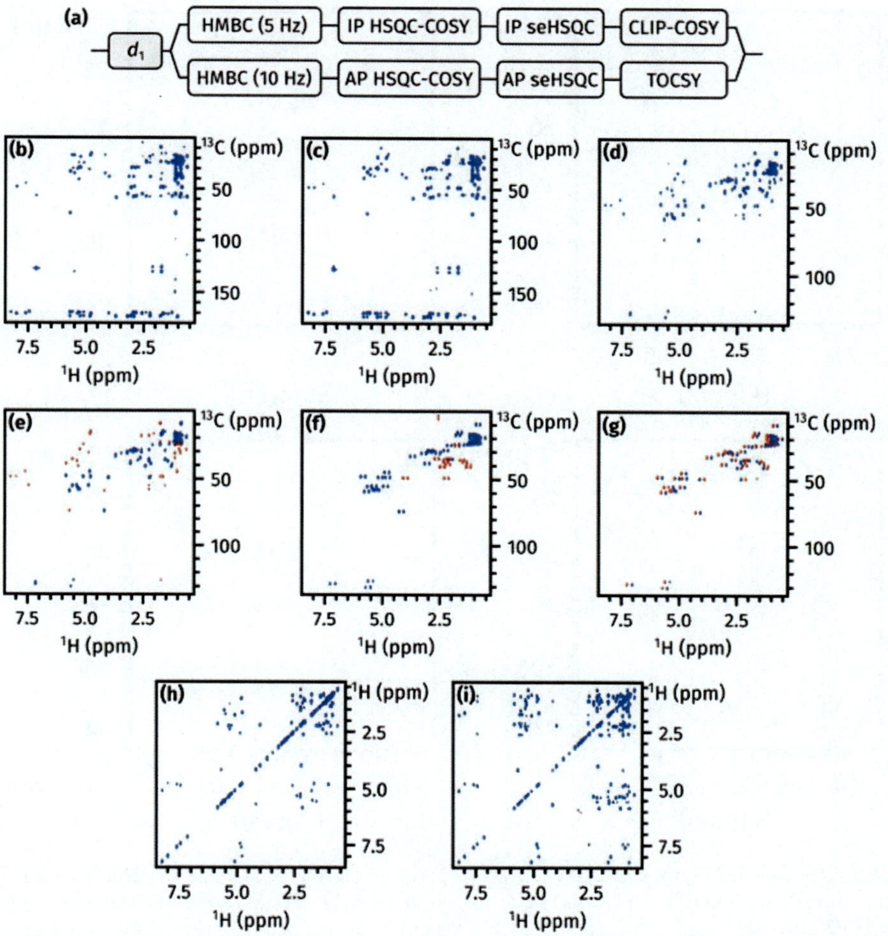

Fig. 4.64 **a** Condensed representation of the supersequence, which may be interpreted in the same way as Fig. 4.63d. **b** HMBC, optimised for $^{n}J_{\mathrm{CH}} = 5\ Hz$. **c** HMBC, optimised for $^{n}J_{\mathrm{CH}} = 10\ Hz$. **d** IP HSQC-CLIP-COSY. **e** AP HSQC-CLIP-COSY. **f** IP multiplicity-edited seHSQC. **g** AP multiplicity-edited seHSQC. **h** CLIP-COSY. **i** TOCSY (60 ms mixing time)

Figure 4.64a shows one example of a more complex parallel supersequence containing 8 modules. In this supersequence, the HMBC module is recorded twice with the delay Δ_{LR} optimised for two different values of $^{n}J_{\mathrm{CH}}$ (Fig. 4.64b and c): these spectra furnish a (slightly) different set of correlations, which allow for more complete structure determination. The HSQC-CLIP-COSY is acquired in an IPAP fashion, where direct/indirect editing is used in one experiment (Fig. 4.64e) and not in the other (Fig. 4.64d). Likewise, the F_2-coupled seHSQC experiment is also performed in an IPAP manner, where this time the IPAP components are the two peaks in each doublet (split by $^{1}J_{\mathrm{CH}}$). Multiplicity editing is additionally used here in both IP and

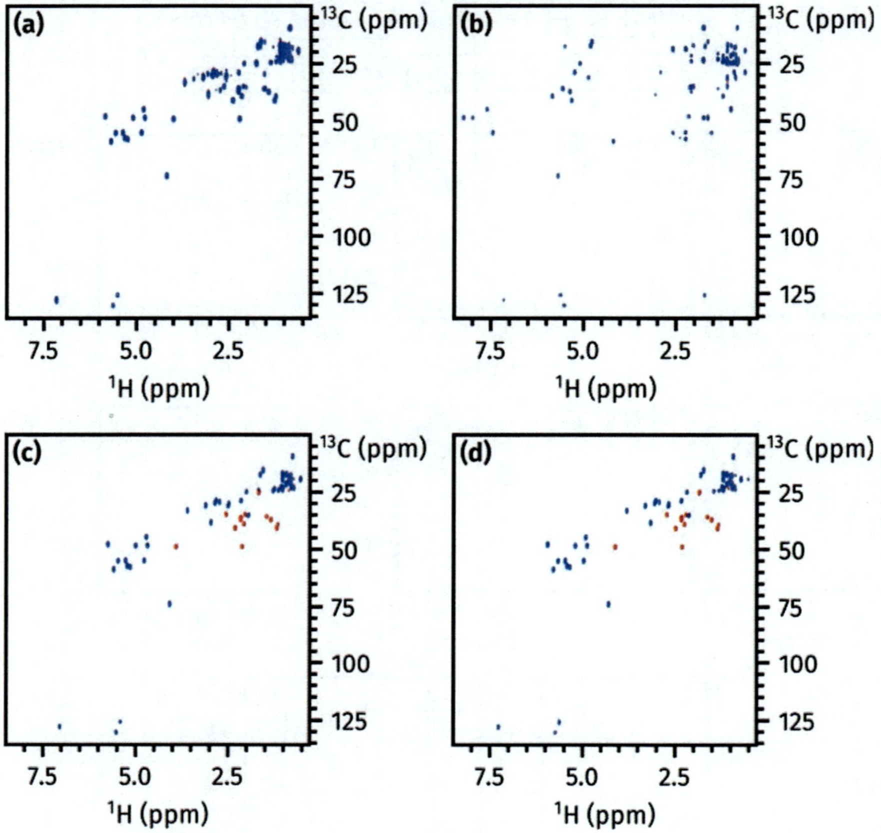

Fig. 4.65 **a–b** IPAP-processed HSQC-COSY spectra (from Fig. 4.64d and e), which separate the 'direct' and 'indirect' HSQC-COSY responses described in Sect. 4.3.5. **c, d** IPAP-processed seHSQC spectra (from Fig. 4.64f and g), each of which contain one peak from every ^{1}H–^{13}C doublet

AP seHSQC spectra. Finally, we have two interleaved homonuclear modules, the CLIP-COSY and TOCSY (Fig. 4.64h and i), which provide complementary information about spin networks. In Fig. 4.64, the heteronuclear spectra are shown before IPAP processing (in other words, these are the IP and AP spectra). The four IPAP-processed spectra (IP − AP and IP + AP) from this supersequence are shown in Fig. 4.65.

4.5.1.1 Time-Shared NMR

In my published work [2], what I refer to as IPAP here is frequently labelled as a TS, or *time-shared* experiment. In TS NMR, pulse sequences are specially designed in order to record two (or more) different signals at once, such as a ^{13}C HSQC and a ^{15}N HSQC: this has found use in both biomolecular [102–107] and small-molecule NMR [5, 26, 27, 108, 109]. Often, the design of TS experiments involves the joint evolution of both heteronuclear chemical shifts in a shared t_1 period: hence the term 'time-shared'.

It should be noted that TS experiments use *simultaneous* acquisition, as opposed to the *sequential* acquisition of NOAH experiments. In other words, the two or more signals recorded in a TS experiment are part of the same FID, meaning that the resulting spectra will contain peaks from both signals. If these can be easily disambiguated (e.g. by the differing chemical shifts of ^{13}C- or ^{15}N-bound protons), then the TS experiment allows more data to be collected in the same amount of time as a non-TS experiment. However, if these signals need to be separated, then the TS experiment must be repeated: once where the two signals have the same phase, and once where they have opposite phases. The underlying signals can then be disentangled through linear combination [102, 110]; this process leads to an improvement in sensitivity, which can be re-converted into time savings by using fewer scans for the TS experiment (if this does not compromise spectral quality too much).

Mathematically, this is exactly the same process as IPAP processing. In my view, what separates the two is the fact that the IPAP technique seeks to disentangle components which are *innate* to the pulse sequence under consideration and cannot usually be isolated—such as doublets in a F_2-coupled HSQC—whereas TS experiments must be *specially crafted* to record these multiple components. Equivalently, the time savings (or sensitivity improvement) from a TS experiment stem from the fact that it is being compared against a pair of 'base experiments' which each yield only one of the two signals detected in the TS sequence. To justly label an experiment as being TS, appropriate choices must be made for these 'base experiments'.

In this case, the IPAP label clearly applies to the pair of F_2-coupled seHSQC modules in Fig. 4.64f and g. However, for the HSQC-COSY modules in Fig. 4.64d and e, the situation is less clear. The two 'signals' which we should consider are the 'direct' and 'indirect' peaks; however, it is not entirely obvious which 'base experiments' correspond to these. The 'direct' peaks correspond to a (^{13}C-decoupled) HSQC experiment.[37] However, for the 'indirect' peaks, there is no real equivalent spectrum to be compared against: the H2BC experiment [79, 80] yields a similar set of peaks, but in this experiment, the 'direct' peaks have in fact been intentionally suppressed through the application of a LPJF. When the LPJF is removed (as in the 2BOB/H2OBC experiments [81]), a similar information content to the HSQC-COSY is obtained. So, it is not entirely accurate to write that the HSQC-COSY combines HSQC plus H2BC spectra. Consequently, I personally prefer to view it as an IPAP experiment.

Of course, the analysis above is only applicable to the modules shown in Fig. 4.64. None of this precludes the application of *true* time-shared experiments to NOAH supersequences: it is completely possible to implement, for example, a ^{13}C/^{15}N TS experiment within the context of a parallel supersequence. However, many TS exper-

[37] It should be noted that the intensities of the direct peaks in an HSQC-COSY spectrum are lower than in the corresponding HSQC spectrum, because the direct peaks in an HSQC-COSY stem only from magnetisation which does *not* evolve under $^{n}J_{\mathrm{HH}}$ during the mixing period.

iments are already designed to detect signals arising from different magnetisation pools (e.g. $^{1}H^{C}$ and $^{1}H^{N}$). This is already something which can be accomplished in a cleaner manner using (linear) NOAH supersequences.

4.5.2 Generalised Supersequences

From the discussion above, it is evident that there is no need to stop at *two* interleaved modules: it is possible to stack an arbitrary number of modules in a vertical manner. Furthermore, different modules may be acquired a different number of times. Thus, in Fig. 4.63b, two supersequences were interleaved, and the TOCSY and NOESY modules were acquired once each in each t_1 increment. However, if M different supersequences are being interleaved, then a set of two different modules may be acquired m_1 and m_2 times, as long as $m_1 + m_2 = M$. For example, a supersequence with $m_1 = 1$ and $m_2 = 2$ is shown in Fig. 4.66. Generally, it is desirable to assign larger m_i values to less sensitive modules, as the FIDs can be summed up to yield improved signal-to-noise.

4.5.2.1 Pulse Programs

Unfortunately, GENESIS is not (at present) advanced enough to create pulse program for generalised supersequences. Thus, the pulse programs for this final section have been written by hand. The overall structure of such a pulse program is shown in Listing 4.4. The parameter `CNST51` corresponds to the number of supersequences being interleaved, i.e., M in the discussion above. The values of m_1 and m_2 are encoded as `CNST52` and `CNST53`; the latter of these can be automatically calculated, since $m_2 = M - m_1$.

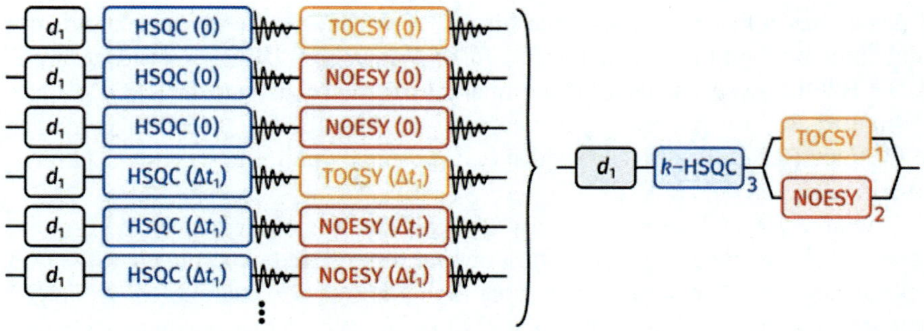

Fig. 4.66 A basic example of a 'generalised' NOAH supersequence, where an arbitrary number of supersequences (here three) can be vertically 'stacked'. For each value of t_1, the TOCSY is acquired once, the NOESY twice and the (k-scaled) HSQC three times, as indicated by the subscripts in the condensed representation on the right

Listing 4.4: The overall structure of the pulse program for the generalised supersequence in Fig. 4.66

```
"l0     = td1/2"               ; TD1/NBL
"cnst53 = cnst51 - cnst52"    ; automatically calculate m_2 = M - m_1
; ...

1 ze
2 30m
4 50u UNBLKGRAD
  d1 st0

  ; HSQC goes here
  ; ...
  goscnp ph30 cpd2:f2
  50u do:f2

  ; check which module to run
  if "l3 % cnst51 < cnst52"
{
  ; TOCSY goes here
  ; ...
  go=2 ph31
}
  else
{
  ; NOESY goes here
  ; ...
  go=2 ph31
}

  ; move to next 'row' in diagram
  1m iu3
  30m wr #0 if #0 zd

  ; check if M 'rows' have passed
  if "l3 % cnst51 == 0"
{
  1m iu1
  1m igrad EA    ; HSQC echo-antiecho gradients
  ; ...
}

  lo to 4 times l0
end
```

This structure provides a clear blueprint for how GENESIS can be adapted to produce generalised supersequences. The only thing which GENESIS needs to detect is how many modules are being interleaved, which determines the number of m_i's. (It does not need to know the *values* of the m_i's, since that is a parameter to be set by the user.) It should be noted that the 'parallel' supersequence of Fig. 4.63c can be obtained by setting $M = 2$ and $m_1 = m_2 = 1$, and standard linear supersequences simply have $M = m_1 = 1$. Thus, if generalised supersequences were to be implemented in GENESIS, all other types of supersequences can be obtained 'for free'.

A small technical detail in Listing 4.4 is that each `go=2` statement loops back to the label `2`, such that the experiment is repeated `NS` times. This means that in the left side of Fig. 4.66, each row is already acquired a total of `NS` times, in addition to any repetition which is explicitly shown. Thus, after summation of the explicitly repeated increments, the HSQC will in fact end up with a total of `3*NS` scans. However, the *phase cycle* is only moved forward within the inner `NS` loop; each time `NS` repetitions are completed, the phase cycle is reset to the beginning. This means that even though each FID in the HSQC is acquired `3*NS` times, only an `NS`-step phase cycle is used. Thus, in order to obtain spectra of the highest possible quality, any common divisors of M and the m_i's should be factorised out and placed into `NS` instead.

4.5.2.2 Examples

Since generalised supersequences allow the effective number of scans to be individually customised for each module, they offer an avenue for modules with (very) different intrinsic sensitivities to be combined. On the very-low-sensitivity end are spectra such as ADEQUATE and ^{15}N HMBC; compared to these, the HSQC experiments may be considered high-sensitivity. It should be reiterated that the time savings to be obtained through this combination (ρ_t) are still on the order of 2, because only two modules are being horizontally concatenated at a time.

Figure 4.67 shows some examples of such experiments, where the relative numbers of scans are chosen with the aim of balancing the sensitivities of each module. The ADEQUATE module used here is the ZIP-ADEQUATE experiment shown in Fig. 4.58; this consumes ^{1}H^{C} magnetisation and preserves the ^{1}H$^{!C}$ magnetisation required by the ^{15}N HMBC, ^{13}C HMBC and ^{15}N seHSQC. However, the ^{13}C HSQC module draws on the same magnetisation pool. Therefore, in order to maximise the intensities in this experiment, a period of isotropic mixing is applied only before the HSQC module in order to effect polarisation transfer from ^{1}H$^{!C}$ spins. The spectra thus obtained are shown in Figs. 4.68 and 4.69.

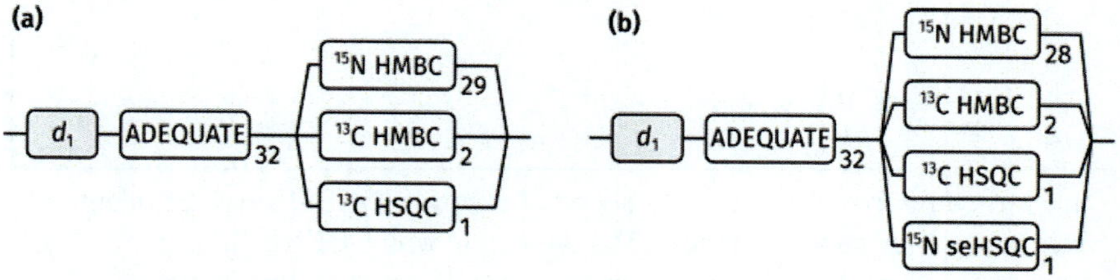

Fig. 4.67 Two examples of generalised NOAH supersequences studied in this work. **a** A NOAH-4 AB$_N$BS supersequence. **b** A NOAH-5 AB$_N$BSS$_N^+$ supersequence

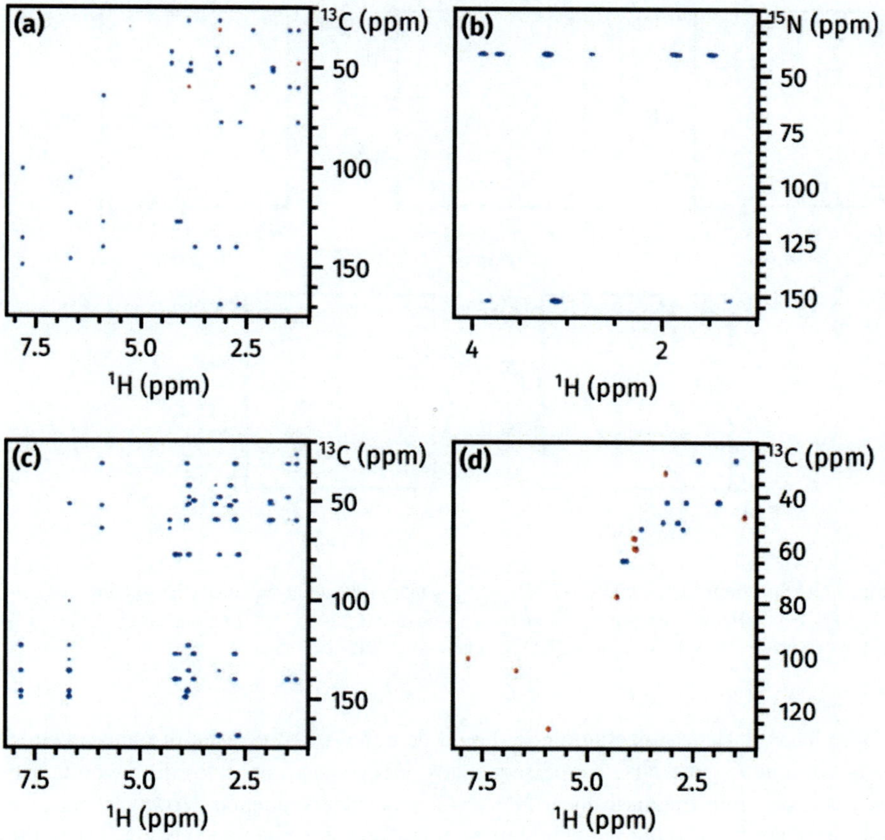

Fig. 4.68 Spectra from the NOAH-4 AB$_N$BS generalised supersequence (Fig. 4.67a). **a** ADEQUATE. **b** ^{15}N HMBC, optimised for an $^nJ_{NH}$ value of 8 Hz. **c** ^{13}C HMBC, optimised for an $^nJ_{CH}$ value of 8 Hz. **d** ^{13}C HSQC. *Data code:* `7X-220604`

Another additional implementation detail, related to the earlier point about phase cycling, is that between each of the 28 or 29 repetitions of the ADEQUATE/^{15}N HMBC increments, the ^{15}N HMBC signal (specifically, the first ^{15}N 90° pulse) and receiver phases are additionally inverted. This can be accomplished by adding the corresponding extra phase incrementation instructions just underneath the `1m iu3` line in Listing 4.4, and can be thought of as a form of 'manual' phase cycling, which is independent of (and complementary to) the usual phase cycle specified in the pulse program. The reason for doing this is to suppress artefacts which appear at $F_1 \approx 0$ in the ^{15}N HMBC, which stem from the ADEQUATE module (when the ^{15}N HMBC is run on its own, these artefacts do not appear). The exact origin of these artefacts is still unclear,[38] but their phases are constant; thus, inverting the ^{15}N HMBC receiver

[38] Their frequencies in F_1 are not exactly zero, so there is *some* kind of modulation in the t_1 period of the ADEQUATE module, but it is not entirely clear where this is happening. My current best guess

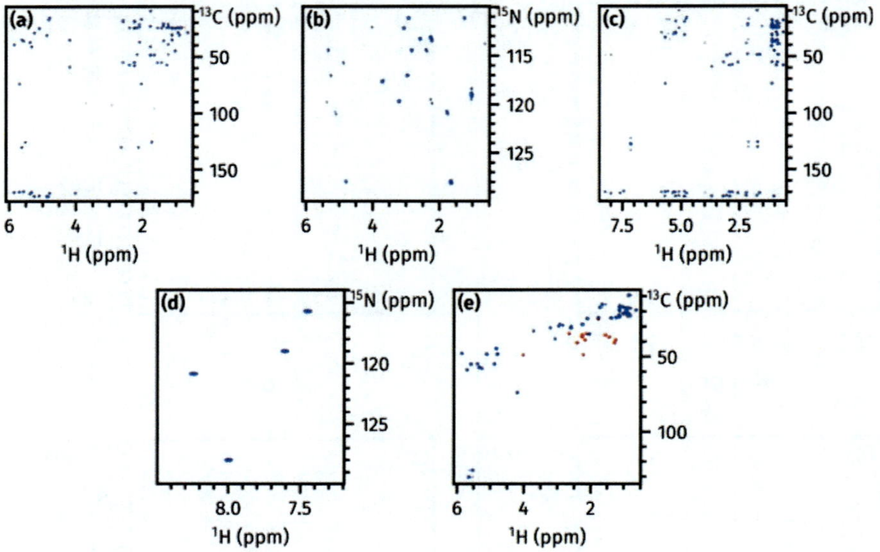

Fig. 4.69 Spectra from the NOAH-5 $AB_NBSS_N^+$ generalised supersequence (Fig. 4.67b). **a** ADEQUATE. **b** ^{15}N HMBC, optimised for an $^nJ_{NH}$ value of 4 Hz. **c** ^{13}C HMBC, optimised for an $^nJ_{CH}$ value of 8 Hz. **e** ^{15}N seHSQC. **d** ^{13}C HSQC. *Data code:* `7C-220722`

phase leads to their cancellation. It should be noted that the issue of phase cycling is not unique to generalised supersequences: indeed, such phase cycling would also be necessary in a standard linear NOAH-2 AB_N supersequence. However, because the degree of built-in phase cycling in a generalised supersequence is lower than the actual number of transients recorded for each module, some of it must be explicitly specified in this way.

4.5.2.3 Covariance Analysis

The generalised supersequences shown in Figs. 4.67 and 4.69 yield very thorough information about heteronuclear connectivity, including one-bond and long-range ^{1}H–^{13}C and ^{1}H–^{15}N couplings. These spectra can be used as inputs for indirect covariance processing [111–113] in order to yield 'unnatural' (but very informative) double heteronuclear correlation spectra, in which each peak represents a correlation between two dilute heteronuclei (^{13}C or ^{15}N). Directly recording such spectra through an NMR experiment would require substantially more time, because of the extremely low probability of finding two heteronuclei in the same molecule.

---

is that the offending magnetisation evolves during the constant-time period of the ADEQUATE, where the double-quantum frequencies are reconverted into single-quantum frequencies.

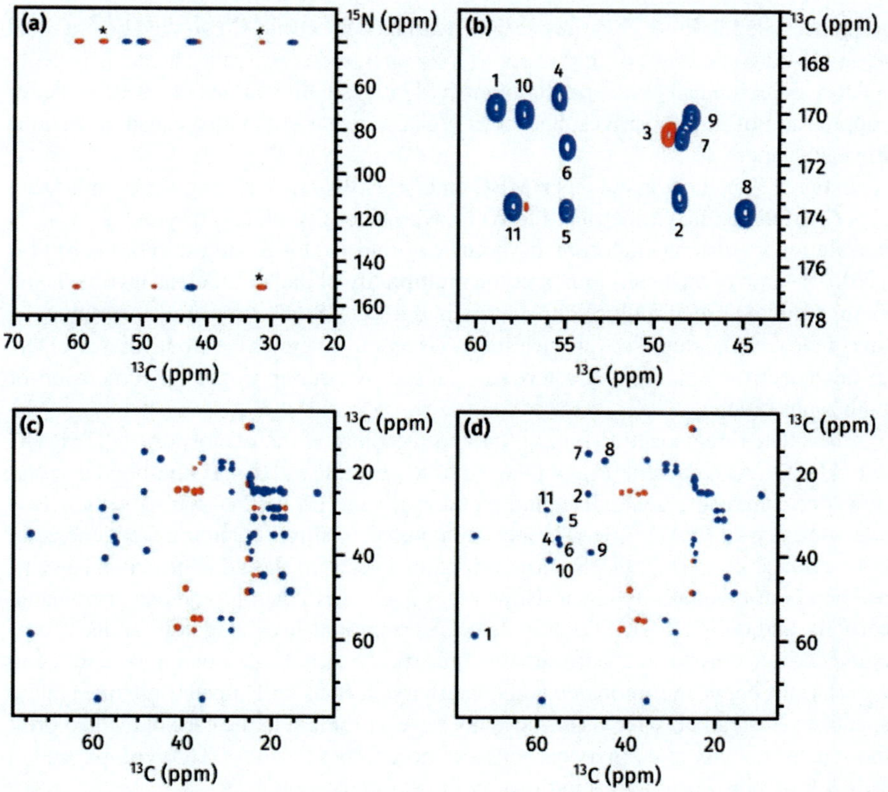

Fig. 4.70 Examples of covariance spectra obtained from generalised NOAH supersequences. **a** ^{13}C–^{15}N correlation spectrum for brucine. Asterisks indicate artefacts arising from peak overlap. **b** Inset of the ^{13}C–^{13}C one-bond correlation spectrum for cyclosporin, obtained by processing the ADEQUATE and ^{13}C HSQC spectra in Fig. 4.69a and d using generalised indirect covariance ($\lambda = 0.5$). The region shown here contains correlations between CO and Cα carbons. **c, d** The same ^{13}C–^{13}C correlation spectrum as in (b), but this time showing correlations in the alkyl region. The spectrum in (c) is untouched, but the spectrum in (d) has been further subject to a sign-preserving symmetrisation process. The CO–Cα and Cα–Cβ correlations for all residues are labelled in (b) and (d), following the residue numbering in Fig. A.2

Some possible covariance spectra obtained from the data in Figs. 4.68 and 4.69 are shown in Fig. 4.70. All covariance calculations were performed in Python, according to the formulae:

$$\mathbf{C} = \mathbf{S}_1\mathbf{S}_2^T \tag{4.23}$$

for unsymmetrical indirect covariance (where $\mathbf{S}_1$ and $\mathbf{S}_2$ are the input spectra and $\mathbf{C}$ the covariance spectrum), or

$$\mathbf{M} = \left[\begin{pmatrix}\mathbf{S}_1 \\ \mathbf{S}_2\end{pmatrix}(\mathbf{S}_1\ \mathbf{S}_2)\right]^\lambda \tag{4.24}$$

for generalised indirect covariance; the covariance spectrum **C** can then be obtained as an off-diagonal block of the matrix **M**. λ is a parameter which is chosen in order to balance peak intensity and artefact intensity [112]: smaller values of λ lead to better suppression of false positives, but lower signal intensity (and also cause distortions in peak shapes).

In Fig. 4.70a, the brucine ^{15}N HMBC and ^{13}C HSQC spectra (Fig. 4.68b and d) are processed using unsymmetrical indirect covariance (Eq. (4.23)) to yield a ^{13}C–^{15}N correlation spectrum which contains both one-bond and long-range correlations [114, 115]. The sign of the peaks indicates the multiplicity of the ^{13}C nucleus involved, and stems from the use of multiplicity editing in the ^{13}C HSQC spectrum. Unfortunately, one of the weaknesses of covariance spectra is on show here: there are several artefacts in this spectrum (marked by asterisks), caused by overlap in the ^{1}H dimension of both input spectra.

The other three spectra (Fig. 4.70b and d) combine the cyclosporin ^{13}C HSQC and ADEQUATE spectra (Fig. 4.69a and d) to generate a ^{13}C–^{13}C one-bond correlation spectrum, using generalised indirect covariance (Eq. (4.24)) with $\lambda = 0.5$. This is essentially an INADEQUATE spectrum, but with single-quantum frequencies in both dimensions [116, 117]. Such a correlation spectrum should in principle be symmetric about the main diagonal. However, since the HSQC only detects protonated carbons, and the ADEQUATE only detects correlations involving at least one protonated carbon, correlations between two quaternary carbons do not appear at all, and correlations between protonated and quaternary carbons will appear only once in the spectrum (instead of twice). This does not pose a problem in the case of cyclosporin, though, as the only quaternary carbons are those of the carbonyl (CO) groups, which will still display correlations to the adjacent Cα carbons.

These Cα–CO correlations are shown in Fig. 4.70b. There are 11 peaks here, one for each amino acid residue in cyclosporin: again, the sign indicates the Cα multiplicity (all are CH, except for the CH_2 of the sarcosine residue). Figures 4.70c and 4.70d show the Cα–Cβ correlations. Figure 4.70c shows the original spectrum, but in Fig. 4.70d, the spectrum has been further subjected to a sign-preserving symmetrisation procedure, where the intensity at each point $p(\Omega_1, \Omega_2)$ is replaced by

$$p(\Omega_1, \Omega_2) \rightarrow \mathrm{sgn}[p(\Omega_1, \Omega_2)] \cdot \min\{|p(\Omega_1, \Omega_2)|, |p(\Omega_2, \Omega_1)|\}. \qquad (4.25)$$

Here, $\mathrm{sgn}\, x$ refers to the sign of x, or equivalently $x/|x|$ (for $x \neq 0$). It should be emphasised that such a procedure can only be safely carried out where peaks on both sides of the diagonals are expected to be seen.[39] Thus, if the symmetrisation procedure were to be applied across the entire spectrum, correlations between quaternary and non-quaternary carbons (such as those in Fig. 4.70b) would be lost. However, since all Cα and Cβ carbons in cyclosporin are protonated, the symmetrisation process is safe to use in this region; in this case, it helps to remove some of the artefacts which tend to plague the interpretation of covariance spectra.

[39] The peak intensities $p(\Omega_1, \Omega_2)$ and $p(\Omega_2, \Omega_1)$ need not be exactly equal, but both peaks should at least be above the noise level.

4.6 Conclusion

In this chapter, I covered a variety of improvements to NOAH supersequences, including the development of new or improved modules (Sect. 4.3), the implementation of basic solvent suppression (Sect. 4.4) and the generalised NOAH sequences obtained through 'vertical stacking' of different supersequences (Sect. 4.5).

As a technique in fast 2D NMR, NOAH is in my view fairly mature despite its relative recency (the first paper was published in 2017 [32]). The underlying idea of separating magnetisation pools, and the tools used to do so, have been around for quite a while—most recently, and notably, in the form of ASAP 2D NMR. Thus, rather than making fundamental conceptual breakthroughs, I would characterise my work here as 'polishing the remaining rough edges', whether by removing artefacts or by incorporating slightly more specialised modules.

Naturally, the possibility of using the same concept to accelerate 3D (or higher) experiments is extremely attractive. Some of the groundwork for this has already been laid, in the form of ASAP-type 3D experiments [14]. However, my opinion is that this research topic would extend far beyond NOAH itself: in particular, 3D experiments are often applied to isotopically labelled substances, which lead to a different set of magnetisation pools which are not so easily separable.

At this point, assuming that there are few fundamental breakthroughs to be made in NOAH itself, we must turn to the question of how to increase its uptake. The accessibility of new techniques is *always* a significant factor, regardless of the merit of the underlying technique. For this reason, I am particularly proud of the GENESIS website (Sect. 4.2) and the associated improvements in the processing pipeline; the feedback from the NMR community has also been greatly encouraging. At the same time, there is still some way to go: although one may consider a 'standard' NOAH experiment easy enough to set up, after taking into account other desirable features such as automation and NUS, the entire process can become rather non-trivial.

Another important, but under-explored, area is the use of NOAH in the sensitivity-limited regime. The results in this chapter are derived from samples which are not particularly dilute and hence belong in the resolution-limited regime. While this may lead to a maximisation of the *relative* (in that $\rho_t > \rho_{t,\text{eff}}$), the *absolute* time savings are small because the standalone 2D experiments would be completed fairly quickly anyway. This is particularly relevant for (e.g.) unstable samples, for which 2D data need to be acquired in as short a time as possible. Conclusively demonstrating the benefits of NOAH experiments on such samples would go a long way towards convincing more synthetic chemists to adopt the NOAH technique.

References

1. Yong JRJ, Hansen AL, Kupče Ē, Claridge TDW (2021) Increasing sensitivity and versatility in NMR supersequences with new HSQC-based modules. J Magn Reson 329:107027. https://doi.org/10.1016/j.jmr.2021.107027
2. Kupče Ē, Yong JRJ, Widmalm G, Claridge TDW (2021) Parallel NMR supersequences: ten spectra in a single measurement. JACS Au 1:1892–1897. https://doi.org/10.1021/jacsau.1c00423
3. Yong JRJ, Kupče Ē, Claridge TDW (2022) Modular pulse program generation for NMR supersequences. Anal Chem 94(2271–2278):4964. https://doi.org/10.1021/acs.analchem.1c0
4. Yong JRJ, Kupče Ē, Claridge TDW (2023) A general scheme for generating NMR supersequences combining high- and low-sensitivity experiments. Chem Commun 59:7827–7830. https://doi.org/10.1039/d3cc01472j
5. Kupče Ē, Frydman L, Webb AG, Yong JRJ, Claridge TDW (2021) Parallel nuclear magnetic resonance spectroscopy. Nat Rev Methods Primers 2021(1):No. 27. https://doi.org/10.1038/s43586-021-00024-3
6. Yong JRJ, Kupče Ē, Claridge TDW (2023) Fast 2D solution-state NMR: concepts and applications, Giraudeau P, Dumez J-N (eds). Royal Society of Chemistry, London. https://doi.org/10.1039/BK9781839168062-00084
7. Barna JCJ, Laue ED, Mayger MR, Skilling J, Worrall SJP (1987) Exponential sampling, an alternative method for sampling in two-dimensional NMR experiments. J Magn Reson 73:69–77. https://doi.org/10.1016/0022-2364(87)90225-3
8. Kazimierczuk K, Stanek J, Zawadzka-Kazimierczuk A, Koźmiński W (2010) Random sampling in multidimensional NMR spectroscopy. Prog Nucl Magn Reson Spectrosc 57:420–434. https://doi.org/10.1016/j.pnmrs.2010.07.002
9. Mobli M, Hoch JC (2014) Nonuniform sampling and non-Fourier signal processing methods in multidimensional NMR. Prog Nucl Magn Reson Spectrosc 83:21–41. https://doi.org/10.1016/j.pnmrs.2014.09.002
10. Kazimierczuk K, Orekhov V (2015) Non-uniform sampling: post-Fourier era of NMR data collection and processing. Magn Reson Chem 53(921–926):4284. https://doi.org/10.1002/mrc
11. Schulze-Sünninghausen D, Becker J, Luy B (2014) Rapid heteronuclear single quantum correlation nmr spectra at natural abundance. J Am Chem Soc 136:1242–1245. https://doi.org/10.1021/ja411588d
12. Schanda P, Van Melckebeke H, Brutscher B (2006) Speeding up three-dimensional protein NMR experiments to a few minutes. J Am Chem Soc 128:9042–9043. https://doi.org/10.1021/ja062025p
13. Kupče Ē, Freeman R (2007) Fast multidimensional NMR by polarization sharing. Magn Reson Chem 45:2–4. https://doi.org/10.1002/mrc.1931
14. Schanda P (2009) Fast-pulsing longitudinal relaxation optimized techniques: enriching the toolbox of fast biomolecular NMR spectroscopy. Prog Nucl Magn Reson Spectrosc 55:238–265. https://doi.org/10.1016/j.pnmrs.2009.05.002
15. Frydman L, Scherf T, Lupulescu A (2002) The acquisition of multidimensional NMR spectra within a single scan. Proc Natl Acad Sci USA 99:15858–15862. https://doi.org/10.1073/pnas.252644399
16. Pelupessy P (2003) Adiabatic single scan two-dimensional NMR spectrocopy. J Am Chem Soc 125:12345–12350. https://doi.org/10.1021/ja034958g
17. Frydman L, Lupulescu A, Scherf T (2003) Principles and features of single-scan two-dimensional NMR spectroscopy. J Am Chem Soc 125:9204–9217. https://doi.org/10.1021/ja030055b
18. Tal A, Frydman L (2010) Single-scan multidimensional magnetic resonance. Prog Nucl Magn Reson Spectrosc 57:241–292. https://doi.org/10.1016/j.pnmrs.2010.04.001
19. Gouilleux B, Rouger L, Giraudeau P (2018) Ultrafast 2D NMR: methods and applications. Annu Rep NMR Spectrosc 2018:75–144. https://doi.org/10.1016/bs.arnmr.2017.08.003

20. Kupče Ē, Freeman R (2003) Two-dimensional Hadamard spectroscopy. J Magn Reson 162:300–310. https://doi.org/10.1016/s1090-7807(02)00196-9
21. Kupče Ē, Nishida T, Freeman R (2003) Hadamard NMR spectroscopy. Prog Nucl Magn Reson Spectrosc 42:95–122. https://doi.org/10.1016/s0079-6565(03)00022-0
22. Jeannerat D (2000) High resolution in the indirectly detected dimension exploiting the processing of folded spectra. Magn. Reson. Chem. 38:415–422. https://doi.org/10.1002/1097-458x(200006)38:6<415::aid-mrc665>3.0.co;2-u
23. Bermel W, Bertini I, Felli IC, Pierattelli R (2009) Speeding up 13C direct detection biomolecular NMR spectroscopy. J Am Chem Soc 131:15339–15345. https://doi.org/10.1021/ja9058525
24. Njock GBB, Pegnyemb DE, Bartholomeusz TA, Christen P, Vitorge B, Nuzillard J-M, Shivapurkar R, Foroozandeh M, Jeannerat D (2010) Spectral aliasing: a super zoom for 2D-NMR spectra. Principles and applications. Chimia 64:235. https://doi.org/10.2533/chimia.2010.235
25. Jeannerat D (2011) In: eMagRes. Wiley. https://doi.org/10.1002/9780470034590.emrstm1187
26. Nolis P, Pérez-Trujillo M, Parella T (2007) Multiple FID acquisition of complementary HMBC data. Angew Chem Int Ed 46(7495–7497):258. https://doi.org/10.1002/anie.200702
27. Parella T, Nolis P (2010) Time-shared NMR experiments. Concepts Magn Reson 36A:1–23. https://doi.org/10.1002/cmr.a.20150
28. Nolis P, Motiram-Corral K, Pérez-Trujillo M, Parella T (2019) Broadband homodecoupled time-shared ^{1}H-^{13}C and ^{1}H-^{15}N HSQC experiments. J Magn Reson 298:23–30. https://doi.org/10.1016/j.jmr.2018.11.005
29. Kupče Ē, Freeman R, John BK (2006) Parallel acquisition of two-dimensional NMR spectra of several nuclear species. J Am Chem Soc 128:9606–9607. https://doi.org/10.1021/ja0634876
30. Kupče Ē, Freeman R (2008) Molecular structure from a single NMR experiment. J Am Chem Soc 130:10788–10792. https://doi.org/10.1021/ja8036492
31. Kovacs H, Kupče Ē (2016) Parallel NMR spectroscopy with simultaneous detection of ^{1}H and ^{19}F nuclei. Magn Reson Chem 54:544–560. https://doi.org/10.1002/mrc.4428
32. Kupče Ē, Claridge TDW (2017) NOAH: NMR supersequences for small molecule analysis and structure elucidation. Angew Chem Int Ed 56:11779–11783. https://doi.org/10.1002/anie.201705506
33. Kupče Ē, Mote KR, Webb A, Madhu PK, Claridge TDW (2021) Multiplexing experiments in NMR and multi-nuclear MRI. Prog Nucl Magn Reson Spectrosc 124–125:1–56. https://doi.org/10.1016/j.pnmrs.2021.03.001
34. Haasnoot CAG, van de Ven FJM, Hilbers CW (1984) COCONOSY. Combination of 2D correlated and 2D nuclear overhauser enhancement spectroscopy in a single experiment. J Magn Reson 56:343–349. https://doi.org/10.1016/0022-2364(84)90114-8
35. Gurevich AZ, Barsukov IL, Arseniev AS, Bystrov VF (1984) Combined COSY-NOESY experiment. J Magn Reson 56:471–478. https://doi.org/10.1016/0022-2364(84)90311-1
36. Motiram-Corral K, Pérez-Trujillo M, Nolis P, Parella T (2018) Implementing one-shot multiple-FID acquisition into homonuclear and heteronuclear NMR experiments. Chem Commun 54:13507–13510. https://doi.org/10.1039/C8CC08065H
37. Nolis P, Parella T (2019) Practical aspects of the simultaneous collection of COSY and TOCSY spectra. Magn Reson Chem 57:S85–S94. https://doi.org/10.1002/mrc.4835
38. Nolis P, Motiram-Corral K, Pérez-Trujillo M, Parella T (2019) Interleaved dual NMR acquisition of equivalent transfer pathways in TOCSY and HSQC experiments. ChemPhysChem 20:356–360. https://doi.org/10.1002/cphc.201801034
39. Nolis P, Motiram-Corral K, Pérez-Trujillo M, Parella T (2019) Simultaneous acquisition of two 2D HSQC spectra with different ^{13}C spectral widths. J Magn Reson 300:1–7. https://doi.org/10.1016/j.jmr.2019.01.004
40. Nagy TM, Gyöngyösi T, Kövér KE, Sørensen OW (2019) BANGO SEA XLOC/HMBC. H2OBC: complete heteronuclear correlation within minutes from one NMR pulse sequence. Chem Commun 55:12208.12211. https://doi.org/10.1039/c9cc06253j.

41. Nagy TM, Kövér KE, Sørensen OW (2020) Double and adiabatic BANGO for concatenating two NMR experiments relying on the same pool of magnetization. J Magn Reson 316:106767. https://doi.org/10.1016/j.jmr.2020.106767
42. Nagy TM, Kövér KE, Sørensen OW (2021) NORD: NO relaxation delay NMR spectroscopy. Angew Chem Int Ed 60:13587–13590. https://doi.org/10.1002/anie.202102487
43. Timári I, Nagy TM, Kövér KE, Sørensen OW (2022) Synergy and sensitivity-balance in concatenating experiments in NO relaxation delay NMR (NORD). Chem Commun 58:2516–2519. https://doi.org/10.1039/d1cc06663c
44. Orts J, Gossert AD (2018) Structure determination of protein-ligand complexes by NMR in solution. Methods 138–139:3–25. https://doi.org/10.1016/j.ymeth.2018.01.019
45. Schulze-Sünninghausen D, Becker J, Koos MRM, Luy B (2017) Improvements, extensions, and practical aspects of rapid ASAP-HSQC and ALSOFAST-HSQC pulse sequences for studying small molecules at natural abundance. J Magn Reson 281:151–161. https://doi.org/10.1016/j.jmr.2017.05.012
46. Vitorge B, Bodenhausen G, Pelupessy P (2010) Speeding up nuclear magnetic resonance spectroscopy by the use of SMAll recovery times. SMART NMR. J Magn Reson 207:149–152. https://doi.org/10.1016/j.jmr.2010.07.017
47. Shaw AA, Salaun C, Dauphin J-F, Ancian B (1996) Artifact-free PFG-enhanced double-quantum-filtered COSY experiments. J Magn Reson Ser A 120:110–115. https://doi.org/10.1006/jmra.1996.0105
48. Howe PWA (2014) Rapid pulsing artefacts in pulsed-field gradient double-quantum filtered COSY spectra. Magn Reson Chem 52:329–332. https://doi.org/10.1002/mrc.4060
49. Claridge TDW, Mayzel M, Kupče Ē (2019) Triplet NOAH supersequences optimised for small molecule structure characterisation. Magn Reson Chem 57:946–952. https://doi.org/10.1002/mrc.4887
50. Kupče Ē, Claridge TDW (2018) Molecular structure from a single NMR supersequence. Chem Commun 54:7139–7142. https://doi.org/10.1039/c8cc03296c
51. Kupče Ē, Claridge TDW (2019) New NOAH modules for structure elucidation at natural isotopic abundance. J Magn Reson 307(106568):6568. https://doi.org/10.1016/j.jmr.2019.10
52. Kontaxis G, Stonehouse J, Laue ED, Keeler J (1994) The sensitivity of experiments which use gradient pulses for coherence-pathway selection. J Magn Reson Ser A 111:70–76. https://doi.org/10.1006/jmra.1994.1227
53. Palmer AG III, Cavanagh J, Wright PE, Rance M (1991) Sensitivity improvement in proton-detected two-dimensional heteronuclear correlation NMR spectroscopy. J Magn Reson 93:151–170. https://doi.org/10.1016/0022-2364(91)90036-s
54. Kay LE, Keifer P, Saarinen T (1992) Pure absorption gradient enhanced heteronuclear single quantum correlation spectroscopy with improved sensitivity. J Am Chem Soc 114:10663–10665. https://doi.org/10.1021/ja00052a088
55. Cavanagh J, Rance M (1993) Sensitivity-enhanced NMR techniques for the study of biomolecules. Annu Rep NMR Spectrosc 27:1–58. https://doi.org/10.1016/s0066-4103(08)60264-1
56. Schleucher J, Schwendinger M, Sattler M, Schmidt P, Schedletzky O, Glaser SJ, Sorensen OW, Griesinger C (1994) A general enhancement scheme in heteronuclear multidimensional NMR employing pulsed field gradients. J Biomol NMR 1994:4. https://doi.org/10.1007/bf00175254
57. Hansen AL, Kupče Ē, Li D-W, Bruschweiler-Li L, Wang C, Bruschweiler R (2021) 2D NMR-based metabolomics with HSQC/TOCSY NOAH supersequences. Anal Chem 93:6112–6119. https://doi.org/10.1021/acs.analchem.0c05205
58. Turner CJ, Connolly PJ, Stern AS (1999) Artifacts in sensitivity-enhanced HSQC. J Magn Reson 137:281–284. https://doi.org/10.1006/jmre.1998.1692
59. Levitt MH (1986) Composite pulses. Prog Nucl Magn Reson Spectrosc 18:61–122. https://doi.org/10.1016/0079-6565(86)80005-x
60. Foroozandeh M, Singh P (2021) Optimal control of spins by analytical lie algebraic derivatives. Automatica 129:109611. https://doi.org/10.1016/j.automatica.2021.109611

61. Wimperis S, Freeman R (1984) An excitation sequence which discriminates between direct and long-range CH coupling. J Magn Reson 58:348–353. https://doi.org/10.1016/0022-2364(84)90227-0
62. Sørensen OW (2022) Selective rotations using non-selective pulses and heteronuclear couplings. Bull Magn Reson 16:49–53. https://ismar.org/wp-content/uploads/2021/09/BMR_16_049-053_1994.pdf (Accessed 25 Sept 2022)
63. Garbow JR, Weitekamp DP, Pines A (1982) Bilinear rotation decoupling of homonuclear scalar interactions. Chem Phys Lett 93:504–509. https://doi.org/10.1016/0009-2614(82)83229-6
64. Uhrin D, Liptaj T, Kover KE (1993) Modified BIRD pulses and design of heteronuclear pulse sequences. J Magn Reson Ser A 101:41–46. https://doi.org/10.1006/jmra.1993.1005
65. Kaltschnee L, Kolmer A, Timári I, Schmidts V, Adams RW, Nilsson M, Kövér KE, Morris GA, Thiele CM (2014) "Perfecting" pure shift HSQC: full homodecoupling for accurate and precise determination of heteronuclear couplings. Chem Commun 50:15702–15705. https://doi.org/10.1039/c4cc04217d
66. Briand J, Sørensen OW (1997) A novel pulse sequence element for biselective and independent rotations with arbitrary flip angles and phases for I and IS spin systems. J Magn Reson 125:202–206. https://doi.org/10.1006/jmre.1996.1095
67. Briand J, Sørensen OW (1998) Simultaneous and independent rotations with arbitrary flip angles and phases for I, ISα, and ISβ spin systems. J Magn Reson 135:44–49. https://doi.org/10.1006/jmre.1998.1556
68. Pérez-Trujillo M, Nolis P, Bermel W, Parella T (2007) Optimizing sensitivity and resolution in time-shared NMR experiments. Magn Reson Chem 45:325–329. https://doi.org/10.1002/mrc.1973
69. Ni F, Scheraga HA (1986) Phase-sensitive spectral analysis by maximum entropy extrapolation. J Magn Reson 70:506–511. https://doi.org/10.1016/0022-2364(86)90145-9
70. Tirendi CF, Martin JF (1989) Fast linear prediction processing in two-dimensional NMR spectroscopy. J Magn Reson 81:577–585. https://doi.org/10.1016/0022-2364(89)90096-6
71. Led JJ, Gesmar H (1991) Application of the linear prediction method to NMR spectroscopy. Chem Rev 91:1413–1426. https://doi.org/10.1021/cr00007a007
72. Koehl P (1999) Linear prediction spectral analysis of NMR data. Prog Nucl Magn Reson Spectrosc 34:257–299. https://doi.org/10.1016/S0079-6565(99)00002-3
73. Donoho DL, Johnstone IM, Stern AS, Hoch JC (1990) Does the maximum entropy method improve sensitivity? Proc Natl Acad Sci USA 87:5066–5068. https://doi.org/10.1073/pnas.87.13.5066
74. Stern AS, Li K-B, Hoch JC (2002) Modern spectrum analysis in multidimensional NMR spectroscopy: comparison of linear-prediction extrapolation and maximum-entropy reconstruction. J Am Chem Soc 124:1982–1993. https://doi.org/10.1021/ja011669o
75. Palmer MR, Suiter CL, Henry GE, Rovnyak J, Hoch JC, Polenova T, Rovnyak D (2015) Sensitivity of nonuniform sampling NMR. J Phys Chem B 119:6502–6515. https://doi.org/10.1021/jp5126415
76. Shaka AJ, Lee CJ, Pines A (1988) Iterative schemes for bilinear operators; application to spin decoupling. J Magn Reson 77:274–293. https://doi.org/10.1016/0022-2364(88)90178-3
77. Enthart A, Freudenberger JC, Furrer J, Kessler H, Luy B (2008) The CLIP/CLAP-HSQC: pure absorptive spectra for the measurement of one-bond couplings. J Magn Reson 192:314–322. https://doi.org/10.1016/j.jmr.2008.03.009
78. Becker J, Koos MRM, Schulze-Sünninghausen D, Luy B (2019) ASAP-HSQC-TOCSY for fast spin system identification and extraction of long-range couplings. J Magn Reson 300:76–83. https://doi.org/10.1016/j.jmr.2018.12.021
79. Nyberg NT, Duus JØ, Sørensen OW (2005) Heteronuclear two-bond correlation: suppressing heteronuclear three-bond or higher NMR correlations while enhancing two-bond correlations even for vanishing $^2J_{CH}$. J Am Chem Soc 127:6154–6155. https://doi.org/10.1021/ja050878w
80. Nyberg NT, Duus JØ, Sørensen OW (2005) Editing of H2BC NMR spectra. Magn Reson Chem 43:971–974. https://doi.org/10.1002/mrc.1698

81. Kupče Ē, Sørensen OW (2017) 2BOB - extracting an H2BC and an HSQC-type spectrum from the same data set, and H2OBC - a fast experiment delineating the protonated ^{13}C backbone. Magn Reson Chem 55:515–518. https://doi.org/10.1002/mrc.4584
82. Hu K, Westler WM, Markley JL (2011) Two-dimensional concurrent HMQC-COSY as an approach for small molecule chemical shift assignment and compound identification. J Biomol NMR 49:291–296. https://doi.org/10.1007/s10858-011-9494-4
83. Thrippleton MJ, Keeler J (2003) Elimination of zero-quantum interference in two-dimensional NMR spectra. Angew Chem Int Ed 42:3938–3941. https://doi.org/10.1002/anie.200351947
84. Gyöngyösi T, Timári I, Haller J, Koos MRM, Luy B, Kövér KE (2018) Boosting the NMR assignment of carbohydrates with clean in-phase correlation experiments. ChemPlusChem 83:53–60. https://doi.org/10.1002/cplu.201700452
85. Gyöngyösi T, Timári I, Sinnaeve D, Luy B, Kövér KE (2021) Expedited nuclear magnetic resonance assignment of small- to medium-sized molecules with improved HSQCCLIP- COSY experiments. Anal Chem 93:3096–3102. https://doi.org/10.1021/acs.analchem.0c04124
86. Aguilar JA, Nilsson M, Bodenhausen G, Morris GA (2012) Spin echo NMR spectra without J modulation. Chem Commun 48:811–813. https://doi.org/10.1039/c1cc16699a
87. Parella T (2019) Towards perfect NMR: Spin-echo versus perfect-echo building blocks. Magn Reson Chem 57:13–29. https://doi.org/10.1002/mrc.4776
88. Koos MRM, Kummerlowe G, Kaltschnee L, Thiele CM, Luy B (2016) CLIP-COSY: a clean in-phase experiment for the rapid acquisition of COSY-type correlations. Angew Chem Int Ed 55:7655–7659. https://doi.org/10.1002/anie.201510938
89. Foroozandeh M, Adams RW, Kiraly P, Nilsson M, Morris GA (2015) Measuring couplings in crowded NMR spectra: pure shift NMR with multiplet analysis. Chem Commun 51:15410–15413. https://doi.org/10.1039/c5cc06293d
90. Foroozandeh M, Adams RW, Meharry NJ, Jeannerat D, Nilsson M, Morris GA (2014) Ultrahigh-resolution NMR spectroscopy. Angew Chem Int Ed 53:6990–6992. https://doi.org/10.1002/anie.201404111
91. Moutzouri P, Chen Y, Foroozandeh M, Kiraly P, Phillips AR, Coombes SR, Nilsson M, Morris GA (2017) Ultraclean pure shift NMR. Chem Commun 53:10188–10191. https://doi.org/10.1039/c7cc04423b
92. Aguilar JA, Faulkner S, Nilsson M, Morris GA (2010) Pure shift ^{1}H NMR: a resolution of the resolution problem? Angew Chem Int Ed 49:3901–3903. https://doi.org/10.1002/anie.201001107
93. Cicero DO, Barbato G, Bazzo R (2001) Sensitivity enhancement of a two-dimensional experiment for the measurement of heteronuclear long-range coupling constants, by a new scheme of coherence selection by gradients. J Magn Reson 148:209–213. https://doi.org/10.1006/jmre.2000.2234
94. Kupče Ē, Schmidt P, Rance M, Wagner G (1998) Adiabatic mixing in the liquid state. J Magn Reson 135:361–367. https://doi.org/10.1006/jmre.1998.1607
95. Reif B, Köck M, Kerssebaum R, Kang H, Fenical W, Griesinger C (1996) ADEQUATE, a new set of experiments to determine the constitution of small molecules at natural abundance. J Magn Reson Ser A 118:282–285. https://doi.org/10.1006/jmra.1996.0038
96. Rao Kakita VM, Hosur RV (2020) All-in-one NMR spectroscopy of small organic molecules: complete chemical shift assignment from a single NMR experiment. RSC Adv 10:21174–21179. https://doi.org/10.1039/d0ra03417g
97. Mareci TH, Freeman R (1982) Echoes and antiechoes in coherence transfer NMR: determining the signs of double-quantum frequencies. J Magn Reson 48:158–163. https://doi.org/10.1016/0022-2364(82)90250-5
98. Hwang TL, Shaka AJ (1995) Water suppression that works. excitation sculpting using arbitrary wave-forms and pulsed-field gradients. J Magn Reson Ser A 112:275–279. https://doi.org/10.1006/jmra.1995.1047
99. Ottiger M, Delaglio F, Bax A (1998) Measurement of J and dipolar couplings from simplified two-dimensional NMR Spectra. J Magn Reson 131:373–378. https://doi.org/10.1006/jmre.1998.1361